W0080967

ENVIRONMENT 2001

A Global Challenge

ENVIRONMENT 2001

A Global Challenge

Edited by

Dr. Ramesh Mathur
M.Sc., Ph.D.
Professor & Head, School of Studies in Zoology,
Jiwaji University, Gwalior

Dr. Sanjay Sharma
M.Sc., Ph.D.
Secretary, Forum For Environment Protection

Dr. Asha Mathur
M.Sc., Ph.D., F.A.Z.
Principal, Navin Girls College, Gwalior

CBS PUBLISHERS & DISTRIBUTORS
4596/1-A, 11 Darya Ganj, New Delhi - 110 002 (India)

ISBN : 81-239-0786-9

First Edition : 2002

Published by :
Satish Kumar Jain for CBS Publishers & Distributors,
4596/1A, 11 Daryaganj, New Delhi - 110002 (India)

Printed at :
Asia Printograph, Shahdara, Delhi-32 (India)

Dedicated

to

Prof. S.Z. Qasim (Padma Bhushan & Padma Shri)
Ex-Member, Planning Commission of India

Preface

Before the turn of the century enormous expectations have been documented for the twenty first century. At the same time enormous environmental problems are expected to be added to the miserable environmental scenario of the world. In the present time environment has not remained a concern of only environmentologists but it is a concern of all. Without contribution of each and everybody condition of the environment will continue to be grave. It is high time that masses are made aware of the environmental issues. The present volume is a holistic approach in this direction with emphasis on matter of worldwide concern. This volume includes papers on agriculture, microbiology, radioactive waste, forestry, mining, pest management, toxicity, wild life and rural upgradation. There is a perennial problem for authors in science as to whether only to summarize the existing knowledge or to deal with the detailed evidence leading to conclusion. Second approach has been preferred in the present volume. The editors have been particular in maintaining conversational writing style so that the matter is conceivable by all including environmental planners and decision-makers at all levels. The editors are very much thankful to authors for contributing useful matter for the readers. They thank Prof. B.G. Kapoor, former Professor and Head, Department of Zoology, J.N. Vyas University, Jodhpur for his constructive criticism and guidance in the preparation of this volume.

The authors hope that the readers will not take this volume as a mere text on environmental issues, but will take cue from the text to dissipate the knowledge to those who can use it for the welfare of themselves and for all. There are bound to be errors in the text which the readers are requested to bring to the notice of the authors so that the future volume may be improved.

Editors

List of Contributors

Anita Pandey
G.B. Pant Institute of Himalayan Environment and Development, Kosi-Katarmal, Almora, UP - 263643

Arif Ali
Gene Expression Laboratory, Centre for Biosciences, Jamia Millia Islamia, New Delhi

Arnulf Seidel
Forschungszentrum Karlsruhe, Institut fur Toxikologie, Postfach 3640, D-76021, Karlsruhe, FRG

Asha Mathur
Navin Girls College, Gwalior

B. Padmanabhamurty
School of Environmental Sciences, Jawaharlal Nehru University, New Delhi - 110067

D.R. Bongirwar
Food Engineering Services Section, Food Technology and Enzyme Engineering Division, Bhabha Atomic Research Centre, BARC, Mumbai - 400085

E. Sharma
G.B. Pant Institute of Himalayan Environment and Development, P.O. Tadong, Sikkim - 737102

Francesco Zilio-Grandii
Faculty of Sciences, Department of Environmental Sciences, University of Venice, Calle Larga Santa Marta, 2137-30123, Venezia, Italy

Gerhard Polzer
Forschungszentrum Karlsruhe, Institut fur Toxikologie, Postfach 3640, D-76021, Karlsruhe, FRG

Girish Sharma
Gene Expression Laboratory, Centre for Biosciences, Jamia Millia Islamia, New Delhi

G.K. Vinci
Central Inland Capture Fisheries Research Institute, Barrackpore - 743101, West Bengal

Gopal Sharma
Environmental Biology Laboratory, Department of Zoology, Patna University, Patna - 800005

G.S. Roonwal
Centre of Georesources, University of Delhi South Campus, Benito Juarej Road, New Delhi - 110021

H.N. Mishra
Timber Engineering Department, Forest Research Institute (ICFRE), Dehradun

Imtiyaz Murtaza
Gene Expression Laboratory, Centre for Biosciences, Jamia Millia Islamia, New Delhi

Ines Lind
Forschungeszentrum Karlsruhe, Institut fur Toxikologie, Postfach 3640, D-76021, Karlsruhe, FRG

Ing L. Szpyrkowicz
Faculty of Sciences, Department of Environmental Sciences, University of Venice, Calle Large Santa Marta, 2137-30123, Venezia, Italy

Ishwar Prakash
INSA Senior Scientist, Desert Regional Station, Zoological Survey of India, Jodhpur - 342009

Jaleel Ahmed
Centralised Waste Management Facility,
Nuclear Waste Management Group, Bhabha Atomic Research Centre, Kalpakkam - 603102

K.B. Lal
Centralised Waste Management Facility,
Nuclear Waste Management Group, Bhabha Atomic Research Centre, Kalpakkam - 603102

K.L. Rai
Department of Applied Geology,
Indian School of Mines, Dhanbad

K. Prasad
Department of Geology, Patna University,
Patna - 800005

K.S. Bhatia
C.S.A. University of Agriculture & Technology,
Kanpur

K.S. Manja
Division of Microbiology, Defence Food Research Laboratory, Siddarth Nagar, Mysore - 570011

L.M.S. Palni
G.B. Pant Institute of Himalayan Environment and Development, Kosi - Katarmal, Almora, UP - 263643

Manika Mishra
DAAD Research Fellow, University of Gottingen,
Germany

M. Wiener-Schmuck
Forschungszentrum Karlsruhe, Institut fur Toxikologie, Postfach 3640, D-76021, Karlsruhe, FRG

Parvin Farshchi
Central Institute of Fisheries Education,
Versova, Mumbai

P.L. Kankane
Desert Regional Station, Zoological Survey of India, 107, Kamla Nehru Nagar, Jodhpur - 342009

P. Sasidhar
Centralised Waste Management Facility,
Nuclear Waste Management Group, Bhabha Atomic Research Centre, Kalpakkam - 603102

Rita Sharma
G.B. Pant Institute of Himalayan Environment and Development, P.O. Tadong, Sikkim - 737102

R.K. Sinha
Environmental Biology Laboratory, Department of Zoology, Patna University, Patna - 800005

R. Mathur
School of Studies in Zoology, Jiwaji University,
Gwalior - 474011

R.N. Sharma
National Chemical Laboratory, Pune - 411008

R. Pitchai
Visiting Professor, Department of Civil Engineering, IIT, Madras - 600036

S.A.H. Abidi
Central Institute of Fisheries Education,
Versova, Bombay

Sanjay Sharma
School of Studies in Zoology, Jiwaji University,
Gwalior - 474011

Subhash Mishra
Gene Expression Laboratory, Centre for Biosciences, Jamia Millia Islamia, New Delhi

Utpal Mishra
Forest Research Institute, Dehradun

V.V. Sugunan
Central Inland Capture Fisheries Research Institute, Barrackpore - 743101, West Bengal

Z.A. Ansari
National Institute of Oceanography,
Dona, Paula, Goa

Contents

Preface vii
List of Contributors ix

1. Biological maintenance of soil fertility in the Himalayas 1
   E. Sharma et al.
2. *In vitro* reactions of alveolar macrophages and bronchial epithelial cells to NO_2 exposure 21
   Arnulf Seidel et al.
3. Performance assessment and its role in radioactive waste disposal—A perspective 35
   P. Sasidhar et al.
4. Utilization of solar energy in agriculture sector 51
   D.R. Bongirwar
5. Water and airborne infections and their control 67
   K.S. Manja
6. Role of agroforestry in conservation of degraded environment 77
   K.S. Bhatia
7. Mining and mineral processing wastes in Indian base metal mines and their environmental management 87
   K.L. Rai
8. Survey of acid rains in the Venice region (Italy) 99
   Francesco Zilio Grandi & Ing. L. Szpyrkowicz
9. Conservation of the desert environment 135
   Ishwar Prakash
10. The Ganges river Dolphin 143
    R.K. Sinha et al.
11. Environmental implications of new pest management concepts : Trends and policies in the new century 153
    R.N. Sharma

12. Sustainable development : Concept and achievement in India 159
G.S. Roonwal

13. Impact of urbanization on climate and air quality—A case study of Delhi 163
B. Padmanabhamurty

14. Problem of oil pollution in the Indian Ocean—An overview 169
Parvin Farschi

15. Molecular mechanisms in mercury detoxification in bacteria 179
Arif Ali et al.

16. Environmental degradation in Indian rivers : A biological perspective 191
V.V. Sugunan & G.K. Vinci

17. Survival of larger mammals in the Thar desert 205
P.L. Kankane

18. Engineered timber housing—An infrastructure need for socio-environmental upgradation of rural India 211
Utpal Mishra et al.

19. Quality assessment of groundwater sources adjoining sewage channel in Gwalior 217
Sanjay Sharma, R. Mathur & Asha Mathur

Chapter 1

Biological Maintenance of Soil Fertility in the Himalaya

E. Sharma*, Amita Pandey, Rita Sharma* and L.M.S. Palni****

INTRODUCTION

Development policies and associative activities in tropical, subtropical and mountainous zones have gradually affected the agriculture production. The increase in agriculture production has largely been due to the use of huge quantities of pesticides and chemical fertilizers, in addition to improved irrigation and introduction of high yielding varieties of crop plants. This policy was highly successful in some privileged sectors but it did not benefit, in a substantial way, the "less favoured" areas such as Himalaya where chemical fertilizers and/or irrigation are not available, and the soils are fragile and deficient in major nutrients such as nitrogen and phosphorus. This policy which resulted in the Green Revolution in some parts of the world including India is not appropriate/adequate and does not even serve to address the problem of maintenance of soil fertility in fragile systems (Subba Rao and Rodriguez-Barrueco 1993). According to Rama Krishnan *et al.* (1993), "soil fertility refers to the syndrome of physical, chemical and biological factors and processes influencing the potential of soil to support crop growth desired by human beings. Nutritional and climatic requirements of crop become as important as the capacity of soil to supply nutrients in determining the production efficiency of agroecosystems. The capacity of soil to supply nutrients and water depends upon the interaction of a whole range of biotic and abiotic factors and processes operating in the soil system. Soil fertility management, in a broad sense, embraces actions or interventions for meeting the ever increasing crop-based demands of fast multiplying human population".

Traditionally, upland farming systems in the Himalaya have developed from the integration of various land-uses such as forest, agriculture, horticulture, agroforestry and animal husbandry. All these components in different ways and to varying degrees, play vital role in maintaining soil organic matter for sustaining soil fertility of the upland farming systems (Sharma, 1995). Most of the farming systems in the Hindu-Kush Himalayan region are at the subsistence level (Misra and Ramakrishnan, 1982; Ralhan *et al.*, 1991; Sharma *et al.*, 1992; and Singh *et al.*, 1984) and have evolved over the years from the experiences and keen observations of farmers to meet the demands for food, fodder, fuelwood, timber and other bio-resources (Sundriyal *et al.*, 1994). In the process, excellent and traditional ways for the maintenance of soil fertility have evolved (Rai *et al.*, 1994). Some of the traditional, biological methods practised (Fig. 1.1) in the upland farming systems for soil conservation and fertility maintenance are :

(a) strong bias towards agroforestry,

(b) use of nurse/shade trees in cash crop-based agroforestry systems (i.e., use of N_2-fixing *Alnus* tree with large cardamom),

(c) plantation of N_2-fixing species and bamboos along contour lines on steep unterraced slopes as soil binders,

(d) terracing and cultivation along edges or bunds (such as *Grewia oppositifolia, Thysanolaena, Arundodonax,* etc.),

(e) mixed cropping,

(f) crop rotation using pulses (N_2-fixers),

(g) supplementation of traditional farmyard manure by collection and application of forest-floor-litter for the cultivation of nutrient exhaustive cash-crops such as ginger (*Zingiber officinale*),

(h) shifting cultivation with longer cycles, etc.

Growth in human population has been exerting pressure, gradually but consistently, leading to the disruption of the above mentioned soil fertility related traditional. practices; this has slowly caused :

(a) the conversion of forests into agriculture land leading to reduction in the availability of soil fertility related resources,

(b) use of steep uplands for crop cultivation, such slopes were earlier under tree cover,

(c) family fragmentation and decrease in land-holding size, and

(d) limitation in the area of operation and sustainable utilization of natural resources.

These have caused perturbations on the intricate linkages that were operative among the different components of upland farming systems. Some of the indicators of such perturbations are :

(a) shortening of shifting cultivation cycle causing immense pressure on resources and soil fertility,

(b) change from stall-feeding to open grazing of livestock.

(c) reduction in the number of households rearing livestock as a result of limitation in fodder resources that directly affect soil fertility as a consequence of unavailability of farmyard manure,

(d) intensive mono-crop cultivation with no crop rotation,

(e) changeover to cash-crops for meeting basic needs from outside through higher cash return, and

(f) increased soil erosion, overland flow and nutrient loss due to intensive cultivation.

Most of the above mentioned changes are difficult to revert to the original sustainable practices for biological maintenance of soil fertility. Therefore, over the years research efforts have focussed also on "biotechnological" inputs for the improvement of soil fertility. These include improvement of

Fig. 1.1. Biological soil fertility enhancement related activities in the Himalaya: **(A)** large cardamom agroforestry with N_2-fixing *Alnus* as a tree associate in a temperate location, **(B)** tea cultivation with N_2-fixing *Albizzia* as a tree associate in a subtropical location, **(C)** forest-floor-litter application in the cultivation of nutrient exhaustive cash-crop *Zingiber officinale*, **(D)** use of forest-floor/litter as livestock bedding along with cowbarn wastes for producing farmyard manure in traditional composting, **(E)** composting using various readily available farm resources, and **(F)** weed (*Eupatorium* sp.) compost ready for use after 32 days of incubation.

agroforestry system, organic composting, biological nitrogen fixation, phosphorus solubilization and beneficial plant-microbe interactions. Some of these aspects relating to the biological maintenance of soil fertility in the context of the Himalayan region are presented in this article. An excellent account of soil fertility status under the Jhum System along with the socio-economic analysis of jhum is presented by Ramakrishnan *et al.* (1993).

AGROFORESTRY SYSTEMS

Agroforestry is practised for soil conservation and this, in its wider sense, includes both the control of erosion and maintenance of soil fertility. The relationship between agroforestry and soil conservation varies with the climate, soil type and landform. In the HImalaya, three types of agroforestry systems are common (a) spatial-mixed practice based, e.g., fuel-fodder trees in the cropland, (b) spatial-mixed practice based, e.g., different horticulture trees in the cropland, and (c) plantation crop combinations, e.g., combinations of tea with *Albizzia* and large cardamom with *Alnus* (Fig. 1.1). There are clear, beneficial effects on soils of some systems of trees on cropland and plantation crop combinations (Young, 1989). There are very few studies on agroforestry systems of the Himalaya that have dealt in greater details on the productivity, the magnitude of nutrient flows and the degree of recycling. Recently, Sharma *et al.* (1994 and 1995) have reported on crop-yield, biomass production and nutrient cycling of two types of a agroforestry system *viz.* (a) spatial-mixed practice where mandarin (*Citrus reticulata*) is grown along with annual crops and N_2-fixing tree associate *Albizzia*, and (b) plantation crop combination where perennial large cardamom (*Amomum subulatum* which is used as spice/condiment) is grown under natural mixed trees or N_2-fixing *Alnus*.

In the large cardamom based agroforestry, productivity was more than double and agronomic yield increased by 2.2 times under the influence of N_2-fixing *Alnus* as compared to natural mixed tree species. Similarly, in the mandarin based agroforestry the mandarin fruit yield increased by 1.2 times under the influence of N_2-fixing *Albizzia* (Table 1.1). Greater biomass accumulation, net primary productivity, yield and higher litter production under the influence of N_2-fixing species in these agroforestry systems are the direct expressions of better performance as a consequence of higher soil fertility.

Soil was acidic in the cardamom agroforestry and slightly acidic to neutral in the mandarin agroforestry system. The C/N ratio was higher in cardamom agroforestry indicating lower N-availability than in the mandarin agroforestry. Cardamom stand under the influence of *Alnus* showed relatively lower C/N ratio characterizing higher nitrogen availability than in the natural forest cardamom stand. The study revealed that N_2-fixing species help in maintenance of soil organic levels with higher mineralization rate as land-use change from natural forest tree mixture to sparsely tree based agroforestry with traditional crops (Sharma, 1995).

The nitrogen and phosphorus release from the soil on per unit area basis was much higher in the *Alnus*-cardamom combination, where higher contribution was recorded from all the litter fractions, especially from the leaf litter of *Alnus*, compared to non-N_2-fixing stand. The nitrogen release from the *Albizzia*-mandarin stand was also higher than in the mandarin stand showing a positive influence of N_2-fixing *Albizzia*. However, this influence was much less compared to the *Alnus*-cardamom system. This is explained by the lower density of *Albizzia* in the mandarin stand. Both the N_2-fixing species, *Alnus* in the cardamom and *Albizzia* in the mandarin system, accelerate nitrogen and phosphorus cycling through above ground litter and residue production and affect greater nutrients release (Sharma, 1995).

Annual uptake and return of nitrogen to the soil in the *Alnus*-cardamom and *Albizzia*-mandarin combinations were higher than when cardamom and mandarin were grown with mixed, non-N_2-fixing trees (Table 1.1). The rates of phosphorus uptake and return through litterfall and decomposition were also high in *Alnus*-cardamom and *Albizzia*-mandarin stands compared to their counterpart control stands, which has probably resulted from an increase in the rate of phosphorus supply attributable to geochemical and biological factors influenced by *Alnus* and *Albizzia*. The mandarin-based agroforestry system is a highly nutrient-exhaustive system as compared to large cardamom-based agroforestry evaluated on the annual nutrient exit from the system through the removal of agronomic yield. Pooled data for the agroforestry system show that mandarin-based agroforestry is 15 times more exhaustive for nitrogen and 11 times for phosphorus than large cardamom-based agroforestry (Table 1.1).

Table 1.1. Stand tree density, productivity and nutrient dynamics in cardamom and mandarin-based agroforestry systems in Sikkim Himalaya

	Agroforestry systems			
	Alnus-cardamom	*Forest-cardamom*	*Albizzia-mandarin*	*Mandarin*
Stand tree density (trees/ha)				
N_2-fixer	517	–	56	–
Mixed-tree species	–	850	144	181
Total	517	850	200	181
Stand biomass (kg/ha)	28422	22237	13879	11137
Productivity (kg/ha/year)	10843	7501	8119	7358
Agronomic yield (kg/ha/year)*	454	205	2356	2194
Nitrogen accretion through atmospheric fixation (kg/ha/year)	65.34	–	3.06	–
Nutrient dynamics				
Soil status (kg/ha upto 30 cm depth)				
Total-N	5880	7590	2880	2860
Total-P	1278	1149	1717	1504
Total uptake (kg/ha/year)				
N	143.83	80.56	130.54	120.83
P	13.18	6.52	16.31	15.40
Total release to soil through decomposition (kg/ha/year)				
N	83.67	29.23	39.51	36.48
P	6.15	2.35	5.19	5.02
Nutrient exit from agronomic yield (kg/ha/year)				
N	4.04	1.78	44.23	42.48
P	0.70	0.33	5.82	5.57

* Cardamom is high-valued cash crop which fetched US $2 per kg in 1993. The agronomic yield in *Albizzia*-mandarin and mandarin stands account for both mandarin-fruit yield and crop yield.

Overland flow (percentage of rainfall) was highest in open cropped fields (9.55%) followed by mandarin agroforestry (4.76%), fallow land (3.77%) and lowest in cardamom agroforestry (2.17%). Similarly, the soil erosion rate (kg/ha/year) was least in cardamom agroforestry (30) followed by fallow land (43), mandarin agrofo\iestry (145) and higher in open-cropped fields (477). Overland flow and soil loss showed positive relationship. This illustrates the value of agroforestry systems in efficient soil conservation and for enhanced soil fertility maintenance compared to cropped area with no trees as observed in an upland watershed of Sikkim (Rai and Sharma, 1995).

ORGANIC MANURING

Soil organic matter and nutrients are important "regulators" of stand productivity. Key processes regulating the sustainability of ecosystems include soil surface hydrology, soil organic matter dynamics and the synchrony between soil processes and plant nutrient demands (Brown *et al.*, 1994). Most natural and undisturbed ecosystems have perfect synchrony in nutrient release and uptake by plants. Most of the land management systems suffer from disruption of such synchrony. This is apparent in the sequence of transformation from forest-cardamom (comprising of mixed natural trees) to *Alnus*-cardamom (having planted *Alnus* trees) and to mandarin-based agroforestry systems that have resulted into sequential reduction in soil organic carbon levels (Fig. 1.2). Mineralization of decomposing residues is a major source of plant nutrients in highly weathered soils with little inherent mineral fertility (Sanchez *et al.*, 1989). The activities of microorganisms and soil fauna serve to promote soil aggregation (Oades, 1984), leading to reduced erosion (Lal, 1986) and greater moisture infiltration (Lavelle, 1988). These can be achieved by increasing the soil organic matter, e.g., by the application of compost or green manure.

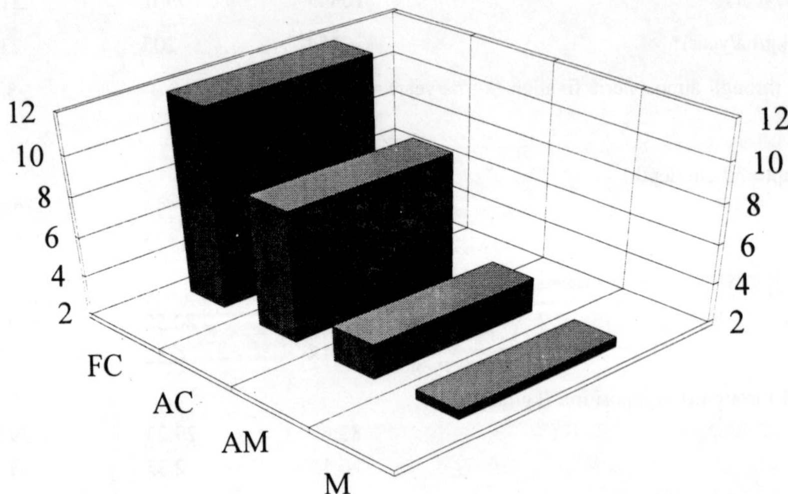

Fig. 1.2. Soil carbon storage (kg/m² upto 30 cm depth) in agroforestry systems (FC = forest-cardamom, AC = *Alnus*-cardamom; AM = *Albizzia*-mandarin; M = mandarin) in the sequence of land-use transformation from natural forest tree mixture to sparsely tree-based agroforestry with traditional crops (after Sharma, 1995).

Some findings on the use of green manure and/or compost are elaborated and discussed below. Synchronization of nutrient release and plant uptake should be established especially on fragile upland farming systems to minimize loss. Both, green manure and compost, have potential for establishing synchrony of nutrient release with crop growth demands. Synchrony applies to both nitrogen and phosphorus. In the upland farming systems of the Himalaya, phosphorus synchrony is very important as it is readily leached out due to high rainfall, and available phosphorus is also converted to unavailable form by secondary fixation.

Green Manures and Mulches

Green manure in terms of *Azolla* and cyanobacteria application in the paddy cultivation has been quite common in the plains of tropical India and less common in the foothills and much less or non-existent in the mid and higher Himalaya. The common practice in the Himalayan upland farming systems has been the cultivation of legume crops such as pulses; the crop residue is applied back to the fields. Cultivation of legumes (*Glysine* spp. and *Phaseolus* spp.) even on the terrace bunds along with paddy is seen in the uplands of Sikkim. The use of leaves of tree legumes such as *Albizzia* and actinorrhizal N_2-fixer, such as *Alnus*, for increasing the organic carbon levels in highly weathered soils and nutrient-exhaustive crop cultivation is practised. Green manures are capable of supplying nitrogen to crops and it is possible that the release pattern may have an inherent synchrony. Short-term green manures of mungbean and cowpea have been effective in supplying nitrogen to paddy in tropical Asia (Morris *et al.*, a, b). Such green manures were found to be highly effective (as effective as equivalent quantities of nitrogen applied as fertilizer) with utilization ranging between 33 and 41%. The response of rice to green manure was directly related to the quantity of nitrogen available through green manure (Morris *et al.*, 1989). It has also been shown that much of the nitrogen fertilizer not utilized by the rice crop is lost to the atmosphere (Fillery *et al.*, 1984) and that gaseous loss from incorporated green manure is relatively small (Nagarajah, 1988). The relatively high efficiency of nitrogen utilization achieved in the above studies with green manures, and the expectation that much of the unutilized nitrogen would remain in the soil organic matter pool, indicates that synchrony was achieved. The rice growing valleys and foothills of the Himalaya have high potential for the use of green manures such as *Azolla* and cyanobacteria, however detailed trials are required. For the rainfed farming systems, use of N_2-fixing legumes and actinorrhizals can be indispensable. Since in most cases these "green manures" are ploughed back into the field and the mulching effect achieved increases soil porosity, organic matter and positively improves infiltration and reduces water loss through evaporation, run off, etc. In most cases weed incidence is also reduced.

Composting

Composting is, in broad terms, the biological reduction of organic wastes to humus and soil-building substances (Martin and Gershung, 1992). In nature, all organic wastes, dead plants and animals are decomposed/reduced to humus slowly. In composting, the process is accelerated and the reduced organic manure is produced relatively fast without losing much of the nutrient elements. These composts provide nutrients and enrich soil organic levels that also modifies the soil physical structure. High organic levels in soil protect it from erosion because of increased infiltration and reduced surface runoff. It also helps in moisture conservation by increased absorption and adsorption. In the upland farming systems composting can supplement considerably the traditional farmyard manure.

In the tropics, farmyard manure (cattle and buffalo), piggery effluent and poultry manure are used for soil amendment. These materials are collected and stored before being applied to the soil. As a result of storage, changes take place which may lead to a loss of nitrogen and changes in the quality of other resources. During 3-6 months storage in cool temperate conditions, 20% or more of the nitrogen may be lost from manure and the content of humic substances may increase (Kirchmann, 1985). Such losses can be reduced by the addition of energy-rich materials, such as straw, to the manure before storage. The first crop grown on soil after amendment with manure generally utilizes 15-35% of the manure nitrogen (Kirchmann, 1985). Lesser amounts are recovered in subsequent crops. Some of the variation between different manures is attributable to identifiable quality factors such as C/N ratio and content of available mineral nitrogen. Other nutrients such as phosphorus are readily available in animal manures (Maraikar and Amarasira, 1989). Therefore, more efficient use of the animal wastes can be made by composting with energy rich materials.

Composting is largely a microbiological process based upôn the activities of bacteria, actinomycetes, and fungi. Basically, any system or design that ensures efficient decomposition of organic matter can constitute a composting method (Bhardwaj, 1995). Conventionally, two methods of composting are known in India:

1. **Indore method :** This method was worked out by Sir Albert Howard, a British agronomist based at Indore (Madhya Pradesh) during 1924-30. It was based on the traditional composting practices being followed in China and India (Howard, 1933). Since this is an aerobic process, a pit (3' deep and 6'-8' wide) is dug near a cattle shed on a site free from water logging. Waste organic materials are laid in the pit in alternate layers with animal manure and soil. Accordingly, 15 cm thick and 1.5-12.0 m layer of organic materials to be composed is placed on a hard upland place. This is followed by a 4-5 cm layer of animal manure, which is in turn covered by a sprinkling of surface soil mixed with a small quantity of lime or wood ashes, about 30 mm in thickness. The layers are repeated until the heap reaches a height of 1.5-2.0 m. The final layer should be of the compostable material covered by a thin layer (about 60 mm) of soil. To provide optimum moisture (60-70%), water is sprinkled over each layer. The heap is turned thoroughly at intervals of 3, 6 and 12 weeks. A good quality compost gets ready in about three months.
2. **Bangalore method :** This method is initially aerobic, followed by anaerobic decomposition later. This method was worked out, firstly to overcome some of the disadvantages of the Indore method, and secondly to process night soil and city refuse. The compostable refuse is dumped into the trench or pit and spread out with rakes forked shorels to make a layer of abôut 1.5 cm thickness. Night soil or dung is then placed over the refuse in a layer of about 5 cm. The process is repeated until the trench on pit is filled up to about 30 cm above the ground level and a final layer of compostable material is placed on the top. At each layering, water is sprinkled over the material to make it optimally moist. The above ground material is made into a dome shape and covered with about 2.5 cm mud plaster. The compost gets ready in about five to six months.

This Institute has tried weed and other organic waste composting in collaboration with St. Alphonsus Social and Agriculture Centre (Kurseong). Twenty-six types of compost compositions were experimented and the nutrient qualities of these composts, on maturity, were examined. Pure *Eupatorium* weed composting required 32 days for maturity and the other twenty-five combinations by using raw ingredients for composting matured between 18 to 33 days. Traditional cowbarn composting takes more than 200 days for maturity. These twenty-six combinations of composts were prepared by using raw materials in different proportions and these resources (such as *Eupatorium* weed, poultry wastes, soil, soaked paper, cowbarn wastes, vegetable wastes, egg shells, kitchen wastes, other plants, *Cryptomeria* needles and old compost) were available in the villages/farms (Fig. 1.1). Organic carbon ranged from 3.02 to 7.23% in above composts and 1.33% was recorded in the control soil, total nitrogen ranged from 0.64 to 1.49% as against 0.20% in the control soil, available phosphorus varied from 0.018 to 0.155% as against 0.002% in the control soil; exchangeable potassium ranged from 0.564 to 0.946% as against 0.191% in the control soil; and the pH ranged from 7.03 to 8.46 as against 6.31 in the control soil. Ratio of C/N is a good indicator of net nitrogen mineralization while N/P for net phosphorus mineralization in soils and composts. In the control soil and different composts, net nitrogen mineralization did not seem to be a constraint but phosphorus mineralization was limiting in the control soils and the same in composts depended solely on plant wastes (Fig. 1.3). The N/P ratio is known to be a better indicator, with phosphorus mineralization at or below N/P ratios of 10. The N/P ratio was below 10 in the experiments, where proportion of poultry waste and egg shells contributed substantially to higher phosphorus values and lower N/P ratios (Fig. 1.3). The composts with lower N/P ratios are expected to provide both nitrogen and phosphorus quite readily for plant uptake. Soil phosphorus in the Himalayan region is secondarily fixed as a result of acidity but as these composts have higher pH the available phosphorus increases significantly thus avoiding phosphorus limitation. Total nitrogen and exchangeable potassium contents also increased considerably in the composts. These composts can be highly beneficial for soil fertility maintenance in the upland farming systems.

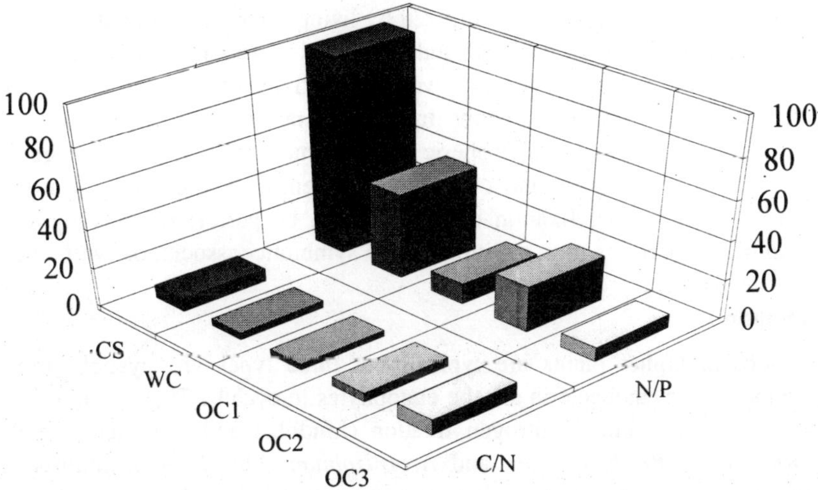

Fig. 1.3. Ratio of C/N and N/P in control soil (CS), weed compost (WC), OC1 (compost with ingredients such as soil, paper, cowbarn wastes, poultry wastes, vegetable wastes, *Eupatorium* and *Cryptomeria*), OC2 (compost with *Eupatorium*, poultry wastes, mixed plants, cowbarn wastes, kitchen wastes), and OC3 (compost with *Eupatorium*, poultry wastes, soil, paper, mixed plants, vegetable wastes and egg shells).

Two sets of pit experiments for demonstrating biocomposting method were also carried out in Haigad watershed (Central Himalaya). The ingredients for the first experiment were (i) chopped farmyard weeds, (ii) cattle dung, (iii) soaked waste paper, and (iv) sieved soil, in 1 : 1 : 1 : 1 ratio. In the second experiment, local weeds were replaced by dried chir-pine needles. Composting was carried out in pits (50 × 50 × 50 cm), near a cattleshed on a site free from waterlogging. The mixture was tightly packed in the pits and the heap was raised upto 25 cm. A polythene cover using wooden framework was made over these pits to provide protection against rain water during decomposition. Sufficient moisture was provided by sprinkling water, and turning the heap at regular intervals. Decomposition of pine needles took longer period than that of weeds. Good quality compost was ready within 40 to 50 days in weed composting and 100 to 110 days in the case of chir-pine composting. Biocomposting demonstrations on larger scale have also been set up to provide compost for protected vegetable cultivation at Haigad and Hawil villages.

Use of earthworms for improving soil fertility is a common practice in North-East India. Vermicomposting is the bioconversion of organic waste materials through the intervention of earthworms, and is increasingly becoming popular in both rural and urban areas. It is a well established fact that the soil fauna, including earthworms, can be manipulated to improve the physical properties of soil and regulate decompositional processes.

BIOLOGICAL NITROGEN FIXATION

Tropospheric atmosphere comprises of metabolically important gases such as oxygen (21%), carbon dioxide (0.03%), nitrogen (78%) and others. Molecular nitrogen is the major constituent of atmosphere but is inert chemically and cannot be used by most forms of life. The great majority of living organisms obtain their nitrogen in some combined forms like nitrate, ammonia or more complex compounds like amino acids. The atmospheric nitrogen can be utilized by living organisms only after fixation (reduction). The atmospheric nitrogen is fixed by physical and biological processes apart from industrial fixation. Ammonium ions arise from industrial burning, volcanic activity and forest fires, while nitrate ions arise from oxidation of N_2 by oxygen or ozone in the presence of lightening or ultraviolet radiation. This small amount of physically fixed atmospheric nitrogen moves to the soil through rain and is then absorbed by plant roots. Nitrogen is the most important nutrient next to carbon and it primarily follows the gaseous cycling. The biological fixation of nitrogen can be performed only by selected group of micro-organisms, either in free living condition or in symbiotic association with higher plants.

Symbiotic association

Symbiotic associations in higher plants are typically of three types *viz.* cyanobacteria, rhizobia and *Frankia* based. Filamentous cyanobacteria are the endophytes in cycads. These cyanobacteria are present in coralloid roots and are capable of nitrogen fixation (Lindblad and Bergman, 1990). Three genera of rhizobia are *Rhizobium, Bradyhizobium* and *Azorhizobium. Rhizobium* symbiosis is characteristic of the family Leguminaceae in which case the plants bear nodules in their roots. In contrast, *Frankia* symbiosis has been recorded from different genera of many families which bear actinorrhizal root nodules. In all the three types of symbioses, nitrogenase enzyme is responsible for reduction of nitrogen to ammonia for assimilation. Nitrogen fixed by various symbiotic plants is recycled in ecosystem in the form of root exudates, plant surface leachates, and mostly through litter production, decomposition and mineralization. This is then made available to other non-nitrogen-fixers for uptake. Some of the fixed nitrogen is returned back to the atmosphere by denitrification in soil by anaerobic bacteria.

Sharma (1995) made extensive inventory investigation for sorting out both leguminous and non-leguminous (actinorrhizal) symbiotic associations in the eastern Himalaya. She reported about 80 species belonging to the family Leguminaceae, including 65 species of sub-family Papilionoideae, 10 species of sub-family Minosoideae and 5 species of sub-family Caesalpinioideae. The altitudinal distribution of these leguminous plants varied from the foothills to the alpine zones. The habits of these species are climbers (10 species), herbs (26 species), shrubs (21 species), lianas (26 species) and trees (17 species). *Albizzia* spp. is the predominant tree legume in the subtropical zone of Himalaya (Sharma, 1995). Genus *Albizzia,* belonging to subfamily Mimosoideae, is most widely distributed in tropical and subtropical belts of the world. In India, *Albizzia* is distributed in tropical and subtropical climate all over the country and the genus has taken an important place in agroforestry systems (Kumar and Toky, 1994). In Sikkim, five species *viz. A. lebbeck, A. lucida, A. odoratissima, A. procera* and *A. stipulata* are found in cultivated as well as in natural areas. Nitrogen fixation rates based on the acetylene reduction technique in the root nodules of *A. stipulata* from *Albizzia*-mandarin agroforestry showed 43 μmol N/g nodule dry weight/day in the growing season. Root nodule biomass was estimated as 206 g/tree and annual tree N_2-fixation with 55 g N/tree/year by *A. stipulata* (Sharma, 1995).

Many angiospermic plants other than legumes also possess N_2-fixing root nodules (Akkermans & Van Dijk, 1981; Becking, 1977). The actinomycetous endophyte of such nodules belong to the genus *Frankia* (Backing, 1970). These plants are called actinorrhizals. *Frankia* symbioses have been recorded in more than 200 plant species belonging to 24 genera within the angiospermeae (Dixon and Wheeler, 1986). *Alnus* is the most prominent actinorrhizal genus in the Himalaya, other actinorrhizal genera are *Elaeagnus, Myrica, Hippophae* and *Coriaria.* Two species of alders are found in the region *viz., Alnus nepalensis* in the eastern Himalaya and *Alnus nitida* in the north-western Himalaya, while both species grow in the central and western Himalaya. *A. nepalensis* grows between 1000 m and 2500 m elevation belt most predominantly in the eastern Himalaya. It is pioneer species on freshly exposed landslide soils. It grows on sandy eroded soils, denuded habits, rocky slopes, landslide affected slopes, steep stream sides and in natural areas. It has been a common species in natural forests and in recent times has also become an important species of plantation forestry in Sikkim Himalaya. It has been considered as a useful species for a social forestry and agroforestry. It is also used as a shade tree in large cardamom (*Amomum subulatum*) based traditional agroforestry systems in the region.

Frankia is the partner responsible for nitrogen fixation in this symbiosis and vesicles in nodule cortical cells are the sites of nitrogenase activity. Pradhan (1993) isolated *Frankia* strains from *A. nepalensis* and these strains showed cushion-like branched septate hyphae ranging from 0.5-1.2 μm in diameter and these produced vesicles.

Sharma (1988) estimated the nitrogenase activity by acetylene reduction technique assuming that the rates of acetylene reduction reflect the enzyme activity. Fig. 1.4 shows the change of this activity with altitude between 1000 m in naturally regenerated *A. nepalensis* seedlings of landslide-affected sites. Measurements made in peak growing season showed that the activity ranged from 5-19 μmol C_2H_4/g nodule dry weight/h. Although there was considerable variation in the values from any one site, it is clear from Fig. 1.4, that activity is highest in the mid zone where the natural regeneration of *A. nepalensis* is most vigorous. The rate of acetylene reduction was significantly related to both the mean soil temperature and the root nodule moisture. *A. nepalensis* at the 1830 m elevation site in the rainy season (July) was most active in terms of nitrogenase activity and thus this site was selected for study of diurnal variation (Sharma, 1988). The experiment was conducted between 5.00-20.00 hours in July 1983 and a marked diurnal variation in the rate of acetylene reduction was recorded. The ecological amplitude of *A. nepalensis* is narrow (1000-2500 m) altitudinally but fairly wide in relation to environmental conditions as there is a large change in environmental conditions with increasing altitude. Nitrogenase activity is high between 1500 m and 2000 m altitude and outside this range *A. nepalensis* performance goes down.

Sharma and Ambasht (1984) made a detailed study on nitrogenase activity which ranged between 3-25 µmol C_2H_4/g nodule dry weight/h in an age series of *A. nepalensis* plantation stands (7-, 17-, 30-, 46- and 56-year-old). They reported that the seasonal variation in nitrogenase activity was quite distinct with the highest activity in growing season and lowest in winter, and the analysis of variance showed both season and nodule age to be highly significant in plantations of different ages. There was a significant interaction between age-class of nodules and age of plantations (Sharma and Ambasht, 1984), but nodule biomass in a hectare area was much greater in the younger plantations (Sharma and Ambasht, 1986). Nitrogenase activity was highest in the young age-class, fairly high in the medium and very low in the old nodules with similar trend of seasonal fluctuations in all the stands of an age series of plantations (Sharma and Ambasht, 1984). Therefore, the difference in the amounts of active nodule biomass and age of the nodules actually caused the difference in total nitrogen fixation in a stand. Nodule nitrogen content when treated as independent variable against nitrogenase

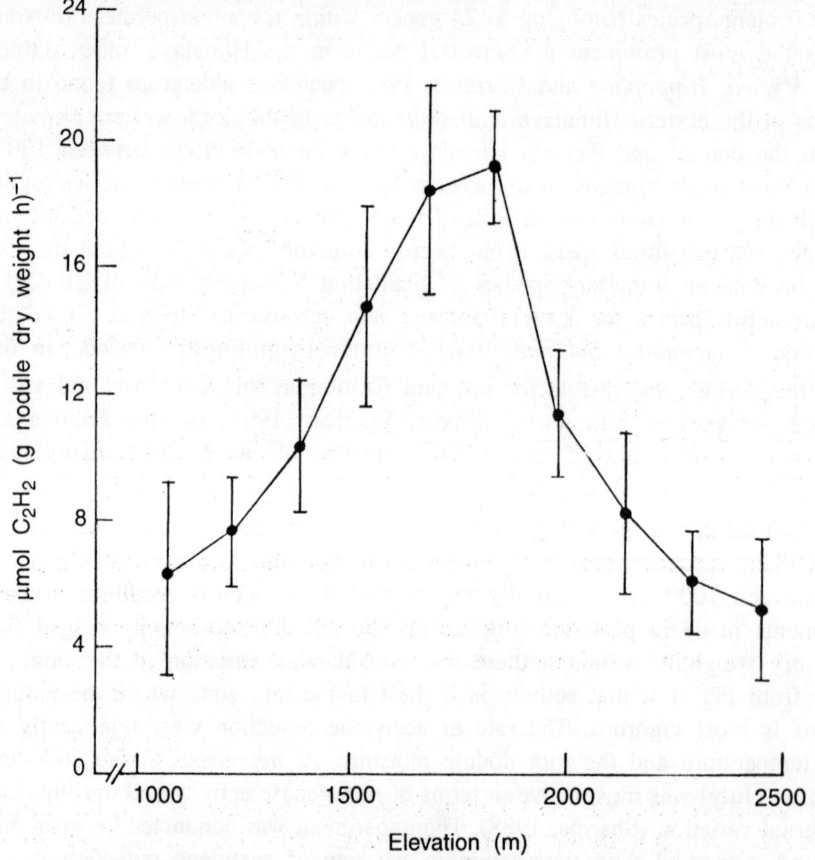

Fig. 1.4. Altitudinal variation in nitrogenase activity of *A. nepalensis* seeding root nodules at ten sites at different elevations in July 1982. These seedlings were naturally regenerated on landslide-affected slopes. Values are means ± SE; *n* = 5 (after Sharma, 1988).

activity as dependent variable gave a highly significant negative correlation in all the months (Sharma and Ambasht, 1984), consistent with Skeffington's hypothesis (1975), which predicts higher nodule activity coupled with lower nodule nitrogen. Annual accretion of nitrogen through biological fixation was highest (117 kg/ha/year) in the 7-year stand of *A. nepalensis* and it decreased with stand age to a lowest (29 kg/ha/year) in the 7-year stand of *A. nepalensis* and it decreased with stand age to a lowest (29 kg/ha/year) in the 56-year stand (Sharma and Ambasht, 1988).

Sharma (1995) estimated nitrogenase activity in *A. nepalensis* (shade tree) in a large cardamom-based agroforestry system in Sikkim. She reported an activity rate of 15 μmol C_2H_4/g nodule dry weight/h in the growing season. *A. nepalensis* not only served as shade trees but fixed atmospheric nitrogen and helped in biological maintenance of soil fertility in such cash-crop based agroforestry systems. She estimated 65 kg/ha/year nitrogen accretion by *A. nepalensis* in a large cardamom-based agroforestry system.

Pradhan (1993) estimated *in vitro* nitrogenase activity in *Frankia* (isolated from *A. nepalensis* host) grown in Defined Proportionate Minimal Medium (DPM) supplemented either with sodium propionate (carbon source) or ammonium chloride (nitrogen source). The activity was recorded with both the supplements showing the higher activity right from the beginning in propionate while in ammonium chloride it is increased with the lapse in time. In propionate supplemented medium, the activity in different *Frankia* strains ranged from 17 to 28 nmol C_2H_4/5 ml suspension/h, while in ammonium chloride, the value ranged from 2 to 21 nmol C_2H_4/5 ml suspension/h.

Associative organisms

Most plants take up their nitrogen from soil, therefore, availability of fixed-nitrogen in the soil is a major determinant of soil fertility and thereby crop production. Some classes of bacteria and cyanobacteria are capable of fixing nitrogen from the atmosphere and therefore can be used as "sources" of fixed nitrogen. A list of bacteria and cyanobacteria capable of fixing atmospheric nitrogen is given in Table 1.2. Amongst free-living nitrogen-fixing bacteria, species of *Azotobacter* and *Azospirillum* have received much attention due to their potential for improving crop productivity (Mishustin and Shilinikova, 1969; Okon, 1985; Okon and Labandera-Gonzalez, 1994; Pandey and Kumar, 1989). Based on the data spanning about 10 years, it was concluded that statistically significant yield increases were reported in up to 60% of the field inoculation trials in India with *Azotobacter chroococcum* showed 115% and that with *Azospirillum brasilense* showed 23% increase in grain yield at a subtropical upland farming system in Sikkim (Pandey *et al.*, 1996).

Family Azotobacteraceae represents four genera, namely *Azotobacter, Azomonas, Beijerinckia* and *Derxia*. *Azotobacter* comprises of the species *A. chroococcum, A. beijerinkii, A. vinelandii* and *A. paspali*. Out of these four species, only *A. chroococcum* has been extensively utilized as biofertilizer. Beneficial effects in terms of higher productivity, following inoculation by *Azotobacter chroococcum* strains, have been demonstrated in major groups of crop plants including cereals, millets, vegetables and fiber, starch and sugar producing commercial crops. Since it is known that the presence of fixed nitrogen inhibits nitrogen fixation in free living bacteria, the beneficial effects of *Azotobacter* inoculation to crops, in the presence of large amounts of inorganic nitrogenous fertilizers, must be due to reasons other than supply of biologically fixed nitrogen.

Four species of nitrogen-fixing azoperilla have been described : *Azospirillum lipoferum, A. brasilense, A. amazonese* and *A. halopraeferens*. Azospirilla, like azotobacters, can extensively colonize the rhizosphere of a diverse variety of plants. In addition they penetrate and develop endosymbiotic association with certain host plants. Thus the association between an *Azospirillum* strain, and the plant genotype sensitive to it, is very intimate. Therefore it is expected that *Azospirillum* should prove to be a better biofertilizer than *Azotobacter*. The beneficial effects obtained due to inoculation with these bacteria are not only in terms of atmospheric nitrogen fixation in the soil, but are also related to the ability of such bacteria to synthesize antibiotics and growth promoting substances, including phytohormones and siderophores, and with respect to phosphate solubilization (Brown and Walker, 1970; Harper and Lynch, 1979; and Pandey and Kumar, 1990).

Table 1.2. List of bacteria and blue green algae capable of fixing atmospheric nitrogen (Alexander, 1961; Roper & Lodha, 1995 and ref. therein)

Group	*Taxa*
Bacteria	
(a) Aerobia	*Azomonas, Azotobacter, Beijerinckia, Derxia*
(b) Facultative anaerobic	*Bacillus, Enterobacter, Klebsiella, Escherichia, Citrobacter, Erwinia*
(c) Anaerobic	*Clostridium, Desulfotomaculum, Desulfovibrio*
(d) Microaerobic	*Azospirillum, Aquaspirillum, Thiobacillus, Pseudomonas, Xanthobacter, Rhizobium, Methylosinus, Methylococcus, Mycobacterium*
(e) Photosynthetic	*Rhodomicrobium, Rhodopseudomonas* and *Rhodospirillum* of the non-sulfur purple bacteria; *Chromatium* and *Ectothiorhodospira* of the purple sulfur bacteria; *Chlorobium* of the green sulfur bacteria
Cyanobacteria	
(a) Heterocystous filamentous forms	*Anabaena, Anabaenopsis, Aulosira, Calothrix, Cylindrospermum, Fischerella, Gloeocapsa, Hapalosiphon, Mastigocladus, Nostoc, Scytonema, Stigonema, Tolypothrix, Trichodesmium, Westiellopsis, Westiella*
(b) Non-heterocystous filamentous forms	*Lyngbya, Phormidium, Plectonema, Oscillatoria, Pseudanabaena*
(c) Unicellular forms	*Gloeothece, Cyanothece, Synechococcus*

PHOSPHATE SOLUBILIZATION

Phosphorus is one of the major plant nutrients and is found in the soil, manure, plant and micro-organisms in both organic and inorganic forms. The organic forms of phosphorus in the soil are compounds of phytins, phospholipids, nucleic acids, phosphorylated sugars and coenzymes which come mainly from vegetation and decaying plant residues. The inorganic forms in the soil include compounds of calcium, iron, aluminium and flourine. Both the above mentioned inorganic compounds and organic forms of phosphorus in the soil are not available to plants.

A variety of micro-organisms have shown the ability to solubilize rock phosphate and thus make available to plants. Several bacterial and fungal species like *Pseudomonas, Bacillus, Micrococcus, Flavobacterium, Aspergillus, Penicillium, Fusarium* and *Sclerotium* possess phosphate solubilizing property. The phosphate solubilizing micro-organisms bring insoluble phosphate present in the soil into soluble forms by secreting organic acids such as formic, acetic, propionic, lactic, glycolic, fumaric and succinic acids. The initial isolation of phosphate solubilizers is made by using a medium containing suspended insoluble phosphates, e.g., tricalcium phosphates. The formation of clearing zones around the colonies of the organisms is an indication of the presence of phosphate-solubilizing organisms (Sundara Rao and Sinha, 1963). In our country various phosphate solubilizing micro-organisms such as *Pseudomonas striata, Bacillus polymyxa, Aspergillus awamori, A. niger* and *Penicillium digitatum* have been used for preparing carrier-based inoculants, named as IARI Microphos Culture (Gaur, 1983).

In addition, the role of vasicular-arbuscular mycorrhiza (VAM) in the P nutrition and early growth of a great variety of plants is well documented, and mycorrhizal inoculates are gaining importance day-by-day.

Biological control of detrimental micro-organisms

Bacterial biocontrol agents improve plant growth by suppressing either major or minor pathogens. Micro-organisms that can grow in the rhizosphere provide the front-line defence for roots against attack by pathogens (Weller, 1988). The modes of interaction between plant pathogens and their antagonists in the rhizosphere can usually be considered as parasitism or predation of one organism by another, antibiosis, or competition where demand exceeds immediate supply material or space (Lynch, 1987).

The only commercially available bacterial biocontrol agent is *Agrobacterium rhizogenes* strain, 84, which controls crown gall caused by *A. tumefaciens* (Kerr, 1972). *Bacillus* spp. and *Pseudomonas* spp. are other two bacteria receiving much attention as biocontrol agents. Two strains of *Pseudomonas flourescens* (one to control bacterial blotch and the other to suppress fungal diseases of seedlings) have also been evolved for commercial use (Powell *et al.*, 1990). The take all disease of cereals and grasses caused by *Gaeumannomyces graminis* has been controlled by *Pseudomonas flourescens* (Weller and Cook, 1983) and also by *Bacillus pumilus* (Capper and Campbell, 1986).

Amongst fungi, *Trichoderma* is the most extensively studied fungal antagonist. *Trichoderma viride* has been exploited commercially on impregnated wood dowels placed in drilled holes of the trunk and major limbs of plum, peach or nectarine trees (Dubos and Ricard, 1974) to control silver leaf disease caused by *Stereum purpureum*. The major interest is in its activity against soil-borne diseases of a broad host range. Various species of *Trichoderma* have been reported to reduce the survival of number of plant pathogens like *Fusarium oxysporum, F. solani, Rhizoctonia solani, Sclerotinia sclerotium* (Lynch, 1987) and many others.

Product of phytohormones

Many bacteria isolated from soil are known to produce plant growth hormones in culture, and are assumed to affect seed germination, seedling development and subsequently the overall plant performance. Production of gibberellin-like substances and indolyl-3-acetic acid by several bacteria which are often used as seed inoculants have been investigated by a number of researchers (Brown and Burlingham, 1968; Janzen *et al.*, 1991; Katznelson and Cole, 1965).

Plant growth promoting rhizobacteria (PGPR) improve plant growth by colonizing the root system. Besides promoting growth, these organisms are effective biocontrol agents of certain soil-borne fungal plant pathogens. Kloepper *et al.* (1980) reported that specific significant yield increases. The bacteria produced an extracellular siderophore and pseudobactin (Teintze *et al.*, 1981) which was claimed to chelate iron and make it unavailable to pathogens and deleterious rhizobacteria. Recently a large number of reports are being placed on record indicating the role of PGPR for better plant growth. More information is required on microbial as well as plant functioning to achieve a clear role of PGPR in plant growth promotion experiments.

In our laboratory, experiments are being conducted to explore the possibilities to use soil microbes for their beneficial properties at higher elevations. Isolations and selections of appropriate microbes are being carried out. A large number of microbes have been isolated and are being studied for their beneficial properties like nitrogen-fixing ability, phosphate solubilization, production of antimicrobial substances, etc. A few rhizobacteria have been selected and are being tested for their root colonization and plant growth promotion ability (Pandey *et al.*, 1996; Fig. 1.5).

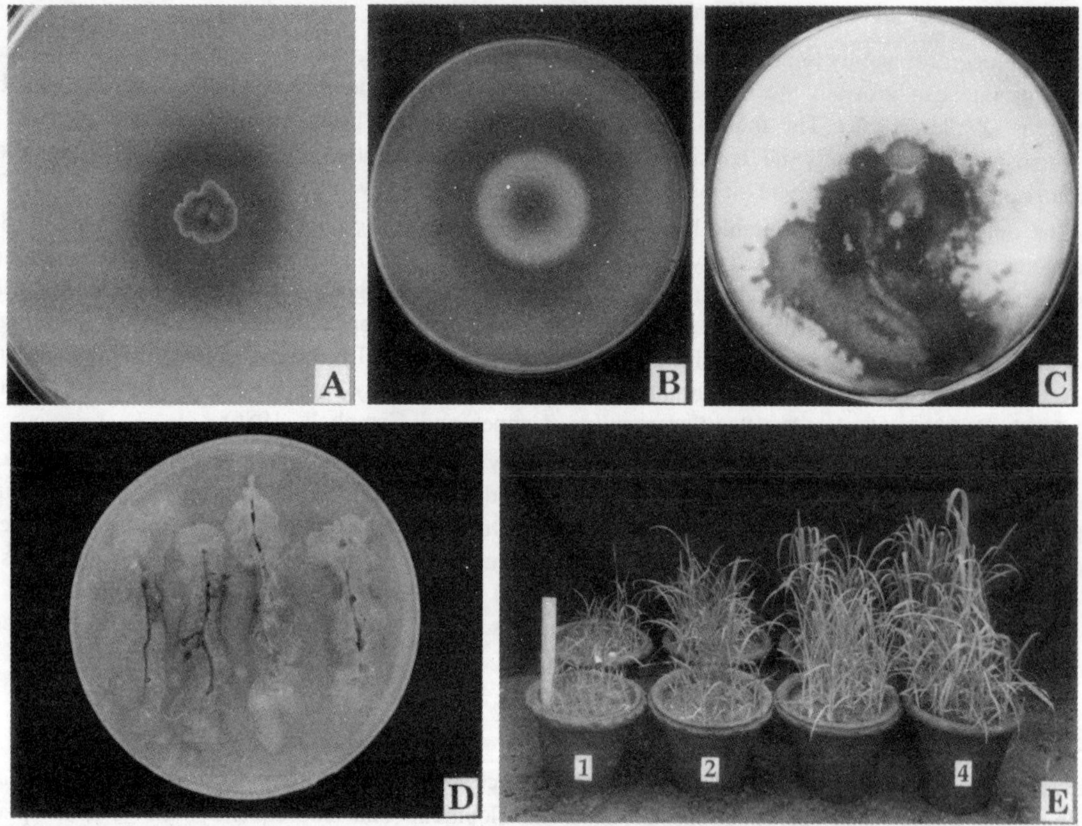

Fig. 1.5. Screening for beneficial microbes from higher elevations in the Himalaya **(a)** phosphate solubilization by a maize associated rhizobacteria, **(b)** phosphate solubilization by *Aspergillus niger* isolated from *Cedrus* rhizosphere, **(c)** production of antifungal compounds by maize associated rhizobacteria, **(d)** root colonization, and **(e)** plant growth promotion of *Eleusine corocana* (1) control (2, 3, and 4) inoculated with different bacteria *i.e.* Rhizobacteria 1, Rhizobacteria 7 and *Azotobacter chroococcum* W_5, respectively.

The above two categories of organisms (biocontrol agents and PGPR) affect plant growth and also plant nutrition in an indirect way and hence have been included here.

CONCLUSION

In conclusion, the issues discussed in this paper have a direct (or, in some cases indirect) bearing on soil fertility and its productive capacity. In the overall assessment of sustainability, at the farming systems' level, maintenance of soil fertility (or increment in the fertility levels) plays a critical role and it affects a number of other important factors in addition to crop yield. A fundamental shift has taken place in agricultural research and world food production. In the past, the major driving force was to increase the yield potential of food crops and to maximize productivity. Now the drive for productivity is increasingly combined with a desire for sustainability. For farming systems to remain productive, and to be sustainable in the long run, it will be necessary to replenish the reserves of nutrients which are removed or lost from the soil. It is widely known that while the system level processes operating in natural and agricultural systems are identical, the rates and consequences of these processes are contrasting (Ramakrishnan *et al.*, 1993). Natural ecosystems operate with internal regulatory mechanisms whereas the modern agricultural systems are regulated from "distal control" centres. The aims of optimizing productivity in the long term, rather than maximizing productivity in the short term, are perhaps the pivotal elements conferring viability and sustainability to traditional systems over modern ones. Thus, as stated by Ramakrishnan *et al.* (1993), there seems considerable scope of developing sustainable farming systems by appropriate mix of traditional and conventional science. In the more developed countries the interest in low-external-input agriculture is driven largely by overproduction and concerns of the environmental effects of intensive agro-chemical use in the less developed countries, interest in low-external-input agriculture is often a necessity, fuelled by the lack of access to high input approaches. In this context maintenance of soil fertility using biological means assumes critical importance particularly in the Himalayan region.

REFERENCES

Akkerman, A.D.L. and C. Van Dijk, 1981 : Non-leguminous root-nodule symbioses with actinomycetes and *Rhizobium.* In : *Nitrogen Fixation I. Ecology*, pp. 57-103, Oxford University Press, Oxford.

Alexander, M., 1961 : *Introduction to Soil Microbiology*, John Wiley & Sons, New York.

Becking, J.H., 1970 : Frankiaceae family mov. (Actinomycetales) with one new combination and six new species of the genus *Frankia*. Brunchorst, 1986. 174. *Int. J. Sys. Bact.* 20 : 201-220.

Becking, J.H., 1977 : Dinitrogen-fixing association in higher plants other than legumes. In : *A Treatise on Dinitrogen Fixation III. Biology*. (Eds. R.W.F. Hardy and Warren S. Silver), John Wiley & Sons, New York.

Bhardwaj, K.K.R., 1995 : Improvements in microbial compost technology : a special reference to microbiology of composting. In : *Wealth from Waste* (Eds. S.K. Khanna and Krishna Mohan), TERI Publication.

Brown, M.E. and S.K. Burlingham, 1968 : Production of plant growth substances by *Azotobacter chroococcum. J. Genetics and Microbiol.*, 53 : 135-144.

Brown, M.E. and N. Walker, 1970 : Indolyl-3-acetic acid formation by *Azotobacter chroococcum. Plant and Soil*, 32 : 250-253.

Brown S, J.M. Anderson, P.L. Woomer, M.J. Swift and E. Borrios, 1994 : Soil biological processes in tropical ecosystems. In : *The Biological Management of Tropical Soil Fertility* (Eds. P.L. Woomer and M.J. Smith), pp. 15-46, John Wiley & Sons, New York.

Capper, A.L. and F. Campbell, 1986 : The effect of artificially inoculated antagonistic bacteria on the prevalence of take-all disease of wheat in field experiments. *J. Appl. Bacteriol.*, 60 : 155-160.

Dixon, R.O.D. and C.T. Wheeler, 1986 : *Nitrogen Fixation in Plants.* Blackie and Sons, Glasgow.

Dubos, B. and J.L. Ricard, 1974 : Curative treatment of peach trees against silver leaf disease (*Stereum purpurem*) with *Trichoderma viride* preparations. *Plant Dis. Rep.*, 58 : 147-150.

Fillery, I.P.R., J.R. Simpson and S.K. De Datta, 1984 : Influence of field environment and fertilizer management on ammonia loss from flooded rice. *Soil Sci. Soc. Am. J.*, 48 : 914-920.

Gaur, A.C., 1983 : Biofertilizers and crop productivity. In : *Advances in Soil Science* (Ed. K.V. Paliwal), pp. 127-178. Books and Periodicals, New Delhi.

Harper H.T. and J.M. Lynch, 1979 : Effects of *Azotobacter chroococcum* on barley seed germination and seedling development. *J. Gen. Microbiol.*, 112 : 45-51.

Howard, A., 1933 : The waste products of agriculture : their utilization as humus. *J. Royal Soc. Arts*, 82 : 84-120.

Janzen, R.A., S.B. Rood, J.F. Dormaar and W.B. McGill, 1992 : *Azospirillum brasiliense* produces gibberellin in pure culture on chemically defined medium and in co-culture on straw. *Soil Biol. & Biochem.*, 24 : 1061-1064.

Katznelson, H. and S.E. Cole, 1965 : Production of gibberellin-like substances by bacteria and actinomycetes. *Can. J. Microbiol.*, 11 : 733-741.

Kerr, A., 1972 : Biological control of crown gall : Seed inoculation. *J. Appl. Bacteriol.*, 35 : 493-497.

Kirchmann, H., 1985 : Losses, plant uptake and utilization of manure nitrogen during a production cycle. *Acta Agri. Scand. Suppl.*, 24 : 1-77.

Kloepper, J.W., J. Leong, M. Teintze and M.N. Schroth, 1980 : Enhanced plant growth by siderophores produced by plant-growth-promoting rhizobacteria. *Nature*, 286 : 885-886.

Kumar, N. and O.P. Toky, 1994 : Variation in chemical contents of seed and foliage in *Albizia lebbeck* (L.) Benth. of different provenance. *Agroforestry System*, 25 : 217-225.

Lal, R., 1986 : Soil surface management in the tropics for intensive land use and high and sustained productivity. In : *Advances in Soil Science*, (Ed. B.A. Steward), Springer-Verlag, New York.

Lavelle, P., 1988 : Earthworm activities and the soil system, *Biol. and Fert. of Soils*, 6 : 237-251.

Lindblad, P. and B. Bergman, 1990 : The cycad-cyanobacterial symbiosis. In : *CRC Handbook of Symbiotic Cyanobacteria*. (Ed. A.N. Rai), pp. 137-159. CRC Press, Boca Raton, Florida.

Lynch, J.M., 1987 : Biological control within microbial communities of the rhizosphere. In : *Ecology of Microbial Communities*. (Eds. M. Fletcher and T.R.G. Gray), Cambridge University Press, Cambridge.

Maraikar, S. and S.L. Amarasira, 1989 : Effect of cattle and poultry drug addition on available P and exchangeable K of a red-yellow podzolic soil. *Trop. Agriculturalist*, 144 : 51-59.

Martin, D.L. and G. Gershung (Eds.), 1992 : *The Rodale Book of Composting*. Rodale Press, Emmaus, Pennsylvania, USA.

Mishustin, E.N. and V.K. Shilinikova, 1969 : Free-living nitrogen fixing bacteria of the genus *Azotobacter. Soil Biology*, Reviews of Research, UNESCO Publication, 72-124.

Misra, B.K. and P.S. Ramakrishnan, 1982 : Energy flow through a village ecosystem with slash and burn agriculture in north-eastern India. *Agri. Syst.*, 9 : 57-72.

Morris, R.A., R.E. Furoc and M.A. Dizon, 1986a : Rice responses to a short-duration green manure. I. Grain yield. *Agron. J.*, 78 : 409-412.

Morris, R.A., R.E. Furoc and M.A. Dizon, 1986b : Rice responses to a short-duration green manure. II. N recovery and utilization. *Agron. J.*, 78 : 413-416.

Morris, R.A., R.E. Furoc, N.K. Rajbhandari, E.P. Marqueses and M.A. Dizon, 1989 : Rice responses to waterlog-tolerant green manures. *Agron. J.*, 81 : 803-809.

Nagarajah, S., 1988 : Transformation of green manure nitrogen in lowland rice soils. In : *Green Manure in Rice Farming*. Los Banos, Philippines : IPRI.

Oades, J.M., 1984 : Soil organic matter and structural stability Mechanisms and implications for management. *Plant and Soil*, 76 : 319-337.

Okon, Y., 1985 : *Azospirillum* as a potential inoculant for agriculture. *Trends in Biotech.*, 3 : 223-228.

Okon, Y. and C.A. Labandera-Gonzalez, 1994 : Agronomic application of *Azospirillum* : an evaluation of 20 years worldwide field inoculation. *Soil Biol. Biochem.*, 26 : 1591-1601.

Pandey, A. and S. Kumar, 1989 : Potential of azotobacters and azospirilla as biofertilizers for upland agriculture : a review. *J. Sci. Ind. Res.*, 48 : 134-144.

Pandey, A. and S. Kumar, 1990 : Inhibitory effects of *Azospirillum brasiliense* on a range of rhizosphere fungi. *Ind. J. Exp. Biol.*, 28 : 52-54.

Pandey, A., A. Durgapal, E. Sharma and L.M.S. Palni, 1996 : Isolation and use of growth promoting rhizobacteria for improved plant performance in the hills. *Proceedings of National Symposium on Frontiers in Applied Environmental Microbiology* SES, CUSAT, Cochin, India. (In Press).

Powell, K.A., J.L. Faull and A. Renwick, 1990 : The commercial and regulatory challenge. In : *Biological Control of Soil Borne Plant Pathogens.* (Ed. Hornby), pp. 445-463 CAB Internationa, Wallingford, U.K.

Pradhan, M., 1993 : *Studies on Interaction between VAM Fungi and Frankia in Non-Leguminous Tree Species* (Alnus *sp.*) *of North-East India.* Ph.D. thesis, North Eastern Hill University, Shillong, India.

Rahlan, P.K., C.G.S. Negi and S.P. Singh, 1991 : Structure and function of the agroforestry system in the Pithoragarh district of Central Himalaya : An ecological viewpoint. *Agri. Ecosys. and Env.* 35 : 283-296.

Rai, S.C. and E. Sharma, 1995 : Land-use change and resource degradation in Sikkim Himalaya : A case study from the Mamlay watershed. In : *Sustainable Reconstruction of Highland and Headwater Regions.* (Eds. R.B. Singh and Martin, J. Haigh), pp. 265-278. Oxford & IBH Publishing Co., New Delhi.

Rai, S.C., E. Sharma and R.C. Sundriyal, 1994 : Conservation in the Sikkim Himalaya : Traditional Knowledge and use of the Mamlay Watershed. *Env. Conser.*, 21 : 30-35.

Ramakrishnan, P.S., K.G. Saxena and K.S. Rao, 1993 : Agro-ecological approaches for soil fertility management. In : *Tropical Soil Biology and Fertility Research : South Asian Context.* (Eds. P.S. Ramakrishnan, K.G. Saxena, M.J. Swift and P.D. Seward), pp. 77-125, Oriental Enterprises, Dehradun, India.

Roper, R.M. and J.K. Ladha, 1995 : Biological N_2 fixation by heterotrophic and phototrophic bacteria in association with straw. *Plant and Soil*, 174 : 211-224.

Sanchez, P.A., C.A. Palm, L.T. Scott, E. Cuevas and R. Lal, 1989 : Organic input management in tropical agroecosystems. In : *Dynamics of Soil Organic Matter in Tropical Ecosystems.* (Eds. D.C. Coleman, J.M. Oades and G. Uehara), University of Hawaii Press, Honolulu, USA.

Sharma, E., 1988 : Altitudinal variation in nitrogenase activity of the Himalayan alder naturally regenerating on landslide-affected sites. *New Phytologist*, 108 : 411-416.

Sharma, E. and R.S. Ambasht, 1984 : Seasonal variation in nitrogen fixation by different ages of root nodules of *Alnus nepalensis* plantations in the Eastern Himalayas. *J. Appl. Ecol.*, 21 : 265-270.

Sharma, E. and R.S. Ambasht, 1986. Root nodule age-class transition, production and decomposition in an age sequence of *Alnus nepalensis* plantation stands in the Eastern Himalayas. *J. Appl. Ecol.*, 23 : 689-701.

Sharma, E. and R.S. Ambasht, 1988 : Nitrogen accretion and its energetics in the Himalayan alder. *Functional Ecol.*, 2 : 229-235.

Sharma, E., R.C. Sundriyal, S.C. Rai, Y.K. Bhatt, L.K. Rai, R. Sharma and Y.K. Rai, 1992 : *Integrated Watershed Management : A Case Study in Sikkim Himalaya.* Gyanodaya Prakashan, Nainital, India, pp. 120.

Sharma, R., 1995 : *Symbiotic Nitrogen Fixation and Maintenance of Soil Fertility in the Sikkim HImalaya.* Ph.D. thesis, H.N.B. Garhwal University, Srinagar (Garhwal), India.

Sharma, R., E. Sharma and A.N. Purohit, 1994 : Dry matter production and nutrient cycling in agroforestry systems of cardamom grown under *Alnus* and natural forest. *Agroforestry System*, 27 : 293-306.

Sharma, R., E. Sharma and A.N. Purohit, 1995 : Dry matter production and nutrient cycling in agroforestry systems of mandarin grown in association with *Albizia* and mixed tree species. *Agroforestry Systems*, 29 : 165-179.

Singh, J.S., U. Pandey and A.K. Tiwari, 1984 : Man and Forests : A Central Himalayan case study. *Ambio.* 12 : 80-87.

Skeffington, R., 1975 : *Study on Nitrogen Fixation in Non-Legume Plants*. Ph.D. thesis, University of Dundee, U.K.

Subba Rao, N.S. and C. Rodriguez-Barrueco (Eds.), 1993 : *Symbioses in Nitrogen-Fixing Trees*. Oxford & IBH Publishing Co., New Delhi.

Sundara Rao, W.V.B. and M.K. Sinha, 1963 : Phosphate dissolving organisms in the soil and rhizosphere. *Ind. J. Agri. Sci.* 33 : 272-278.

Sundriyal, R.C., S.C. Rai, E. Sharma and Y.K. Rai, 1994 : Hill agroforestry systems in south Sikkim, India. *Agroforestry Systems*, 26 : 215-235.

Teintze, M., M.B. Hossain, C.L. Barnes, J. Leong and D. Van der Helm, 1981 : Structure of ferric pseudobactin, a siderophore from a plant growth promoting *Pseudomonas. Biochemistry*, 20 : 6446-6457.

Wani, S.P., 1990 : Inoculation with associative nitrogen fixing bacteria : role in cereal grain production improvement. *Ind. J. Microbiol.*, 30 : 363-393.

Weller, D.M., 1988 : Biological control of soil borne plant pathogens in the rhizosphere with bacteria. *Ann. Rev. Phytopathology*, 26 : 379-407.

Weller, D.M. and R.J. Cook, 1983 : Suppression to take all of wheat by seed treatments with fluorescent pseudomonads. *Phytopathology*, 73 : 463-469.

Young, A., 1989 : *Agroforestry for Soil Conservation*. CAB International, Wallingford, U.K.

Chapter 2

In Vitro Reactions of Alveolar Macrophages and Bronchial Epithelial Cells to NO_2 Exposure

Arnulf Seidel, Gerhard Polzer, Ines Lind and M. Wiener-Schmuck

The paper summarizes the available information in the literature regarding the response of alveolar macrophages and bronchial epithelial cells to NO_2 exposure and describes some own results of the authors. Only *in vitro* experiments are considered. Some recent results indicate that the stimulus-induced release of mediators by alveolar macrophages is impaired with higher NO_2 concentrations whereas lower concentrations might have priming effects, which causes a higher than normal response to immunostimulants. There is also evidence that alveolar macrophages and bronchial epithelial cells can release mediators after short-term exposure to concentrations of NO_2, which might still occur in large cities of developing countries.

INTRODUCTION

Nitrogen dioxide (NO_2) is a common component of outdoor and indoor air pollutants and is generated in various combustion processes, with traffic as one of the major sources. Due to its poor solubility, NO_2 penetrates into the lower respiratory tract, where it acts on cells according to its oxidative potential.

The effects of NO_2 on pulmonary tissue have been reviewed frequently [the references concerning clinical and epidemiological studies with NO_2 are given in detail in one of our previous papers (Polzer *et al.*, 1994), in this paper only those referring to *in vitro* or *ex vivo* results are quoted]. It has been demonstrated that the inhalation of NO_2 may result in an increased reactivity of the respiratory tract and in changes in the epithelium and interstitium of the lung as well as in "emphysema like" diseases. In the bronchiolar regions, the first morphological changes noticed occur in ciliated cells, where a loss of cilia is detectable. In the alveolar region, especially near the openings of the terminal bronchioles, the Type I cells exhibit swelling. In some areas of alveoli, the covering of Type I cells is lost, and some interstitial and alveolar edema becomes apparent.

In addition, animal studies provide strong evidence for a relationship to exist between NO_2 inhalation and enhanced susceptibility to bacterial and viral respiratory infections. It was shown that NO_2 exposure increases the streptococcal-induced mortality of mice in a time - as well as in a concentration-dependent manner and that this effect is strongly influenced by the pattern of exposure and by the time of streptococcal challenge after the exposure. In epidemiological studies a higher incidence of lower respiratory tract illness was found in children living in homes where gas stoves were used for cooking as compared to homes with electric stoves. Gas cooking may enhance drastically the indoor levels of NO_2, and for this reason NO_2 was assumed to cause an increase in the number of respiratory diseases. In school children with asthmatic symptoms, a reduction of lung function parameters dependent on the environmental concentration of NO_2 has been observed (Moseler *et al.*, 1994).

The evaluation of the effects of NO_2 on the immunologic system of the lung is essential to understand this phenomenon. The alveolar macrophages (AM) play a key role in nonspecific immunological processes, such as phagocytosis, as well as in specific immunological defence mechanisms. Depending on the physiological requirement, the AM produce and release a number of diverse products, including cytokines and reactive oxygen species, which can participate in these defence mechanisms.

Numerous investigations of the effect of NO_2 have been carried out with *in situ* exposed AM. These experiments allow alterations in the cell population to be determined and phagocytic and metabolic activities to be analyzed, but the infiltration of cells from the interstitium and the bloodstream into the airways and the impossibility of determining exact exposure concentrations are a drawback. For this reason, various methods have been developed that allow direct contact to be achieved between a single cell population and the atmosphere and permit an *in vitro* evaluation of the effects of gases on cells. In this paper, the current status of *in vitro* research with AM as well as with bronchial epithelial cells have been reviewed and some own results have been shown.

A compilation of *in vitro* studies with NO_2 and AM is presented in Table 2.1. Generally, results obtained with concentrations above 1 ppm are more of experimental interest. The data for bronchial epithelial cells and endothelium are summarized in Table 2.2. Despite the fact that there are inevitably conflicting results between different laboratories, the reasons of which are usually not very clear, it can be seen that NO_2 has effects on cells also *in vitro*. This is confirmed by our own results, which have been presented in the following sections and discussed together with those from the above tables.

MATERIALS AND METHODS

The experiments were performed with HL-60-cells differentiated to macrophage like cells by calcitriol (HL-60-M). For the morphological studies with the Atomic Force Microscope, a human bronchial epithelial cell line was used. All cells were exposed to air or NO_2 in a gas phase system, allowing direct contact between cells and the gases. The NO_2 concentration varied between 0.2 and 3 ppm,

Table 2.1. *In vitro* studies with NO_2 and alveolar macrophages*

Authors	*NO_2 concentration (ppm)*	*Exposure period (min.)*	*Observations*	*Species*
Alink and Rietjens (1990)	b	b	Protective effect of Vit. C and E concerning impairment of yeast phagocytosis	Rat
Bart *et al.* (1988)	0.2	30	Increased release of O_2 and of chemokines for neutrophils	Human
Kienast *et al.* (1993; 1994a, b; 1996)	0.1-5 ppm	15 or 30	Non-stimulated cells; No change of IL-I-IL-6 TNF- and TGF beta-release. With 1 or 5 ppm decreased chemotactic migration to C5a. LPS-stimulated cells: Impaired secretion of IL-6, IL-8 and TNF after higher NO_2 concentration, increase of TNF production with 0.1 ppm	Human
Kouzan *et al.* (1989)	0.2-1 ppm	30	No effects on AA metabolism	Rat
Pinkston *et al.* (1988)	5-15 ppm	180	No effect on spontaneous or Zymosan stimulated release of chemokines for neutrophils or influenza virus-induced release of IL-1	Human
Rietjens *et al.* (1986)	b	b	Concerning yeast phagocytosis O_3 is ten times more toxic than NO_2. Protective effect of Vit. C and E	Rat
Robison and Forman (1987)	0.1-50 ppm	"short period"	Significant alteration of AA metabolite profile depending on time and concentration	Rat
Robison *et al.* (1990)	1-20	60	Increased release of LTB_4 and chemokines for neutrophils; concentration-dependent decrease of PMA-induced O_2 - production	Rat
Robison and Forman (1991; 1993)	1-5	60-240	Increase of cyclooxygenase and lipoxygenase products only when NO_2 was followed by ionophore A 23187	Rat
Sone *et al.* (1983)	10-40	60-180	Increased tumor cell killing capacity	Rat
Tu *et al.* (1995b)	2-20	10-20	No effects on mRNA expression for TNF, IL-1, MIP-1-alpha or MIP-1-beta	IC 21mouse macrophage line
Vasallo *et al.* (1973)	b	b	Inhibition of phagocytosis and intracellular killing of bacteria	Rabbit
Voisin *et al.* (1977)	0.1-2	30	Decrease of bactericidal capacity	Guinea pig
Voisin *et al.* (1987a, b)	0.1-0.5	16-24 h	Positive effects of autoxidative enzymes and glutation on NO_2-induced cytotoxicity	Guinea pig
Voisin *et al.* (1989)	0.2	30	Increased release of O_2 and chemokines for neutrophils	Human

* Due to the limited space schematic information is given. For details, the original literature should be considered.

Table 2.2. *In vitro* studies with NO_2 and epithelial and endothelial cells*

Authors	*NO_2 concentration (ppm)*	*Exposure period (min.)*	*Observations*	*Species and cell type*
Cheek *et al.* (1986; 1988)	2-50	20	Inhibition of "dome" formation	Rat Type II pneumocytes
Davies *et al.* (1992)	0-0.8	15	Expression of mRNA for IL-1, IL-8, TNF, release of TNF	Human bronchial epithelial cells
Devalia *et al.* (1993a, b)	0.1-0.8	360	With 0.4 ppm: Increased release of IL-8, TNF, GM-CSF. Increased permeability for serum albumin	Human bronchial epithelial cells
Ebert *et al.* (1988a, b)	1-5	30	Decrease of viability, morphological changes	Endothelial cells and human Type II pneumocyte line (A549)
Li *et al.* (1994)	5	4-48 h	Increased phosphatidylserin content in cells and in their membranes	Pig pulmonal arteria endothelial cells
Patel *et al.* (1988)	5	3-24 h	After 24 h: Increased phosphatidylserin and ethanolamin content in lipid extracts. Decrease in insulin receptor density	Pig pulmonal arteria endothelial cells
Sapsford *et al.* (1991)	0.4-2	15 min	Increased Cr-51 release, increase of PGE_2, LTC_4 and 15-HETE. With 0.8 and 2 ppm reduction of ciliary beat frequency	Human bronchial epithelial cells
Tu *et al.* (1995a)	2-20	10-20	With 2 ppm: Reduced colony forming efficiency; Above 5 ppm: Increased LDH release, decreased glutathion content	Human umb. vein endothelial cells and Type II line (A549): Mouse macrophage line (IC-21)

* Due to the limited space schematic information is given. For details, the original literature should be considered.

the duration of exposure was one or two hours. The production of O_2 was assayed immediately after exposure, that of TNF and IL-8 during a period of three hours after exposure. In one series of experiments, the cells were treated with dipalmitoyl-lecithin (DPL, 100 μg/ml/medium) for one hour prior to exposure. In order to analyze possible combination effects, cells were incubated with latex beads (1.2 μm, Sigma) or quartz (AMAD1.7 μm, Sikron F 600) for one hour and thereafter exposed to air or NO_2; the mass concentration of the particles was 100 μg/ml. Details concerning experiments with HL-60-cells have been described previously (Polzer *et al.*, 1994).

It was possible to visualize living bronchial epithelial cells by Atomic Force Microscopy and to detect changes of the cell membranes. The human bronchial epithelial cell line BEAS 2B was exposed to 0.2 ppm NO_2 for two hours in the same arrangement as the macrophages. The Atomic Force Microscope was a "AFM 2010", Topometrix, Darmstadt, FRG. For the analysis the fluid chamber of Topometrix was used. It was performed in "constant force mode" with a force of <10nN and with two lines per second. Other methodological informations have been reported elsewhere (Polzer and Seidel, 1996).

RESULT AND DISCUSSION

As shown in Fig. 2.1, the lowest NO_2 concentration used (0.2 ppm) has no influence on the viability of the cells. Above this level, the effects are dependent on concentration and time. Regarding the results presented for O_2 in Fig. 2.2, it should first be mentioned that the response of HL-60-M towards Zymosan as standard stimulus varies from one experiment to the other already for "normal", air exposed cells. However, since the aim of the study is the comparison between air and NO_2-exposed cells for each concentration level separately, this discrepancy is not essential. For the 3 ppm concentration, there is a clear inhibition of the Zymosan-induced O_2-release, which might also occur for 1.5 and 1.0 ppm, but is not statistically significant. The 0.2 ppm exposure has clearly no statistically significant effect. In order to elucidate a possible stimulus specificity we have also used TPA instead of Zymosan to trigger the O_2 release. Interestingly, the TPA-elicited superoxide anion production was not affected by NO_2 (Fig. 2.3).

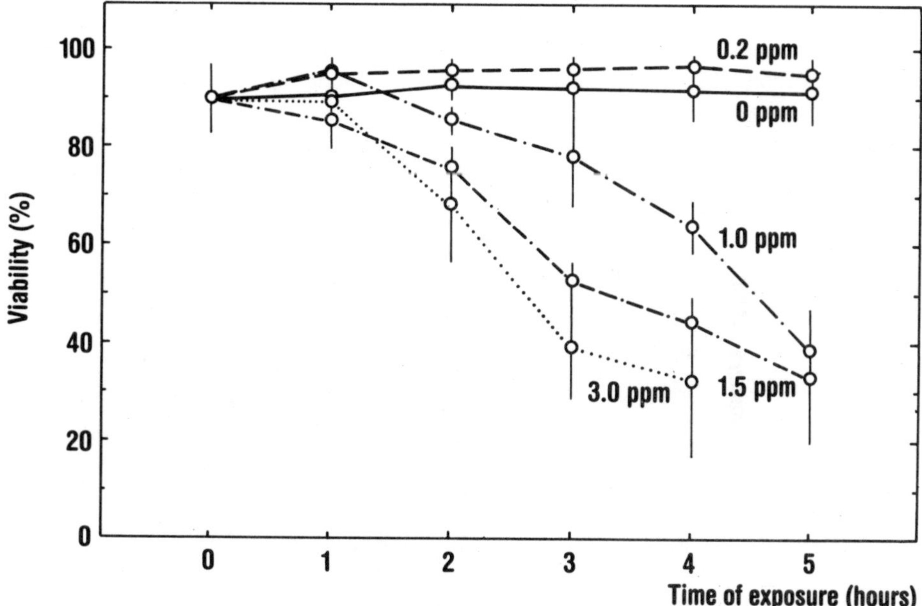

Fig. 2.1. Viability of HL-60-M as dependent on NO_2 concentration and time of exposure. Arithmetic mean ± S.E., n = 4.

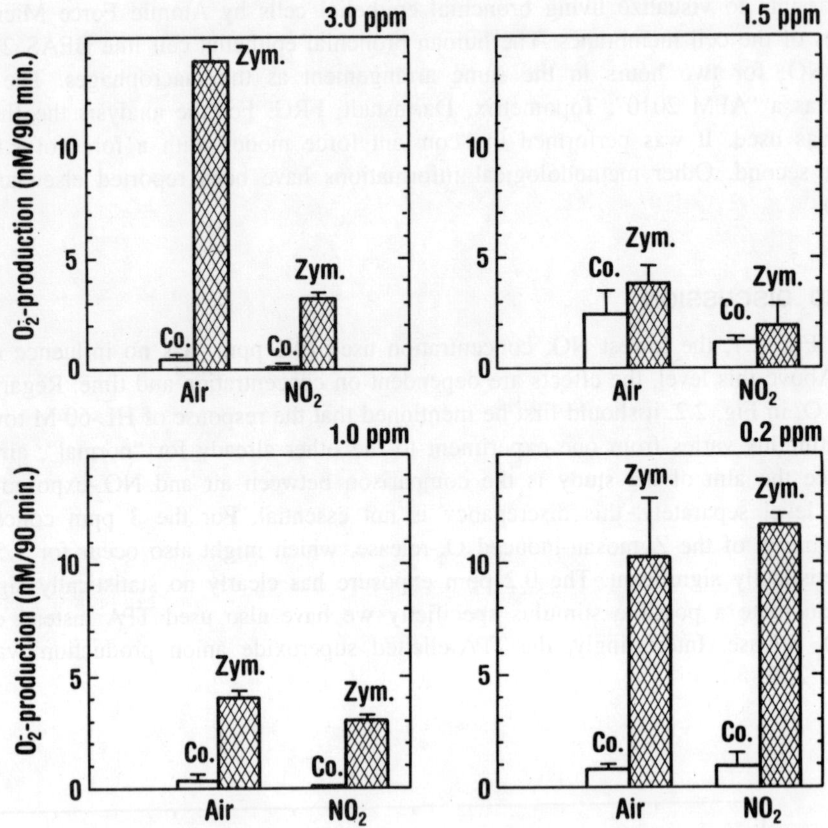

Fig. 2.2. Effect of NO_2 on the release of O_2 by HL-60-M as determined with or without Zymosan stimulation (Zym.) immediately after exposure (1 h). Arithmetic mean ± S.E., n = 3 (one representative experiment out of three is presented).

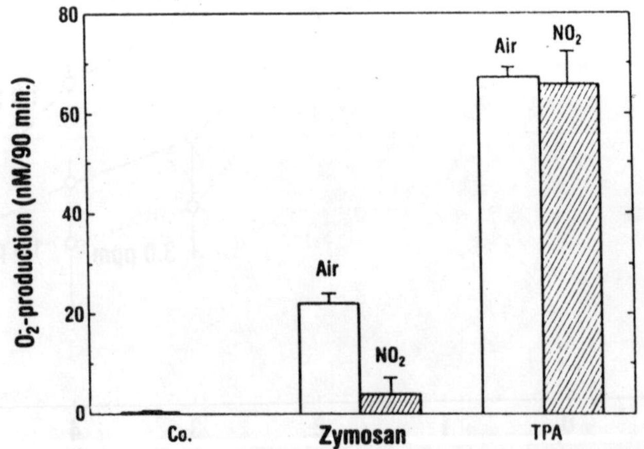

Fig. 2 3. Effect of NO_2 on the release of O_2 by HL-60-M (1.5 ppm, 2h). Immediately after exposure the release of O_2 with or without stimulation with Zymosan or tPA was measured. Arithmetic mean ± S.E., n = 3.

The impairment of the stimulus-induced reaction by higher concentrations of NO_2 is also visible in Fig. 2.4, which concerns the spontaneous and LPS-stimulated release of TNF. For the 3.0 and 1.5 ppm values, there is a clear inhibition of TNF production, the 1.0 ppm concentration has no effects. A very interesting phenomenon is the distinctly higher TNF production by the LPS-treated cells, which had been exposed previously to 0.2 ppm NO_2. We have investigated also for TNF, whether the inhibitory effects of NO_2 are dependent on the stimulus. As can be seen in Fig. 2.5 also, the TPA-triggered TNF production reduced by NO_2 is obviously a more general phenomenon, since it can also be demonstrated for the release of IL-8 (Fig. 2.6).

The influence of the surfactant component DPL on the results is shown in Fig. 2.7. DPL by itself reduces the measurable TNF content already in the air exposed controls or LPS-stimulated cells (left side of Fig. 2.7). Without DPL, NO_2 has its already described inhibitory effect (compare the LPS-columns of the "no DPL" data). When the cells were exposed to NO_2 after pretreatment with DPL prior to gas exposure there is no difference to the results obtained without DPL pretreatment.

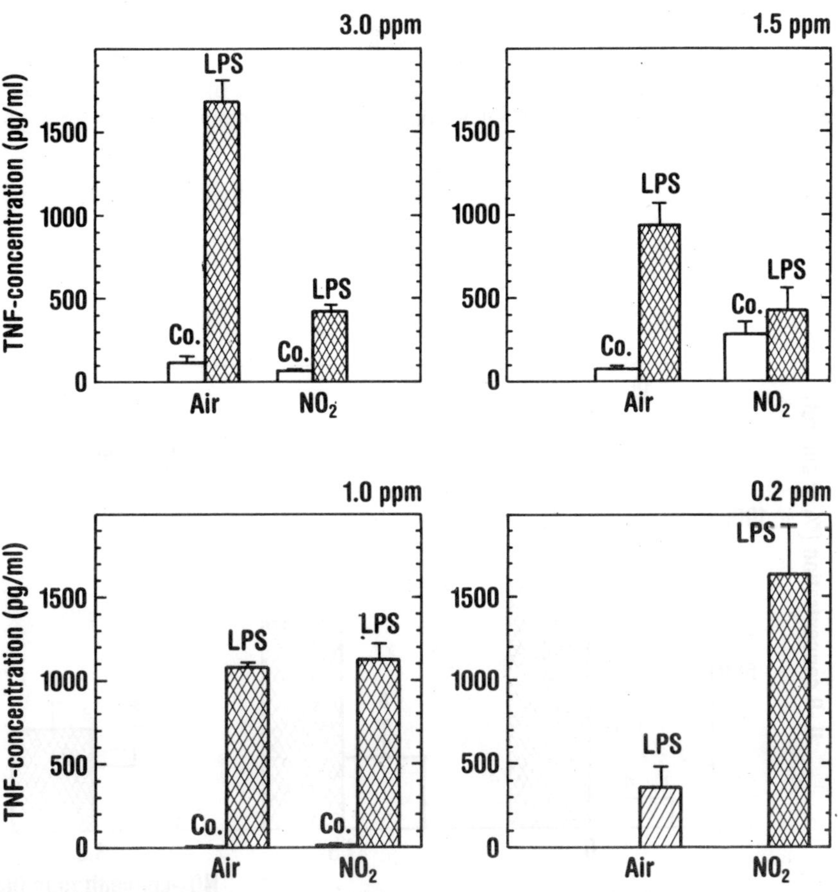

Fig. 2.4. Effect of NO_2 on the release of TNF by HL-60-M with or without stimulation with LPS. The TNF production was determined after 1 h exposure to NO_2 followed by 3 h incubation. Arithmetic mean ± S.E., n = 3 (one representative experiment out of three is presented).

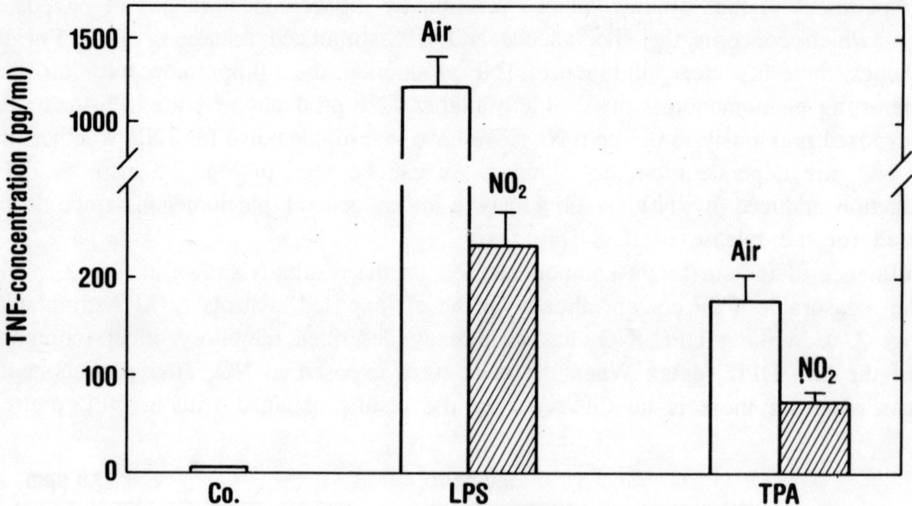

Fig. 2.5. Effect of NO_2 on the release of TNF by HL-60-M (1.5 ppm, 2 h). After NO_2 exposure, the cells were incubated for further 3 h with or without stimulation with LPS or TPA. Arithmetic mean ± S.E., n = 3.

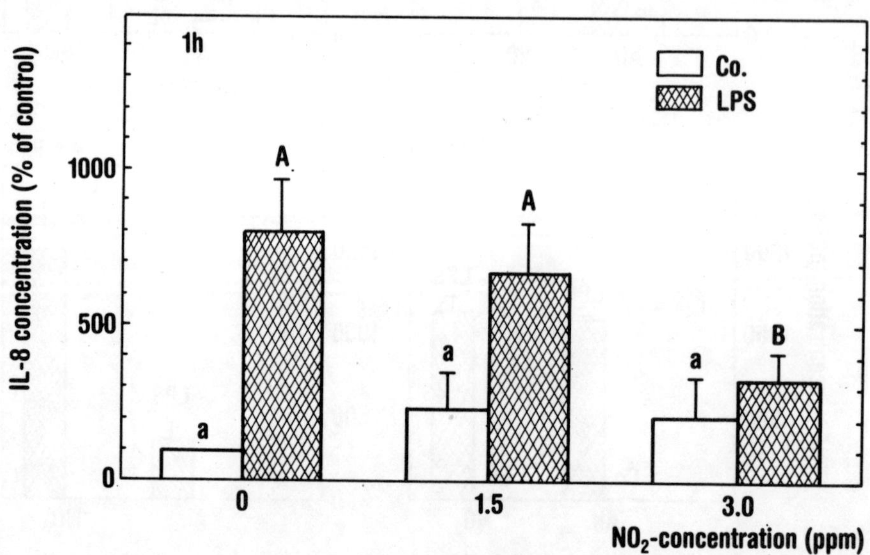

Fig. 2.6. Effect of NO_2 on the release of IL-8 by HL-60-M. The cells were exposed for 1 h. Thereafter, the release of IL-8 during a period of 2 h with or without LPS stimulation was determined. Arithmetic mean ± S.E., n = 3.

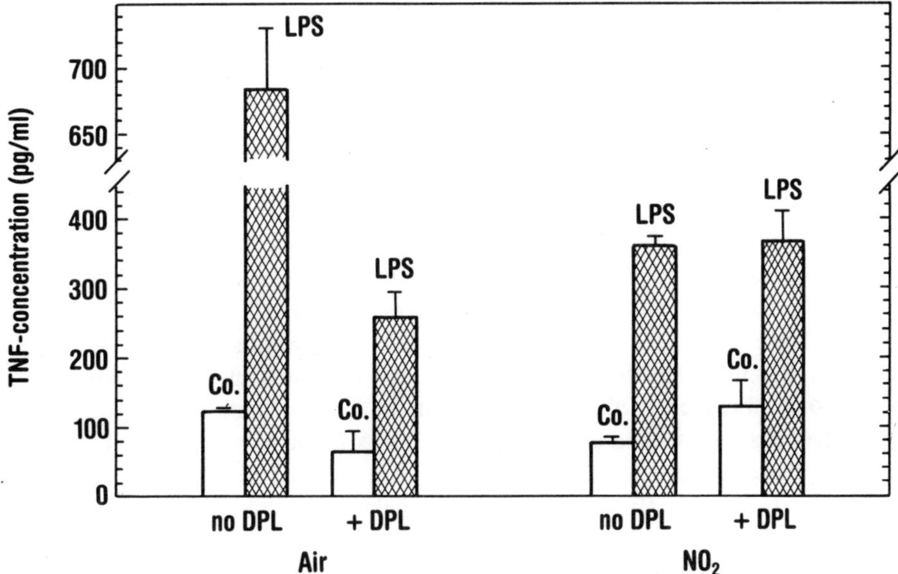

Fig. 2.7. Influence of DPL on effects of NO_2. HL-60-M were preincubated with DPL for 1 h. The release of TNF by the cells was determined after exposure to air or NO_2 (1.5 ppm) followed by 3 h incubation with or without LPS stimulation. Arithmetic mean ± S.E., n = 3.

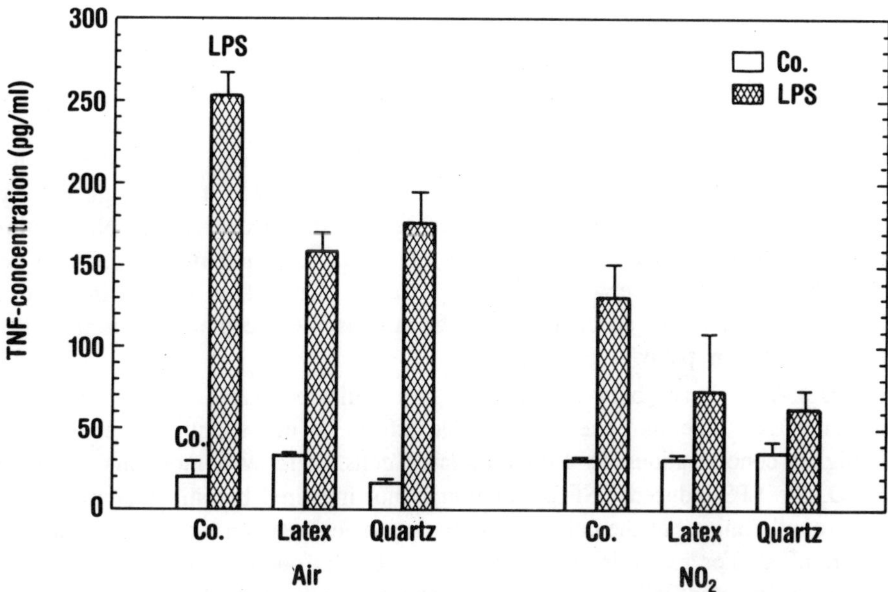

Fig. 2.8. Effect of particles combined with NO_2 on the release of TNF by HL-60-M. The cells were preincubated with latex or quartz (100 mg/ml) for 1 h. The release of TNF by the cells was determined after exposure to air or NO_2 followed by 3 h of incubation with or without LPS stimulation. Arithmetic mean ± S.E., n = 3.

In the Fig. 2.8, the results of the studies on the interaction of NO_2 and particles concerning TNF are presented. With non-stimulated cells, the latex- or quartz-particles had not much effect on the TNF-production, the same is true when they are combined with NO_2. The inhibitory effect of NO_2 on the LPS-stimulated TNF release is not changed after preincubation of the cells with the particles, which are inhibitory by themselves. The combination "particle plus NO_2 plus LPS" yields only 20-25% of the value for LPS alone.

The Atomic Force Microscope (AFM) was used to visualize the surface of alveolar macrophages and bronchial epithelial cells (BEAS 2B) under physiological conditions. In addition, the surface morphology as influenced by NO_2 of these cells was examined. Living AM as well as bronchial epithelial cells appeared to be more flat and smooth as compared to fixed cells. In AM exposed to NO_2 "bleb"-like protrusions of the membrane became demonstrable. The results obtained during the incubation with Cytochalasin B and Colchizine of BEAS 2B cells indicate that the component responsible for the submembranous structures is the actin skeleton, microtubules do not play a role. The surface morphology of the cells seems to be uninfluenced by the exposure to air compared to submerse cultured cells. In contrast, the exposure to 0.2 ppm NO_2 for 2 h induced the formation of small bulges and caves in the apical membrane (Fig. 2.9) which are only detectable with the AFM but not in SEM studies. Exposure to 0.1 ppm ozone for 2 h had no detectable effects. The data show that the AFM can be used to image living lung cells at high spatial resolution, to observe dynamic events at the cell surface and to demonstrate morphologic effects of low concentrations of NO_2 on plasma membranes.

Some general conclusions can be drawn from our original results (presented in the figures) together with the results obtained from the *in vitro* experiments summarized in the Table 2.1 and some additional *ex vivo* data. Phagocytosis and antimicrobicidal capacity are impaired, a finding which is corroborated also by *ex vivo* studies (Frampton *et al.*, 1989; Katz and Laskin, 1976). While our results confirm the *in vitro* data of Robinson *et al.* (1990) with respect to a decreased O_2-production with concentration of >1 ppm, and the *ex vivo* data of Suzuki *et al.* (1986), there was no altered O_2-release by human AM from volunteers. However, the product of NO_2-concentration *vs.* time of exposure was much smaller than in the experiments with rats performed by Suzuki *et al.* Below 1 ppm, we neither observed an effect on O_2-production by HL-60-M nor by AM (Hockelė and Seidel, unpublished). The finding that NO_2 inhibited the O_2-release induced by Zymosan but not that which was triggered by TPA needs further elucidation, since it might be helpful in explaining the mechanism of NO_2 action. As shown by others and ourselves, the surface morphology of AM is changed by NO_2, which certainly must influence membrane-linked or receptor-dependent processes. Since there was no influence on the TPA-mediated O_2-release, one might speculate that NO_2 affects the receptors rather than the other parts of the signal transduction pathway.

A striking similarity exists between some of the results published by Kienast *et al.* (Table 2.1) and ours insofar as NO_2 does not exert remarkable effects on non-stimulated cells and has inhibitory effects with higher concentrations on LPS-stimulated cells. It is even more interesting that with 0.1 or 0.2 ppm NO_2 the LPS-induced TNF production is not impaired but still increased. Obviously, NO_2 exposure leads to a "priming" of the cells. Some discrepancies between the results from different groups cannot be overlooked. For example, Pinkston *et al.* (1988) found no gas effects on the release of mediators by non-stimulated cells, which is in agreement with some data from Kiensat *et al.*, Tu *et al.* (same table) and our ones, but in their experiments there was no influence on stimulated cells. It should also be noted, that there are reports on an increased O_2 and chemokine production induced by a short-time exposure to only 0.2 ppm. We could not confirm this findings, neither with HL-60-

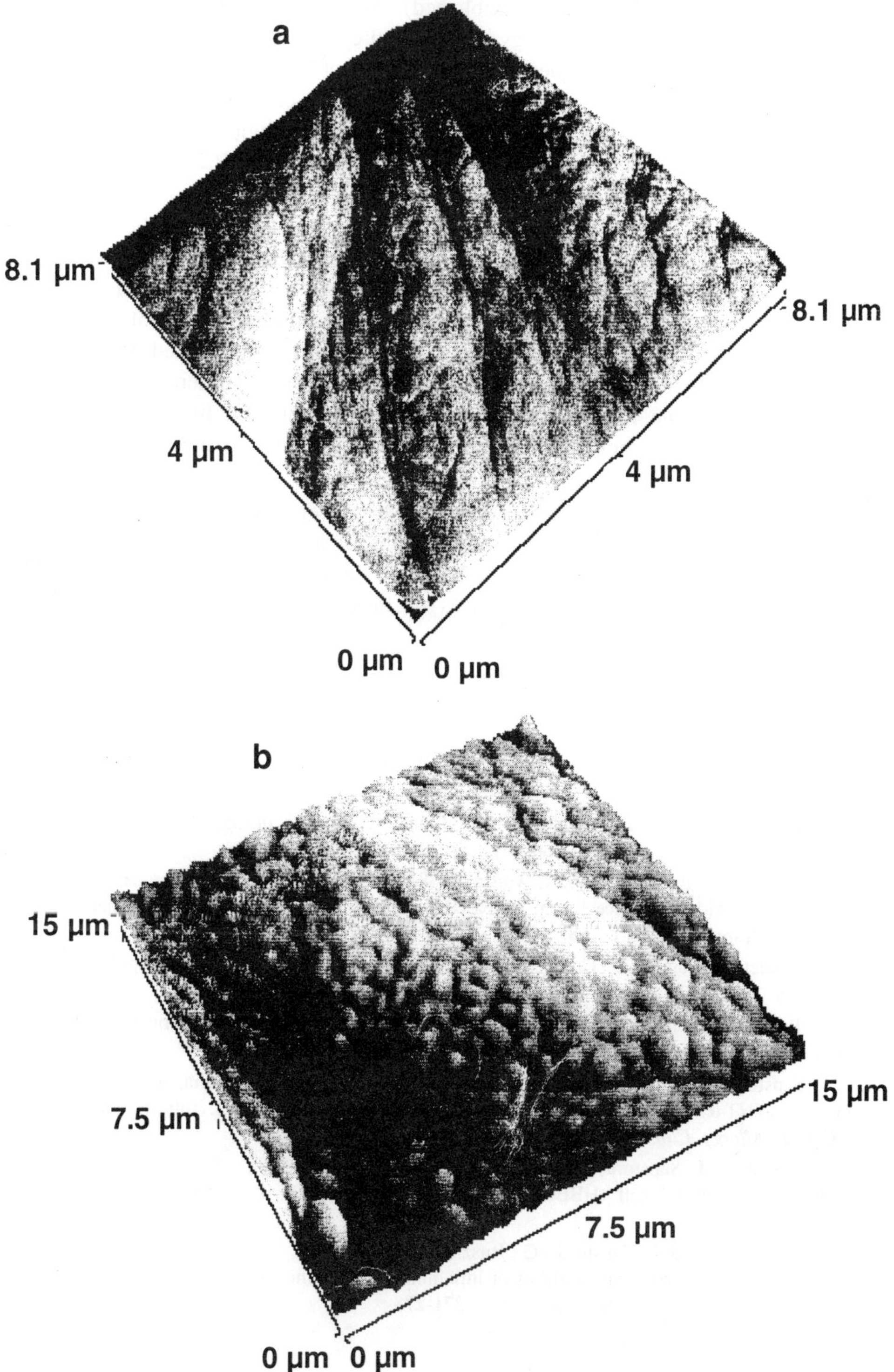

Fig. 2.9. Living BEAS 2B cells as visualized by Atomic Force Microscopy. **(a)** Cells after exposure to air (2 h). **(b)** Cells after exposure to 0.2 ppm NO_2 (2 h).

M nor with AM (Hockele and Seidel, unpublished). With Atomic Force Microscopy we could demonstrate changes in the surface morphology of bronchial epithelial cells. Admittedly, the mechanisms leading to these changes are not understood. It is probable that they reflect changes in the submembranous cytoskeleton (Polzer and Seidel, 1996). In any case this is the first time that NO_2 effects have been observed with living cells. The finding of morphological changes with 0.2 ppm fully corresponds to biochemical results published by other authors (Table 2.2).

When the cells had been coated with the surfactant component DPL prior to NO_2 exposure, there was no general change of their reaction and also no protective effect concerning the inhibitory influence of NO_2. This is in some contrast to our results with ozone, which had been performed under identical experimental conditions (Mosbach *et al.*, 1996b). The reason is not clear but both gases differ from much in their action, since ozone alone is very effective stimulus for the release of cytokines (Mosbach *et al.*, 1996a, b). There was not much interaction between particles phagocytized prior to NO_2 exposure and the gas. Also after particle phagocytosis, NO_2 had not much more influence than without particles. The extent of inhibition of the LPS-induced response was similar in cells with or without previous particle phagocytosis. A similar lack of interaction between gas and particles was also noted in our previous studies with ozone (Mosbach *et al.*, 1996a).

Summing up, there is sufficient evidence that NO_2 at concentrations, which can probably still be found in large third world cities, affects immunofunctions of airway cells, either by impairing or by exaggerating pathophysiological reactions. Disturbance of lung development or functions and aggravation of infectious diseases can be consequences of these cellular effects.

REFERENCES

Alink, G.M. and I.M.C.M. Rietjens, 1988 : Mechanisms of O_3 and NO_2 toxicity in lung cells *in vitro*. In : *Environmental Hygiene* (Eds. N.H. Seemayer and W. Hadnagy), pp. 7-10, Springer, Berlin.

Bart, F., C. Aerts, C. Deroubaix, B. Wallaert and C. Voisin, 1988 : *In vitro* short exposure of human alveolar macrophages to 0.2 ppm nitrogen dioxide. Evidence for an increased generation of superoxide anion and neutrophil chemotactic activity. *Am. Rev. Respir. Dis.*, 4 : 168.

Cheek, J.M., E.M. Postlethwait and E.D. Crandall, 1986 : Sensitivity of cultured alveolar epithelial cell monolayers to NO_2 *Fed. Proc.*, 45 : 915.

Cheek, J.M., E.M. Postlethwait and E.D. Crandall, 1988 : Effects of culture conditions on susceptibility of alveolar epithelial cell monolayers to NO_2. *Toxicol. Lett.*, 40 : 247-255.

Davies, R.J., R.J. Sapsford, C. Rusznak, A.M. Campbell, D. Quint and J.L. Devalia, 1992 : Synthesis of tumor necrosis factor-α (TNT-α) by cultured human bronchial epithelial cells (HBE), following exposure to nitrogen dioxide (NO_2). *J. Allergy Clin. Immunol.*, 89 : 211.

Devalia, J.L., C. Rusznak, R.J. Sapsford, M. Calderon and R.J. Davies, 1993a : Effect of nitrogen dioxide (NO_2) on human bronchial epithelial cell (HBEC) permeability and synthesis of GM-CSF and IL-8 *in vitro*. *Thrax*, 48 : 456.

Devalia, J.L., A.M. Campbell, R.J. Sapsford, C. Rusznak, D. Quint, P. Godard, J. Bousquet and R.J. Davies, 1993b : Effect of nitrogen dioxide on synthesis of inflammatory cytokines expressed by human bronchial epithelial cell *in vitro*. *Am. J. Respir. Cell Mol. Biol.*, 9 : 271-278.

Ebert, W., E. Moll, K. Kayser and D. Komitowski, 1988a : Die Wirkung von Luftschadstoffen auf kultivierte Lungenzellen. *Prax. Klin. Pneumol.*, 42 : 259-262.

Ebert, W., E. Moll, D. Komitowski and I. Vogt-Moykopf, 1988b : Die Wirkung von Luftschadstoffen auf kultivierte Lungenzellen. *Prax. Klin. Pneumol.*, 42 : 259-262.

Ebert, W., E. Moll, D. Komitowski and I. Vogt-Moykopf, 1988b : Biochemical and morphological studies on ozone and nitrogen dioxide treated cultured lung cells. *Zentralbl. Bakteriol. Mikrobiol. Hyg.*, 185 : 472-473.

Frampton, M.W., A.M. Smeglin, N.J. Roberts, J.N. Finkelstein, P.E. Morrow and M.J. Utell, 1989 : Nitrogen dioxide exposure *in vivo* and human alveolar macrophages inactivation of influenza virus *in vitro*. *Environ. Res.*, 48 : 179-192.

Katz, G.V. and S. Laskin, 1976 : Pulmonary macrophages response to irritant gases. In : *Air pollution and the lung*, Symp. Proc., Ma'alot, Israel, 83-100, John Wiley & Sons, New York.

Kienast, K., M. Knorst, J. Muller-Quernheim and R. Ferlinz, 1993 : Modulation der TNFa-Sekretion von Alveolarmakrophagen unter NO_2-Exposition in umweltrelevanten Konzentrationen. *Atemwegs and Lungenkrk.*, 19 : 427-428.

Kienast, K., M. Knorst, J. Muller-Quernheim and R. Ferlinz, 1994a : *In vitro* study of human alveolar macrophage IL-6, IL-8 and TGF-β release induced by nitrogen dioxide in indoor relevant concentrations. *Tubercle Lung Dis.*, 75 : 105-106.

Kienast, K., M. Knorst, R. Neuwirth, B. Fries, S. Grob, J. Muller-Quernheim and R. Ferlinz, 1994b : Das chemotaktische Verhalten von Alveolarmakrophagen und Blutmonozyten nach Exposition mit unterschiedlichen NO_2 Konzentrationen. *Dtsch. Med. Wschr.*, 199 : 899-903.

Kienast, K., M. Knorst, J. Muller-Quernheim and R. Ferlinz, 1996 : Modulation of IL-1-β, IL-6, IL-8, TNF-α, and TGF-β secretions by alveolar macrophages and NO_2 exposure. *Lung*, 174 : 57-67.

Kouzan, S., T. Fournier, C. Voisin, M.C. Jaurand and J. Bignon, 1989 : Arachidonic acid metabolite production by alveolar macrophages cultured in gaseous phase. Effects of NO_2 and diesel exhaust. In : NATO ASI Series, Vol. H3O. *Effects of Mineral dusts on cells*. (Eds. B.T. Mossman and R.O. Begin). pp. 215-222, Springer, Berlin.

Li, Y.D., J.M. Patel and E.R. Block, 1994 : Nitrogen dioxide-induced phosphatidylserine biosynthesis and subcellular translocation in cultured pulmonary artery endothelial cells. *Toxicol. Appl. Pharmacol.*, 129 : 114-120.

Moseler, M., A. Hendel-Kramer, W. Karmaus, J. Forster, K. Weiss, R. Urbanek and J. Kuehr, 1994 : Effect of moderate NO_2 air pollution on the lung function of children with asthmatic symptoms. *Environ. Res.*, 67 : 109-124.

Mosbach, M., M. Wiener-Schmuck and A. Seidel, 1996a : Influence of coexposure of Ozone with Quartz, Latex, Albumin and LPS on TNF-α and chemotactic factor release by bovine alveolar macrophages *in vitro*. *Inhalation Toxicol.*, 8 : 625-638.

Mosbach, M., M. Wiener-Schmuck and A. Seidel, 1996b : Influence of surfactant on cytokine release from ozone exposed human and bovine alveolar macrophages. *Inhalation Toxicol.*, 8 (In press).

Patel, J.M., D.A. Edwards, E.R. Block and M.K. Raizada, 1988 : Effect of nitrogen dioxide on surface membrane fluidity and insulin receptor binding of pulmonary endothelial cells. *Biochem. Pharmacol.*, 37 : 1497-1507.

Pinkston, P., A. Smeglin, N.J. Roberts, F.R. Gibb, P.E. Morrow and M.J. Utell, 1988 : Effects of *in vitro* exposure to nitrogen dioxide on human alveolar macrophage release of neutrophil chemotactic factor and interleukin-1. *Environ. Res.*, 47 : 48-58.

Polzer, G., I. Lind, E. Kruger and A. Seidel, 1994. *In vitro* effects of nitrogen dioxide on the release of superoxide anions. TNF-α and IL-8 by HL-60-macrophages and bovine alveolar macrophages, *Inhalation Toxicol.*, 6 : 359-377.

Polzer, G. and A. Seidel, 1996 : Zellmembranveranderungen and Lungenzellen durch umweltrelevante Konzentrationen von Ozon and NO_2 : Untersuchungen mittels Atom-Kraft-Mikroskopie. Report of Forschungszentrum Karlsruhe, *FZKA-PUG* 24 (In Press).

Rietjens, I.M.C.M., M.C.M. Poelen, R.A. Hempenius, M.J.J. Gijbels and G.M. Alink, 1986 : Toxicity of ozone and nitrogen dioxide to alveolar macrophages : comparative study revealing differences in their mechanism of toxic action. *J. Toxicol. Environ. Health*, 19 : 555-568.

Robinson, T.W. and H.J. Forman, 1987 : Alterations of prostaglandin synthesis by alveolar macrophages exposed to nitrogen dioxide. *Fed. Proc.*, 46 : 872.

Robinson, T.W., M.J. Thomas, M. Samuel and H.J. Forman, 1990 : Generation of aldehydes from rat alveolar macrophages exposed to nitrogen dioxide. *Free Rad. Biol. Med.*, 9 : 115.

Robinson, T.W. and H.J. Forman, 1991 : Changes in pulmonary alveolar macrophage arachidonate metabolism in response to *in vitro* and *in vivo* exposure to nitrogen dioxide. *Am. Rev. Respir. Dis.*, 143 : A640.

Robinson, T.W. and H.J. Forman, 1993 : Dual effect of nitrogen dioxide on rat alveolar macrophage arachidonate metabolism. *Exp. Lung Res.*, 19 : 21-36.

Robinson, T.W., D.P. Duncan and H.J. Forman, 1990 : Chemoattractant and leukotriene B_4 production from rat alveolar macrophages exposed to nitrogen dioxide. *Am. J. Respir. Cell Mol. Biol.*, 3: 21-26.

Sapsford, R.J., D.T. McCloskey, J.L. Devalia, D.R. Cundell and R. Davies, 1991 : Nitrogen dioxide : effects of human bronchial epithelial cell function *in vitro*. *Thorax*, 46 : 306.

Sone, S., L.M. Brennan and D.A. Creasia, 1983 : *In vivo* and *in vitro* NO_2 exposure enhance phagocytic and tumoricidal activities of rat alveolar macrophages. *J. Toxicol. Environ. Health*, 11 : 151-163.

Suzuki, T., S. Ikeda, T. Kanoh and I. Mizoguchi, 1986 : Decreased phagocytosis and superoxide anion production in alveolar macrophages of rats exposed to nitrogen dioxide. *Arch. Environ. Contam. Toxicol.*, 15 : 733-739.

Tu, B., A. Wallin, P. Moldeus and I.A. Cotgreave, 1995a : Cytotoxicity of NO_2 gas to cultured human and murine cells in an inverted monolayer exposure system. *Toxicology*, 96 : 7-18.

Tu, B., A. Wallin, P. Moldeus and I.A. Cotgreave, 1995b : Direct exposure to nitrogen dioxide fails to induce the expression of some inflammatory cytokines in an IC-21 murine macrophage cell model. *Toxicology*, 104 : 159-164.

Vassalo, C.L., B.M. Domm, R.H. Poe, M.L. Duncombe and J.B.L. Gree, 1973 : NO_2 gas and NO_2 effects on alveolar macrophage phagocytosis and metabolism. *Arch. Environ. Hlth.*, 26 : 270-274.

Voisin, C., C. Aerts, E. Jakubezak, J.L. Houdret and A.B. Tonnel, 1977 : Effects du bioxyde d'azote sur les macrophages alveolaires en survie en phase gazeuse. *Bull. Eur. Physiopath. Resp.*, 13 : 137-144.

Voisin, C., C. Aerts and B. Wallaert, 1987a : Prevention of *in vitro* oxidant-mediated alevolar macrophage injury by cellular glutathione and precursors. *Bull. Eur. Physiopath. Resp.*, 23 : 309-313.

Voisin, C., C. Aerts and B. Wallaert, 1987b : Controlled *in vitro* study of low concentrations nitrogen dioxide (NO_2) on alveolar macrophages in gas phase. *Chest*, 91 : 313.

Voisin, C., C. Aerts, C. Deroubaix and B. Wallaert, 1989 : Controlled *in vitro* approach of low concentration NO_2. Effects on human alveolar macrophage function. In : *Environmental Hygiene II*, (Eds. N.H. Seemayer and W. Hadnagy), 187-190, Springer, Berlin.

Chapter 3 Performance Assessment and its Role in Radioactive Waste Disposal — A Perspective

P. Sasidhar, K.B. Lal, Jaleel Ahmed and R. Pitchai*

The problem of managing radioactive wastes, having various chemical and physical forms, is expected to grow with the development of nuclear technology. Underground disposal of low-level radioactive wastes, appropriately immobilized and/or packaged, is generally agreed to be an adequate way of providing the necessary protection for humans and the environment. With growing public concern and awareness, performance assessment of disposal facilities would assume increasing importance and significance in the years to come.

This paper has the objective of providing an overview on Performance Assessment and presents a case study on quantitative forecasting of the performance of the ground disposal system for nuclear wastes located at Centralised Waste Management Facility (CWMF), Kalpakkam. The case study compares the forecast results with regulatory limits stipulated by the Code of Federal Regulations of USA 10 CFR 61 and national regulatory requirements stipulated by the Atomic Energy Regulatory Board (AERB) in India.

INTRODUCTION

Safe management including disposal of radioactive wastes from the various parts of the nuclear fuel cycle is an important aspect of nuclear technology development. The problem of managing radioactive wastes, having various chemical and physical forms, is expected to grow with the development of nuclear technology. Underground disposal of low-level radioactive wastes, appropriately immobilized and/or packaged, is generally agreed to be an adequate way of providing the necessary protection for humans and the environment (IAEA, 1981). One of the most important problems associated with this method of waste management has been the possibility of radioactive material becoming mobile. Some of the earliest studies of shallow radioactive waste disposal sites at Hanford, USA and Chalk River, Canada have clearly demonstrated the possibility of mobility of dissolved contaminant species. In order to ensure that radioactive waste management facilities are located and designed in such a way as to minimize any possible hazard, it is necessary to understand the mechanism of contaminant transport and the influence of site characteristics on the transport mechanism. Although all of these transport problems are currently being studied, concern over movement of contaminants has led to considerable attention being focussed (Anderson, 1979) on physical and chemical processes that tend to retard or dilute the pollutants (Mercer and Charles, 1980). With growing public concern and awareness, performance assessment of disposal facilities would assume increasing importance and significance in the years to come. This could result in performance assessment becoming a regulatory requirement in every country.

Overview of performance assessment

Performance assessment (IAEA, 1983 and 1984) of a shallow ground repository is necessary to determine the expected performance of the repository system and to compare it with acceptability criteria. The overall performance of the disposal system as a whole must satisfy all the regulatory or desired environmental protection requirements. The stipulation of the International Commission on Radiation Protection (ICRP) for dose limitations has been incorporated in the IAEA Basic Performance Standards (IAEA, 1984) and has been accepted by many national authorities. Performance assessment takes into account all the phenomena which may lead to a release of radionuclides from a repository or influence the rate at which release occurs and includes quantitative predictions of doses received by the general public. Safety assessment and performance assessment are interchangeably used in the literature and this paper also uses these terms interchangeably.

It appears that not many countries have really carried out extensive performance assessments even now. In various countries, concerted efforts are being made to ensure compliance with regulatory requirements. The main components of performance assessment include scenario analysis and consequence analysis. These components are illustrated in Fig. 3.1, which also indicates that as a relevant procedure, probabilistic and deterministic techniques are employed in the performance assessment.

Various methods are available for considering how the probabilistic variations in components of a system acting together cause variation in the system as a whole. These include fault/event tree and Monte Carlo Analyses.

Deterministic techniques are also applicable in both scenario and consequence analyses. They form an essential part of consequence analysis, because the rates of transport of radionuclides through the environment cannot be predicted by the use of purely probabilistic techniques. Consequence analysis involves following the progress of the released radionuclides *via* different pathways. Pathways include drinking water, consumption of aquatic foods, bathing and swimming; indirect pathways include transport by plants. The transport in each of the above processes has to be forecasted by deterministic methods but when the parameters used in each of the pathways have distribution functions, consequence analysis introduces probabilistic components. Deterministic techniques also have a role in scenario analysis, where they may be used to predict the effects of various natural phenomena or human intrusion

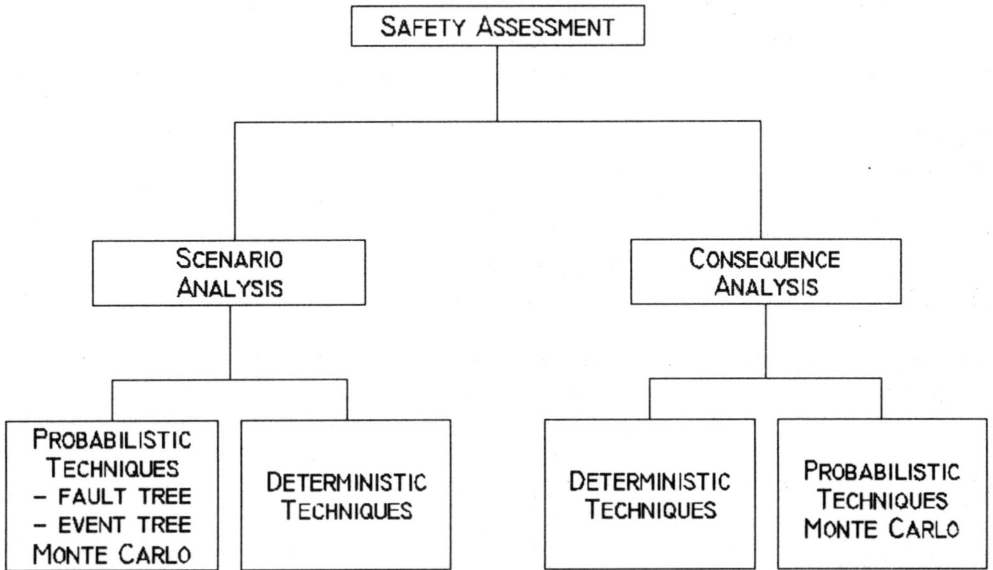

Fig. 3.1. Components of safety assessment.

on the repository system. Deterministic techniques are particularly useful in modelling the effects of continuous processes and can be applied to both steady-state conditions and dynamic conditions. The main components of performance assessment are scenario analysis and consequence analysis.

Scenario analysis

Scenario analysis involves the identification and quantitative definition of phenomena which could initiate the release of radionuclides from a shallow ground repository and/or influence the rate at which releases and transport occur.

Table 3.1 gives a list of phenomena which are potentially relevant to scenario analysis for repositories in shallow ground. These phenomena fall into three broad categories : (1) human activities; (2) natural processes and events, and (3) waste and repository processes. These broad areas are further classified into various processes which emphasize the need for probabilistic and deterministic techniques.

The approach to analysis of scenarios involving phenomena in the various categories in Table 3.1 would require determination of the phenomena or combination of phenomena which are most relevant leading to logical framework for estimation of the probabilities of occurrence of scenarios and for the quantitative definition of release and transport parameters. Scenario analysis involves also the use of submodels to preduct the effects of particular phenomena and thus providing detailed inputs (source terms, boundary conditions, etc.) for subsequent consequence analysis. These submodels are often deterministic in nature.

Consequence analysis

The consequence analysis consists of several steps. As a first step, a forecast of radionuclide release rates from the repository is made, followed by estimates of the radionuclide concentrations in the various

Table 3.1. Phenomena potentially relevant to scenario analysis for shallow ground repositories

1.0 HUMAN ACTIVITIES
1.1 Improper design and operation
Chemical liquid waste disposal
Draining system obstruction
Improper waste emplacement
Top cover failure
1.2 Future intrusion
Construction activities
Farming
Groundwater exploitation
Habitation
Salvage
Re-use of disposal material
Archaeology
2.0 NATURAL PROCESSES AND EVENTS
2.1 Biological intrusion
Animals
Plants
2.2 Faulting/seismicity
2.3 Fluid interactions
Erosion
Flooding
Fluctuations in the water-table
Groundwater flow
Seepage water
2.4 Weathering
Deterioration with time
Freezing/thawing
Wetting/drying
3.0 WASTE AND REPOSITORY PROCESSES
Gas generation
Waste and soil compaction
Waste/soil interaction

compartments of the environment. The second step consists of a prediction of transport rates of released radionuclides between various compartments and man. The third step involves estimates of radionuclude interaction with man resulting in calculation of doses.

It will therefore be seen that consequence analysis involves following the progress of the released radionuclides *via* different pathways. Direct pathways include drinking water, consumption of aquatic foods, water contact through bathing and swimming; indirect pathways include transport by plants and domestic animals.

Acceptability criteria

In order to compare the results of performance assessment of the disposal facility with the regulatory requirements, the Code of Federal Regulations (CFR) of USA formulated by US Nuclear Regulatory Commission (USNRC). 10 CFR 61 (USRC, 1990) defines objectives and minimum requirements for a performance assessment. The important regulation regarding annual dose equivalents is stated as :

* *Protection of the general population from releases of radioactive material by requiring that annual dose equivalents do not exceed 25 mrem/a (0.25 mSv/a) to a member of the public [10 CFR, 61.41].*

The basic requirement for all radiation protection is that radiation exposure of man be kept below an acceptable limit. Guidance on this is found in the recommendations of the ICRP (1983). This ICRP system for dose limitation has been incorporated in the IAEA Basic Performance Standards which have been accepted by many national authorities.

State-of-the-art approach in performance assessment

In order to develop a full performance assessment analysis that can be meaningfully compared with the performance objectives set forth in 10 CFR Part 61.41, the performance assessment begins by identifying appropriate pathways that may be important at the specific site. These pathways should then be ranked in order of importance to the overall performance of the site. The next step should be to develop a conceptual model of the site that appropriately accounts for transport of radionuclides through the crucial pathways previously identified. Once a conceptual model has been developed, the process can be analyzed using appropriate mathematical models. The results of model, including associated uncertainties, can then be compared with the performance objectives. In practice, this approach should be iterative; the choice and ranking of pathways, conceptual model, and choices of mathematical models are all candidates to be tested and revised on successive iterations, which should converge to a satisfactory and defensible performance assessment of the site.

The overall procedure and principles in developing a conceptual model and implementing the methodology are similar for most performance assessment analyses. In this context, it may be noted that the groundwater pathway is expected to be the crucial pathway for many low-level waste facilities (Shipers and Harlan, 1989).

One of the important steps in modelling waste transport in a groundwater system is then the development of the conceptual model. This development calls for abstraction of site characterization data into a form that is capable of being modelled. Such a process generally involves imposing a number of simplifying assumptions, including simplification of the appropriate governing equations to reflect the physical situation. Simplifications are usually made about the geometry of the system, spatial and temporal variability of parameters, isotropy of the system, and also about the influence of the surroundings on the behaviour of the system. In a nutshell, to understand the physical behaviour of the system, cause-effect relationships are determined and a conceptual model of how the system operates is formulated.

A conceptual model can be considered adequate if (1) it accounts for the most important physical, chemical, and geological characteristics of the system, and (2) it adequately represents the response of the system to changes in stresses. Naturally, there can be no rigid rules in the development of conceptual models.

The linkage between model development and field/laboratory observations might best be described as the iterative feedback between site and system characterization with a view to develop, improve, and support appropriate conceptual models representing system behaviour. For generating confidence that performance assessments are reliable and realistic, performance assessment process must be pursued in close association with site characterization and laboratory and field experiments. The criteria for determining when and whether sufficient characterization has been achieved will definitely use uncertainty analysis. Current site and system characterization data, conceptual models, computer models, and any appropriate performance assessment tools could be used to improve characterization and understanding. The above aspects are brought out clearly in Fig. 3.2.

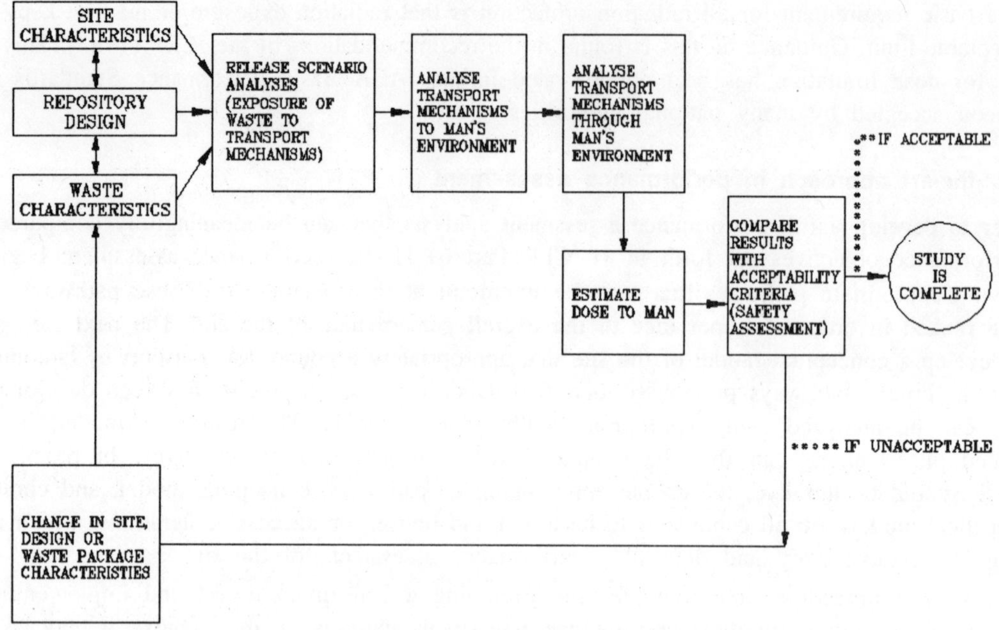

Fig. 3.2. Relationship of safety assessment activities.

Code verification, validation, sensitivity and uncertainty analysis are discussed elsewhere (Sasidhar, 1993) as a means to achieve the required degree of confidence in working of the models for performance assessments of repositories for nuclear waste. The two international projects INTRACOIN and HYDROCOIN projects, are dealing with these issues. Full cooperation between laboratory and field experimentalists, experts in natural analogues, geologists, model developers and users is necessary to achieve a fair degree of validation of models.

Radioactive waste repositories contain usually a mixture of radionuclides emanating from different operations or stages. Since, source-term models generally describe the release of a single radionuclide only, the source-term analysis must be repeated for each radionuclide in the waste inventory. The result is a set of release-rate histories that provide the input to groundwater transport analyses. These analyses include transport from the facility to the aquifer, and from the entry point in the aquifer to a nearby well through groundwater flow and to nearby surface water, if any. If the water table is shallow (near the ground surface), one can neglect the transport in the unsaturated zone and assume that source-term releases occur directly into the aquifer. For sites at which the water table is deep, the methodology employed in this study treats the unsaturated zone as an interval causing a delay time between release from the disposal units and entry into the aquifer.

Once the concentration of each radionuclide is determined as a function of time in well water and surface water (if necessary), the dose history can be generated by applying environmental pathway and dosimetry analyses for each time of interest. Pathway and dosimetry models consist of simple, linear, multiplication factors that convert environmental concentration to an annual committed dose.

A careful review of literature and the above inferences have clearly pointed out the need for carrying out site-specific performance assessment studies for nuclear waste repositories in ground. Accordingly, the present case study has been conceived and executed: it lays emphasis mainly on the performance assessment related to the disposal facility at Centralised Waste Management Facility (CWMF), Kalpakkam. Detailed characterization of the site with respect to geology, hydrology, geochemistry, water balance studies etc., have been carried out in addition to the development of mathematical models and computer codes relevant to the topic.

MODELLING OF RADIONUCLIDE MIGRATION

Mathematical modelling has rapidly become a powerful tool and supplement to conventional forecasting procedures and is extensively used (Broyd *et al.*, 1984) in performance assessment studies or analysis of subsurface contaminant migration. Numerical methods are commonly used in groundwater and solute transport applications. A one-dimensional (1-D) numerical code of radionuclide migration using finite-difference method (FDM) has been developed as a part of this study to understand the transport and migration of radionuclide in the saturated region, based on the IAEA, 1984 governing advention-dispersion equation (1) :

$$(\partial C/\partial t)_t = a(x, t)\,(\partial^2 C/\partial x^2)_t + b(x, t)\,(\partial C/\partial x)_t + c(x, t)\,C_t + d(x, t) \qquad \ldots (1)$$

C represents the concentration of radionuclide or pollutant, a, b, c and d are the coefficients represented by $a = D_c/R_t$; $b = -GWV/R_t c = RDC$; $d = 0$ (or source term as the case may be)

where :

D_c = Dispersion coefficient (L^2T^{-1})
GWV = Groundwater velocity (LT^{-1})
R_t = Retardation factor (Dimensionless)
= 1+(1-porosity*K_d*TD/(porosity)

where :

K_d = Distribution coefficient (L^3M^{-1})
TD = True density (ML^{-3})
RDC = Radioactive decay constant (T^{-1})

The code has been developed in Fortran-77 language and is presently operable in SINTRAN operating system of Norsk Data 570 super-mini platform available at Indira Gandhi Centre for Atomic Research (IGCAR), Kalpakkam.

The 1-D Numerical Code has been employed to predict the transport of several radionuclides relevant in the context of waste disposal at Kalpakkam to a hypothetical well 500 m (nearest distance of possible future source of potable water) from the disposal facility in the direction of the flow field.

The dose received by an individual from the predicted concentrations of radionuclides at the well by consumption of 2 L/d of well water has been obtained by using the dose conversion factors of each radionuclide given by ICRP-30 (ICRP, 1983) which are also incorporated in the code.

With this methodology, the performance of the disposal facility has been evaluated by comparing the results from the above code with 10 CFR 61 (USNRC, 1990) for acceptance criteria.

Approach to numerical code development

The code is developed based on the finite difference scheme expressed in equation (1). The process commences with declaration of radionuclide of interest in the main program which makes it possible to call subroutine ISOTOPES (Fig. 3.3). This subroutine on return provides information on half-life (HL), distribution coefficient (DC), dose conversion factor (DGF), and time of containment (TC) in the unsaturated zone for the radionuclide of interest. This information is used while calling subsequently the subroutine PARAMETERS, which on return provides data on retardation factor (R_f) for the radionuclide under consideration, effective porosity, true density of soil and annual water intake by an individual. The time step DT, the space boundaries and number of mesh points on X-axis (NPT), option for higher order correction terms, accuracy required (ACC) and boundary conditions are declared in the main program.

After the data initialisation, the main program calls the subroutine INVALSOL at $t = t_0$ which on return contains the values of C at the stated number of (NPT) points at time $t_0 + \delta t$. The main program also calls subroutine COEFFTS which provides details of coefficients, a, b, c and d. The source-term is defined through the coefficient d in equation (1). INVALSOL calls internally other subroutines BOVALSOL, COEFFTS, CDIFFVAL, DERIVAL and LINEQSOL. Hence these subroutines have to be loaded along with main program. LINEQSOL is a system subroutine which solves set of linear equations of the form A*X = B where A is a square matrix of order NO, X and B are matrices each with NO rows and NR columns. The solution X on return overwrites B. On return from INVALSOL, the main program calls subroutine DOSIMETRY to evaluate the dose received (Sv/y) by an individual at specified time intervals and the same is written in a file. INVALSOL needs to be called as many times as required by the radionuclide mix, to obtain the peak doses for each nuclide. The flow chart of 1-D finite difference code programmed accordingly for radionuclide migration in this study is shown in Fig. 3.3.

METHODOLOGY OF PERFORMANCE ASSESSMENT IN THE CASE STUDY

The disposal site at CWMF, Kalpakkam is located at about 800 m NW of Madras Atomic Power stations (MAPs) spread over 10 hectares. The disposal area is gently sloping towards east and lies between the contours of 11.58 m and 14.02 m above sea level. About 1 km to the west of the site flows Buckingham Canal and eastern side borders the Bay of Bengal. The area is devoid of any natural drainage.

The present case study has the objective of quantitative forecasting of the performance of the ground disposal system for nuclear wastes (*i.e.* the evaluation of the radiological impact of the disposal practice at the disposal facility) located at Kalpakkam, Chengai-MGR District, Tamil Nadu, India. The study attempts to compare the forecast results with regulatory limits stipulated by the Code of Federal Regulations of USA 10 CFR 61 (USNRC, 1990) since national regulatory requirements have not yet been stipulated by the Atomic Energy Regulatory Board (AERB) in India.

The present case study does not attempt detailed modelling of transport in the unsaturated zone which is a thin layer. However, the travel times are computed based on the groundwater velocities in the unsaturated region to arrive at appropriate source-term and arrival times of radionuclides in the saturated region of the aquifer. Attempt has also been made to take into account the attenuation caused by appropriate physical processes namely retardation and radioactive decay, in order to appreciate the role played by each in the overall migration. The following inputs outline certain aspects of the performance assessment at this site :

(i) The assessment deals exclusively with radionuclides having half-lives greater than 5 years and which are projected to contribute over 90% of effective dose equivalent from all pathways during any time period of concern. The present study is mainly concerned, therefore, with assessment in respect of ^{3}H, ^{14}C, ^{137}Cs, ^{90}Sr, and ^{129}I.

(ii) The effective dose equivalents of the above isotopes to man from exposure *via* well water use at 500 m from the disposal facility have been computed from the dose conversion factors provided in the ICRP 30 (ICRP, 1983) assuming a water consumption from the well amounting to 2 L/day, per person.

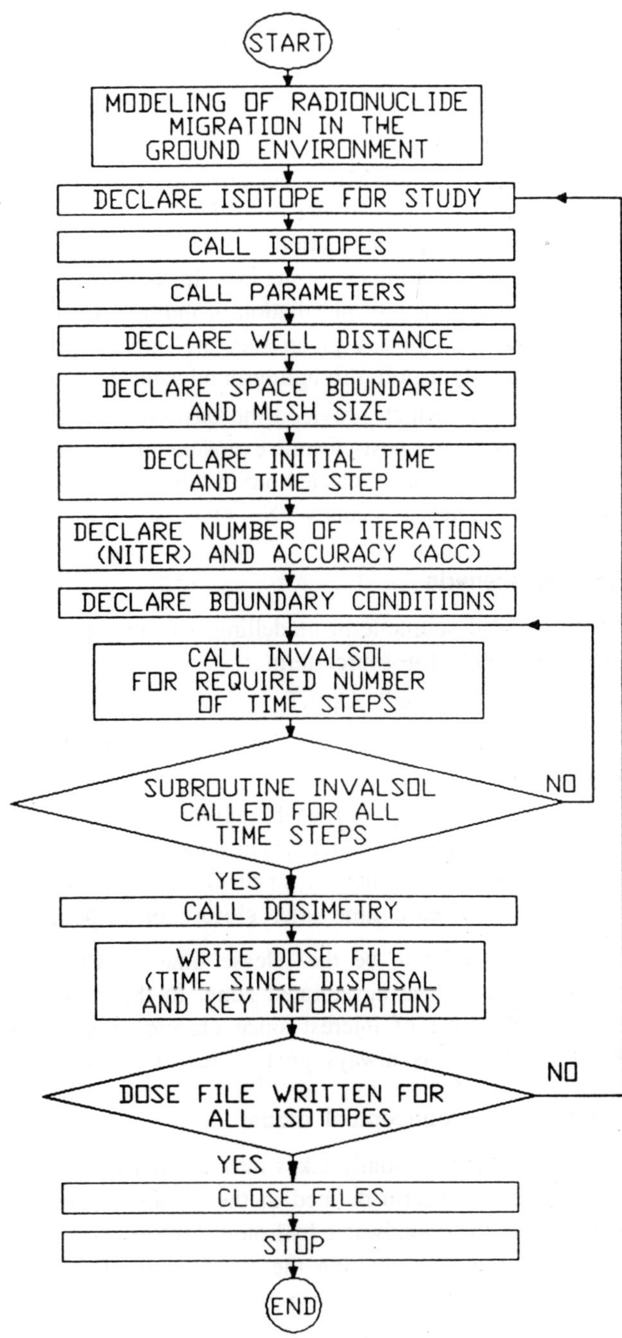

Fig. 3.3. Flowchart of 1-D finite difference code.

The assessment approach involves :

* Specification of waste inventory and characteristics
* Enumeration of geo-hydrological features of the sites
* Selection or development of release/exposure scenarios

* Formulation of conceptual model to describe the transport of radionuclides *via* groundwater
* Prediction of annual dose equivalents received by the general public, based on the 1-D finite difference code.
* Evaluation of results by comparing them with performance objectives.

The facility at CWMF is designed to segregate, collect, treat and dispose of low and intermediate level radioactive waste arising from the operation of 2 × 230 MW(e) Pressurised Heavy Water Reactors (PHWRs) and other installations. Over 95% of the waste generated is stored in a retrievable manner in engineered containment structures. In view of the moderate nuclear programme in the country and the national approach of installation of multiple facilities like reactors, reprocessing facilities and nuclear research facilities at any given site, the impact from intrusion scenario is likely to be less compared to that from transport through groundwater because of the extended institutional control available.

The studies which provided the basis for the 10 CFR Part 61 promulgation demonstrated that the major pathway for post-closure exposure of the public is the groundwater pathway. Accordingly, the performance assessment carried out in this study at Kalpakkam site has also concentrated on groundwater pathway for release scenario. No other scenario appeared more relevant for the site conditions.

Release scenario

As no detailed source-term modelling was attempted in the study, the following equation (2) was used (NSARS, 1992) for source-term :

$$S(t) = (Q/T_s)\, e^{\lambda * t},\ O\mu t\mu T_s \quad \ldots (2)$$

$$= O,\ T > Ts \quad \ldots (3)$$

where :

O is the inventory of each radionuclide (Bq)

λ is the radioactive decay constant (T^{-1})

t is the time of interest since closure of disposal operations (T)

T_s is the release time after closure of disposal operations (T)

T_s is assumed to be the release time in which the inventory of a given radionuclide (Q) from repository enters the unsaturated zone. In the present study T_4 is assumed as 20 years. The term 't' refers to the time of interest since closure of disposal operations. The relation between the 't' and the T_s is that 't' is always greater than T_s after the complete release of inventory has taken place.

Outline of the conceptual model

The overall model initially takes into account the input of radionuclide inventory into unsaturated zone as per the source-term defined in the equation (2) and equation (3). Based on the site characteristics with respect to groundwater hydrology, sub-surface geology and repository design, a conceptual model was developed incorporating the source-term. The conceptual model is shown in Fig. 3.4. The step-wise methodology followed in radionuclide transport modelling is represented in Table 3.2.

A procedure to solve the equation (1) has been developed employing numerical solution techniques and these details have been discussed in Section 2.1.

Given the initial value C_0, at $t = t_0$ the finite difference scheme (FDS) used, finds the solution at $t = t_0 + \delta t$. The equation is solved by the Crank-Nicolson approach, and a code has been developed for the same.

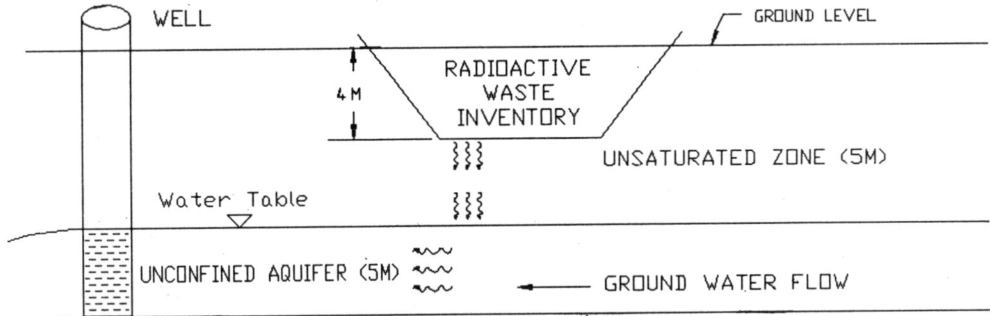

Fig. 3.4. Conceptual model for release of radionuclides at Kalpakkam.

Table 3.2. Steps involved in the radionuclide transport modelling

I.	**Groundwater flow and nuclide transport in unsaturated region**
	Step 1: Evaluation of velocity of groundwater in the unsaturated region (V_u)
	Step 2: Evaluation of retardation factor (R_f)
	Step 3: Computation of time taken for each radionuclide to travel 5 m of unsaturated zone
	Step 4: Evaluation of input concentration into the saturated region
II.	**Groundwater flow and nuclide transport in saturated region**
	Step 1: Determination of groundwater velocity
	Step 2: Assumption of longitudinal dispersivity
	Step 3: Modelling of transport in groundwater of individual radionuclides by 1-D numerical code
III.	**Evaluation of exposures**
	Step 1: Estimation of radionuclide concentration at well (assuming daily intake of 2 L/d and based on adult ingestion rates of ICRP).
	Step 2: Log-log plots of dose received (Sv/year) against time (year)

This code developed at CWMF (BARC), Kalpakkam (Sasidhar *et al.*, 1992) has been verified with varied cases where analytical solutions were available, thereby validating the correctness of the numerical algorithm..

The code has been applied in the present study for migration of the five selected nuclides of concern individually, for band release of these radionuclides, with release time of 20 years into the saturated region. The code was run with following boundary condition:

$C_t = 0$ at $X \pm s$ for all t . O

The code has been developed in Fortran-77 language and is presently operable in SINTRAN operating system of Norsk Data 570 super-mini platform.

Evaluation of results

The safety assessment of the disposal facility has been carried out with the assumptions listed in Table 3.3. Some of the data were derived from the actual determinations and others were assumed reasonably for the purpose of computation, based on prior studies (Sasidhar, 1993) and literature references.

A typical representation of the results obtained from the application of 1-D numerical code after evaluation of exposures is presented in Table 3.4 as per the radionuclide inventories provided by Krishnamoorthy (1992) for total inventories of 1 TBq and 10 TBq, respectively. The time in years for maximum dose to be experienced at an observation well and the maximum dose (Sv/year) for each nuclide can be visualised from the log-log plot in Fig. 3.5 for total inventory of 1 TBq.

The time for experiencing the maximum dose, calculated for the nuclides ^{3}H, ^{14}C, ^{80}Sr, ^{137}Cs and ^{129}I, were 29 years, 35 years, 164 years, 981 years and 35 years, respectively. The time of maximum dose followed the order : $^{137}Cs > ^{90}Sr > ^{129}I > ^{14}C > ^{3}H$. The time of maximum dose was experienced only after a very long time (981 years) for ^{137}Cs which is due to its high retardation factor. The retardation factor is a function of its distribution coefficient (see equation 1) which is characteristic for the particular nuclide, influenced by the soil environment. The above trend clearly shows that the time of maximum dose decreases with decreasing retardation factor. The trend also confirms that the dose received is a function of half-life of the radionuclide.

Table 3.4 brings out the influence of waste composition in terms of the five nuclides considered from the nuclear power plants. It can be seen that for a total inventory of 1 TBq the doses received meet the 10 CFR Part 61 regulatory requirement for all isotopes. A similar trend was noticed for a total inventory of 10 TBq also. These results bring out the salient point that even not considering radioactive decay, the regulatory requirements of 10 CFR Part 61 are met quite adequately. This does indicate that Kalpakkam site provides adequate attention to the nuclide concentration during the transport phenomenon. At Kalpakkam, 10 TBq per year is permitted for disposal and this limit has not been utilised fully in the past 10 years of disposal operation. This would further suggest that the site disposal operations meet the 10 CFR Part 61 regulatory requirements quite satisfactorily. The trends discussed above are graphically shown for inventory of 1 TBq in Fig. 3.5. The comparison of results with AERB limits indicate that trends of compliance are similar to that obtained from 10 CFR Part 61.

With the waste composition which included the effect of decay process, the 1-D numerical code was run and the results are presented in Table 3.5. The effect of radioactive decay on doses received has been presented in Fig. 3.6 for total inventory of 1 TBq after provoding necessary decay corrections.

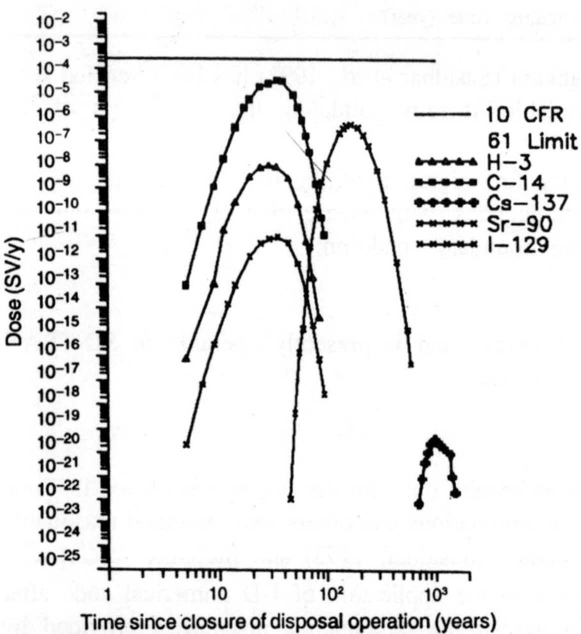

Fig. 3.5. Radionuclide transport at Kalpakkam (without considering radioactive decay) with total inventory of 1 TBq.

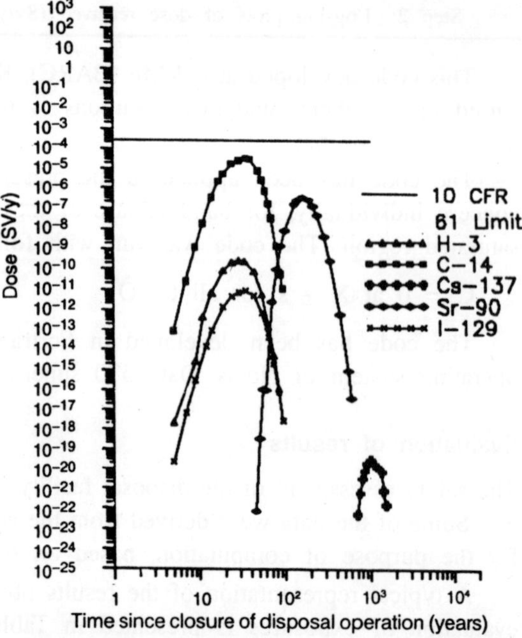

Fig. 3.6. Radionuclides transport at Kalpakkam (considering radioactive decay) with total inventory of 1 TBq.

Table 3.3. Assumptions made for 1-D radionuclide migration in the ground environment

Source and general characteristics		
(1)	Radionuclide inventory for 30 years	1-10 TBq
(2)	Release area of earth trench	90000 m^2
(3)	Depth of the disposal facility	4 m
(4)	Water percolation rate	0.36 m^3/year
(5)	True density of the soil (ρ)	2.5 g/cc
(6)	Distance between the well and the centre of release area at southeast of the facility boundary	500 m
Unsaturated zone		
(7)	Thickness of unsaturated zone (d_{uz})	5 m
(8)	Porosity (x)	0.4
(9)	Moisture content (x_u)	0.3
Saturated zone		
(10)	Saturated thickness of the aquifer	5 m
(11)	Effective porosity (x_s)	0.35
(12)	Groundwater pore velocity (29.2 m/year)	8 cm/day
(13)	The longitudinal dipersivity is assumed as 20 m as per Pickens and Grisak (1981)	
Nuclide characteristics		
(14)	Radionuclides modelled	^{3}H, ^{14}C, ^{129}I, ^{137}Cs and ^{90}Sr
(15)	Half-lives (year)	^{3}H-12.95, ^{14}C-5730, ^{137}Cs-30.0 ^{90}Sr-28.8 and ^{129}I-1.6E+07
(16)	Distribution coefficients (m^3/kg)	^{3}H, ^{14}C and ^{129}I = 0.0 ^{137}Cs = 0.02 and ^{90}Sr = 0.0015
(17)	Dose conversion factors (Sv/Bq) (ICRP 30, 1983)	^{3}H-1.7E-11, ^{14}C-5.7E-10, ^{137}Cs-I.4E-08, ^{90}Sr-3.6E-08 and ^{129}I-7.4E-08

Table 3.4. Maximum exposure doses and times for individual radionuclides without considering radioactive decay*

Total inventory		*10 TBq*	*1 TBq*	*Acceptance criteria*
Radionuclide	*Time of maximum/ (years)*	*Maximum dose (Sv/year)*	*Maximum dose (Sv/year)*	*10 CFR 61 limit (Sv/year)*
^{3}H	29	5.10E-07	5.10E-08	2.5E-04
^{14}C	35	1.72E-04	1.72E-05	2.5E-04
^{137}Cs	981	2.17E-19	2.17E-20	2.5E-04
^{90}Sr	164	2.49E-06	2.49E-07	2.5E-04

*Radionuclide inventories according to Krishnamoorthy (1992).

The regulatory compliance with AERB limits has also shown nearly same trends in the range of 1 to 10 TBq total inventory.

Table 3.5. Maximum exposure doses and times for individual radionuclides considering radioactive decay

Total inventory		*10 TBq*	*1 TBq*	*Acceptance criteria*
Radionuclide	*Time of maximum dose (years)*	*Maximum dose (Sv/year)*	*Maximum dose (Sv/year)*	*10 CFR 61 limit (Sv/year)*
^{3}H	29	1.36E-08	1.36E-09	2.5E-04
^{14}C	35	1.72E-04	1.72E-05	2.5E-04
^{137}Cs	981	1.57E-19	1.57E-20	2.5E-04
^{90}Sr	164	1.80E-06	1.80E-07	2.5E-04
^{129}I	35	4.49E-11	4.49E-12	2.5E-04

*Radionuclide inventories according to Krishnamoorthy (1992).

As a conclusion it may be stated that the Kalpakkam site has favourable characteristics to attenuate the radionuclide migration and also provide performance assurance that the parameters from disposal operations meet the internationally accepted 10 CFR Part 61 regulations provided the inventories of ^{14}C and ^{129}I have to be carefully monitored and controlled. This is necessary in order to ascertain and control the impact of disposal on human population. The other three isotopes *viz.* ^{3}H, ^{137}Cs and ^{90}Sr would meet the regulatory requirements quite well even under worst case scenario. Even ^{129}I and ^{14}C could meet the stringent regulatory requirements despite having a very low retardation factor of 1. This study has shown that stringent surveillance has to be carried out for ^{14}C and ^{129}I at the Kalpakkam site and advanced immobilisation methods have to be adopted for better containment. Above all, appropriate assay methods have to be introduced to control the actual disposal of these two isotopes into the repository. Field experiments should be carried out at least in the unsaturated zone using lysimeters to understand the transport of radionuclide migration.

Summary of results and discussion

The results obtained from the safety assessment studies compared with the regulations 10 CFR 61 (USNRC, 1990) showed that the maximum doses of pertinent radionuclides to which a consumer of well water in the neighbourhood of the disposal site might be exposed under the most unfavourable conditions were well below the regulatory limits in spite of the fact that the source-terms assumed were of relatively larger magnitude and most other parameters were conservative. If the safety assessments had been carried out for actual inventories, the operating experience of waste disposal at Kalpakkam would then have indicated that the quantity of waste disposed so far is far below the limits specified by AERB and that the site meets the regulatory requirements.

It was observed that for a total waste inventory of 1 TBq, the doses received meet the regulatory requirement in respect of all isotopes considered. A similar trend was noticed for a total inventory of 10 TBq also. These results clearly demonstrate that despite not considering the radioactive decay, the regulatory requirements of 10 CFR Part 61 are met quite adequately. It can reasonably be concluded that, subject to close monitoring and control of ^{14}C, Kalpakkam site provides adequate attenuation to the nuclide concentration during the transport phenomenon. At Kalpakkam, 10 TBq per year is permitted for disposal and this limit has not been utilised fully in the past 10 years of disposal operation. This further suggests that the site disposal operations can meet the 10 CFR Part 61 regulatory requirements quite satisfactorily.

The significant results obtained from the safety assessment study are recounted as under :

1. The time of experiencing maximum dose by ingestion of water from a well, 500 m downstream from the disposal site, in respect of the nuclides ^{3}H, ^{14}C, ^{90}Sr, ^{137}Cs and ^{129}I is 29 years, 35 years, 164 years, 981 years and 35 years, respectively. The time of maximum dose follows the order : $^{137}Cs > ^{90}Sr > ^{129}I > ^{14}C > ^{3}H$.
2. The maximum dose received itself follows the order $^{14}C > ^{129}I > ^{90}Sr > ^{3}H > ^{137}Cs$.
3. The maximum dose received itself from exposure to ^{14}C appears to be high and the time for maximum dose is also of the order of a few tens of years; hence this isotope requires special attention in the disposal operations from the long-term hazard considerations.

CONCLUSION

This study has provided a methodology and performance assessment was carried out using simple mathematical models of the radioactive waste disposal operations at CWMF, Kalpakkam. Performance assessment contributes to and facilitates many aspects of the licensing process. The most common perception is that performance assessment simply involves making compliance calculations to ensure that exposure to the general public from a low-level radioactive waste disposal facility is maintained below regulatory limits. This is too simplistic a perception on two accounts : first, it is important to recognize that performance assessment calculations are not simply a compliance tool; they can also be used to reduce the cost of the facility by appropriate design and site characterization activities during the licensing process. Engineering and design of a disposal system should use performance assessment in the same manner as analysis is normally used in the design of any other system. Secondly, performance assessment calculations should not be considered as absolute predictions. Results should be used to verify whether the expected doses within a range of uncertainity will be below specified tolerance limits, as distinct from predicting that the dose will be a specific value.

It is well recognised that it is an exceedingly difficult exercise to evaluate simultaneously all the factors controlling transport and attenuation of radionuclides in the ground environment. With the added complexity of a real geologic situation, it appears that the predictive mathematical approach only could offer some advantage. The model provides a convenient framework in which to incorporate field determined parameters and would seem to be the most suitable tool for quantitatively examining the features of performance assessment.

SCOPE FOR FURTHER STUDIES

A major and important task of the present study has been to produce a model capable of predicting the rate of radionuclide release from a low-level radioactive waste shallow land disposal facility. In this context and based on the experience gained in testing of the model developed, four major areas have been identified for future work :

(a) quantification of low-level waste inventories and disposal practices;

(b) coupling of the process models to form a more unified structure;

(c) further studies on the development of the source-term models, the container degradation models, waste form leaching models, and

(d) model testing and validation.

REFERENCES

Anderson, M.P., 1979 : Using models to simulate the movement of contaminants through groundwater flow system. *Critical Rev. Env. Con.*, 9 : 97-156.

Broyd, T.W., R.B. Dean, G.D. Hobbs, N.C. Knowles, J.M. Putney and J. Wrigley, 1984 : A directory of computer programs for assessment of radioactive waste disposal in geological formations. EUR 8669 EN.

IAEA, International Atomic Energy Agency, 1981 : Shallow ground disposal for radioactive wastes : A Guide book. *Safety Series No. 53, IAEA*, Vienna.

IAEA, International Atomic Energy Agency, 1983 : Concepts and examples of safety analyses for radioactive waste repositories in continental geological formations. *Safety Series No. 58. IAEA*, Vietnam.

IAEA, International Atomic Energy Agency, 1984 : Safety analysis methodologies for radioactive waste repositories in shallow ground. *Safety Series No. 64, IAEA*, Vienna.

ICRP, International Commission on Radiation Protection, 1983 : Limits for intakes of radionuclides by workers. *ICRP publication No. 30*, Pergamon Press, Oxford.

Krishnamoorthy, T.M., 1992 : Radionuclide migration from near-surface burial facilities. *Proc. of Seminar on Transport and Dispersion of Pollutants in Aqueous and Atmospheric Media*, Indira Gandhi Centre for Atomic Research (IGCAR), Kalpakkam.

Mercer, J.W. and R.F. Charles, 1980 : Groundwater modeling : An overview. *Groundwater*, 18 : 2.

NSARS, 1992 : *Near-surface safety assessment of radioactive waste disposal facilities*. Specifications for Test Case 2A of IAEA-CRP on NSARS, Vienna, IAEA.

Pickens, J.F. and G.E. Grisak, 1981 : Modeling of scale-dependent dispersion in hydrogeologic system. *Water Resources Research*, 17 : 1701-1711.

Sasidhar, P., S. Krishnan, K.B. Lal, R. Pitchai and R.V. Amalraj, 1992 : Half yearly progress report (Jan.-June, 1992). Anna University, Madras.

Sasidhar, P., 1993 : Safety assessment of low-level radioactive waste disposal facility at Kalpakkam. 252 pp. Anna University, Madras.

Shipers, L.R. and P.C. Harlon, 1989 : Background information for the development of a low-level waste performance assessment methodology - Identification of potential exposure pathways. *NUREG/CR-5453, SAND89-2509, Vol. 1*, Sandia National Laboratories, USA.

USNRC, 1990 : *Code of Federal Regulations* (CFR), 10 CFR Part 61, USNRC, Washington, D.C.

Chapter 4

Utilization of Solar Energy in Agriculture Sector

D.R. Bongirwar

Solar energy is a heat source and has been used in many applications pertaining to developing countries. One of the main applications, especially in rural areas, is crop drying. The paper examines various aspects of solar crop drying. The evaluation of solar energy and its utilization in various types of dryers in agriculture sector are discussed. Factor affecting the design of these dryers, which are mainly used in developing countries are outlined. The main objective of this communication is to understand the primary factors that inhibit widespread application of solar energy for drying purposes as well as to suggest a course of action that can enhance the viability and implementation of solar drying technology in the developing world.

INTRODUCTION

Solar energy is a very large inexhaustible source of energy. The power from the Sun intercepted by the Earth is approximately 1.8×10^{11} MW, which is many thousands of times larger than the present consumption rate on the Earth of all commercial energy sources. Thus, in principle, solar energy could supply all the present and future energy needs of the world on a continuing basis. This makes it one of the most promising of the unconventional energy sources. It is an environmentally clean and freely available source of energy. India is blessed with plenty of sunshine. The daily average solar energy incident in the country varies from 5 to 7 KWh per square metre. Approximately 5×10^{15} KWh of energy from the Sun is received in India per annum. However, there are a few problems associated with its use as well. The main problem is that it is a dilute source of energy. Even in the hottest regions of Earth, the solar radiation flux available rarely exceeds 1 KW/m^2, which is a low value for technological utilization. Consequently, large collecting areas are required in many applications and these result in excessive costs. A second problem associated with the use of solar energy is that its availability varies widely with time, the variation in availability occurs daily because of the changing day-night cycle, with season and location. Consequently, the energy collected when the Sun is shining must be stored for use during those periods when it is not available. The need for storage also adds significantly to the cost of any system.

Thus the challenge in utilizing solar energy as an energy alternative is of economic nature. One has to strive for the development of cheaper methods for collection and storage so that the large initial investments required at present in most applications are reduced.

Methods of utilizing solar energy for drying

Solar energy has been the oldest and most widely used method of drying crops in developing countries. The methods used to date have largely been based on open-air drying. However, to better utilize this abundant source of energy effectively, systems have to be developed based on specific needs.

Agricultural crop dryers can be designed to use solar energy in various ways. There are two main types of dryers : active dryers and passive dryers. Active dryers use an external device operated, for example, by means of a fan to circulate the air, but passive dryers do not.

Although, passive systems tends to be more realistic for application in developing countries, because of the relatively low initial capital and operating costs, it is possible to use active systems for relatively large scale applications. Their specific uses, therefore, depend on various factors such as availability of a power source, scale of application, location, design and available materials. There are various models in which solar energy can be used for drying.

Open-air drying

In this method the dryers use insulation, wind velocities, ambient air temperatures, and the relative humidity of the air to reduce the moisture content of the crops. This method, which has many variations, has been used extensively because it is cheap to implement.

In one application, the crop is spread on the ground on an area that has been cleared of leaves and grass. The solar energy incident on the thinly spread crop provides heat which is required to evaporate water. The mechanisms by which this is achieved are well understood and have been reported in the literature. The mechanism by which the moisture leaves the crops are : conduction, convection and radiation.

Radiation from the Sun heats the ground and the surrounding air. Heat is transferred to the crop by conduction from the ground, by conduction and convention from the air close to the crop and by radiation from the Sun and the ambient air. The moisture at or near the surface of the crop is thus heated and is vaporized which causes movement of moisture to the surface. The heat transferred to the drop may also be transmitted to the inner core by conduction, which will in turn liberate further vapour. Thus, the rate of drying depends on the available radiation and the ground temperature.

During open-air drying, efforts must be made to ensure that the ground temperature is not so high that it will damage the crop. Sometimes crops are spread out on cement or asphalt pavement. This method is more expensive than ground drying because of the initial cost of constructing the pavement. It does, however, have the advantage of providing a higher surface temperature compared to the ground.

Because there is little or no vertical circulation of air through the crops, they have to be spread thinly when dried by open air method which makes it necessary to have a large area of land. Although, this method is relatively low cost, it has many disadvantages as mentioned earlier.

Another method of open-air drying involves raising the crop above the ground to dry by placing it in a tray and resting it on an open raised platform. This arrangement reduces the problems encountered when drying on the ground. Rodents and insects, such as ants, find it more difficult to infest the crops. To better utilize the solar energy, the base of the tray should be made of wire mesh and the tray should be painted black to attain higher temperature for heat transfer. The depth to which the crops are packed in this instance be higher compared to when there is no wire mesh because of the increased circulation of air through the crops.

Open-air drying on trays or racks can be used for beans, coffee, and cocoa etc. For other crops such as grapes, the trays are stacked on top of each other with a roof overhead that protects the crops from dew or rain (this method is used in Australia). The method is cheap and can be used to process large quantities of grapes.

Direct drying

The discussion on open-air drying has shown that the air used for taking moisture from the crops is effective at the ambient temperature and relative humidity (RH). However, it is well known that air at a given temperature and relative humidity when heated experiences a decrease in relative humidity. Thus heated air takes away more moisture from the crops than unheated air. This fact has been used in various solar crop dryers that use direct and indirect methods of heating the ambient air.

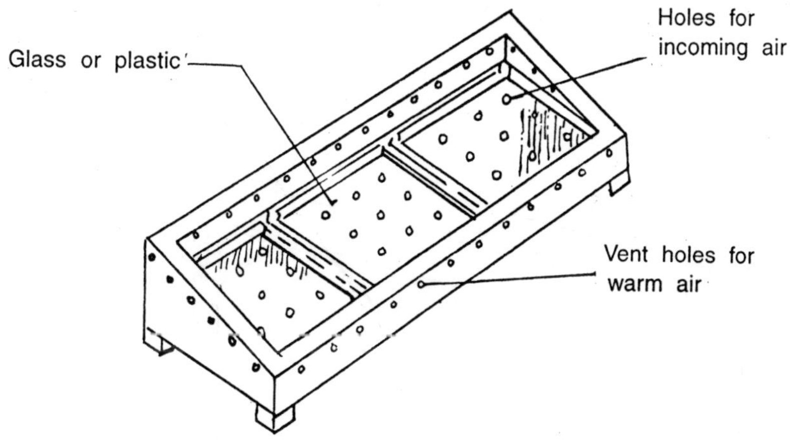

Fig. 4.1. Diagram of a typical direct dryer.

Direct dryers consist of an enclosure with a transparent cover. The crops are placed on trays in the enclosure and the solar energy is absorbed by both the crops and internal mass of the dryer. The elevated temperature causes evaporation of water from the crops. The warm moist air escapes through vent holes usually located on the side of the dryer and fresh air is drawn in through holes at its base. Therefore, there is a continuous flow of ambient air through the dryer.

Cheaper dryers, which use solar energy by the direct method, do not use insulation. The top and sides of the dryers are covered with a transparent cover that serves to collect solar radiation and protect the crop from dirt, bad weather and insects. The design of direct dryers is such that the crops are directly beneath the transparent top covers that are sloped at the appropriate angle to collect the optimum solar radiation. The magnitude of this angle can be calculated for various locations. However, because it changes throughout the year, a recommended value is α = latitude + 10°, although this inclination will not give the best performance when the dryer is used on a year-round basis. It should be noted that tracking the sun would be an expensive exercise making the application very uneconomical.

1. Air inlet mainfold
2. Solar collector
3 to 6. Drying chamber
7. Chimney
8. Stands

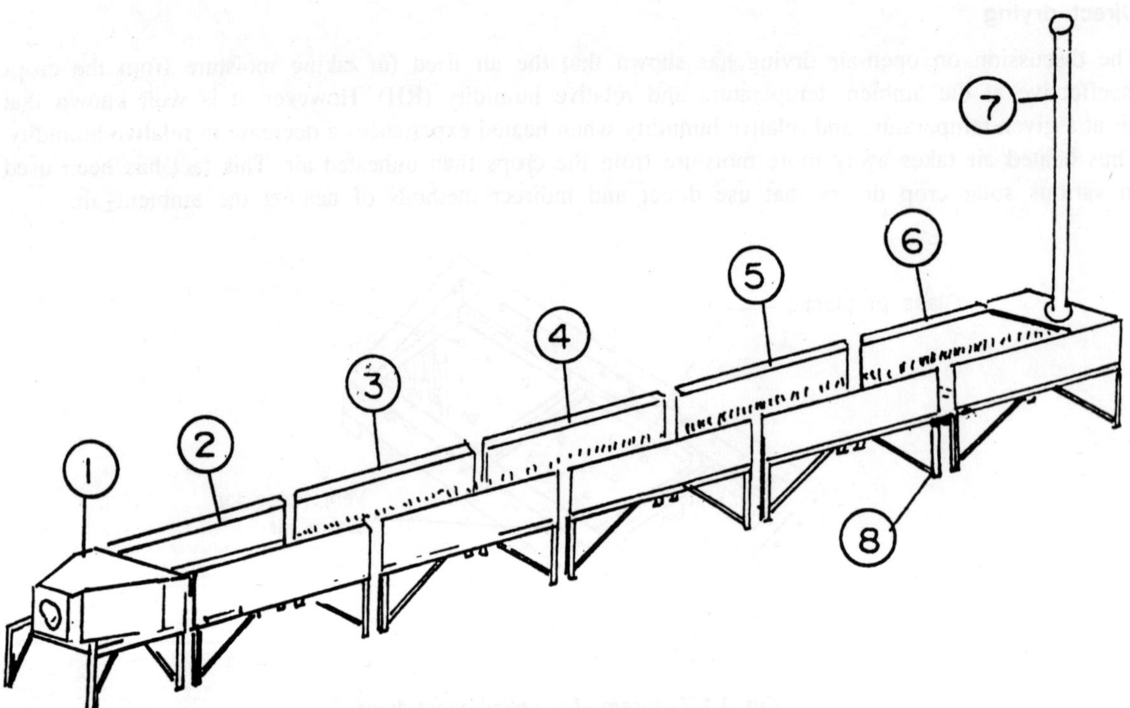

Fig. 4.2. Natural convection solar crop dryer.

Design considerations of direct heating systems involve maximizing the temperatures in the dryer as it determines the load. As mentioned earlier, the insulation must be effective and the materials used can be chosen from sawdust, wood shavings, fibre glass, coconut fibre, straw and others. The thickness of insulation will depend on the material used and the temperature difference between the ambient air and that in the chamber. Wind speeds should also be considered in the design of heating dryer as this causes losses by convection.

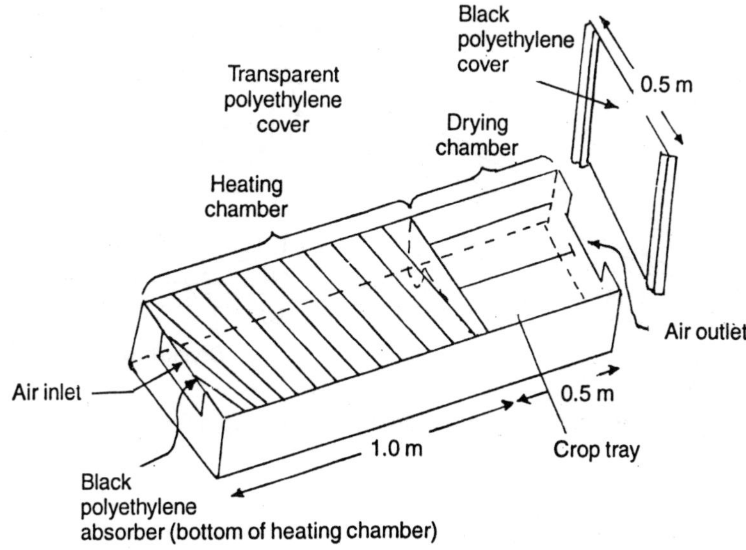

Fig. 4.3. Portable indirect solar dryer.

Indirect drying

In indirect drying, solar energy does not come into direct contact with the crops. The air used for drying is heated in a solar air collector and then circulated through the crop to reduce its moisture content. It is possible to use a fan for circulating the air or just natural convection. The solar energy is collected in the air heater by means of the greenhouse effect using the transparent cover and absorber. Because of buoyancy effect, the warm air rises through the sloped collector and into the drying chamber where the crops are placed. Many designs are possible depending on the mode of circulating the air.

The main objective in the design of indirect dryers is to produce adequate temperature that can be used to dry the crops. The design of air heaters for developing countries must be aimed at minimizing the material requirements for the dryer to be economically attractive. The agent for moving the air

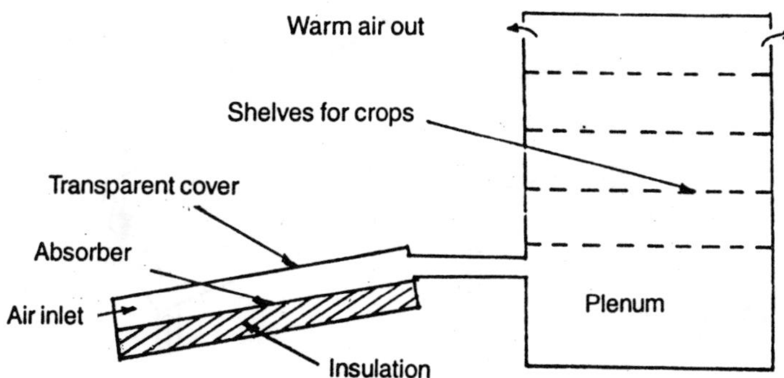

Fig. 4.4. Schematic diagram showing the typical features of an indirect crop dryer.

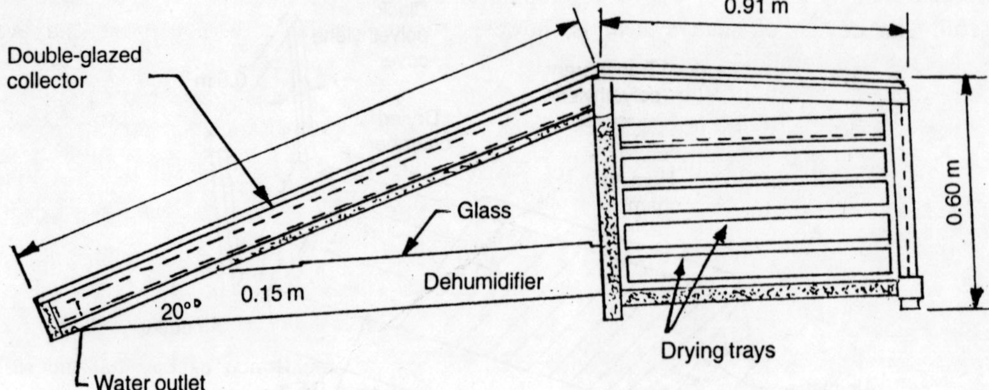

Fig. 4.5. Schematic diagram of an indirect crop dryer developed by Headly and Springer (1973).

in most of the designs available is mechanical, such as a fan. With the restrictions mentioned earlier that exist in developing countries, these designs may be expensive in many situations. The effect of buoyancy can be used by incorporating a chimney. This creates a draft that can cause an adequate mass flow rate of air to pass through the collector and then through the crops. There does not appear a great deal of interest in this method of circulating the air in crop dryers using solar energy, although it is very important in some parts of the world. There is, however, work in progress by some researchers, as previously mentioned, to use natural convection.

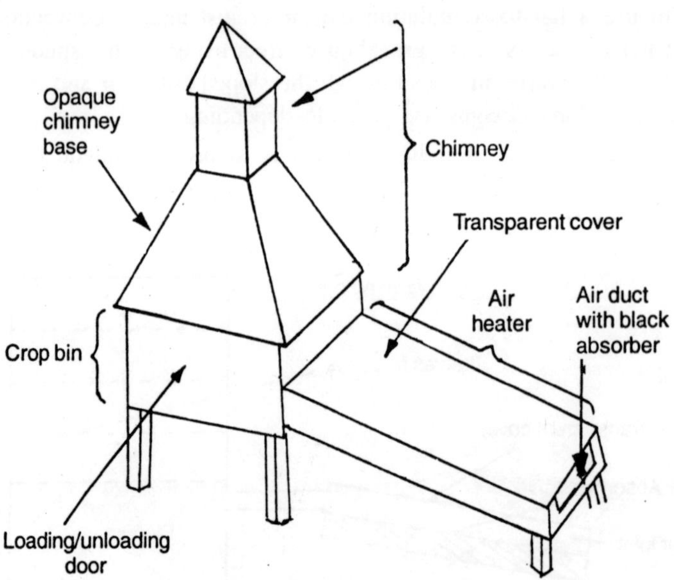

Fig. 4.6. Indirect natural convection solar dryer.

The application of indirect dryers in developing countries has been shown in many studies to be quite useful. The design of such system is, however, dependent on local conditions and available facilities. The relative humidity and temperature of the ambient air, the maximum allowable temperature for the crop and its quantity determine the best design of the system. The design of the chamber depends on whether grains or fruits, for example, are being dried. Another consideration is the versatility of a particular system. It is desirable, although not always feasible, to have solar dryer that can process various types of crops as this is a more economical proposition.

Hybrid and mixed mode drying

The use of solar energy in a dryer so that both indirect and direct heating can be carried out is called mixed mode drying. This method makes use of greenhouse effect in both the drying and in the collector. These dryers are not substantially different from the other types mentioned except that the sidewalls or top use a transparent material such as glass or plastic.

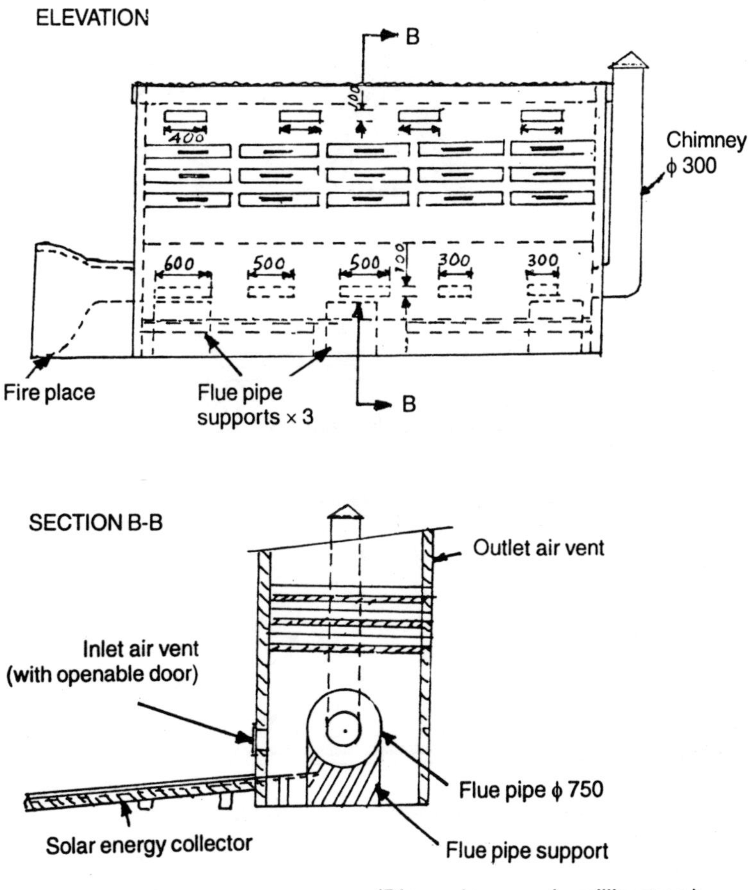

(Dimensions are in millimetres.)

Fig. 4.7. Hybrid solar-wood pyrethrum dryer.

Dryers that use solar energy as well as supplementary heating are called hybrid dryers. The heat source may be fossil fuel, electricity, agricultural waste material etc. The design and operation of such systems is useful in some locations where solar energy is not available to dry crops for a few months of the year. It is also not possible to use solar dryers at night. Therefore, systems that can utilize supplementary heat have possibilities for application in developing countries.

MATERIALS FOR CONSTRUCTING SOLAR DRYERS

Construction of solar dryers for use in developing countries very often poses problems for various reasons. For example, the materials used must be available locally, and the cost of the material would be another factor because this determines the investment that would have to be made by the owner. The expected life-span of the dryer is another factor that must be taken into consideration.

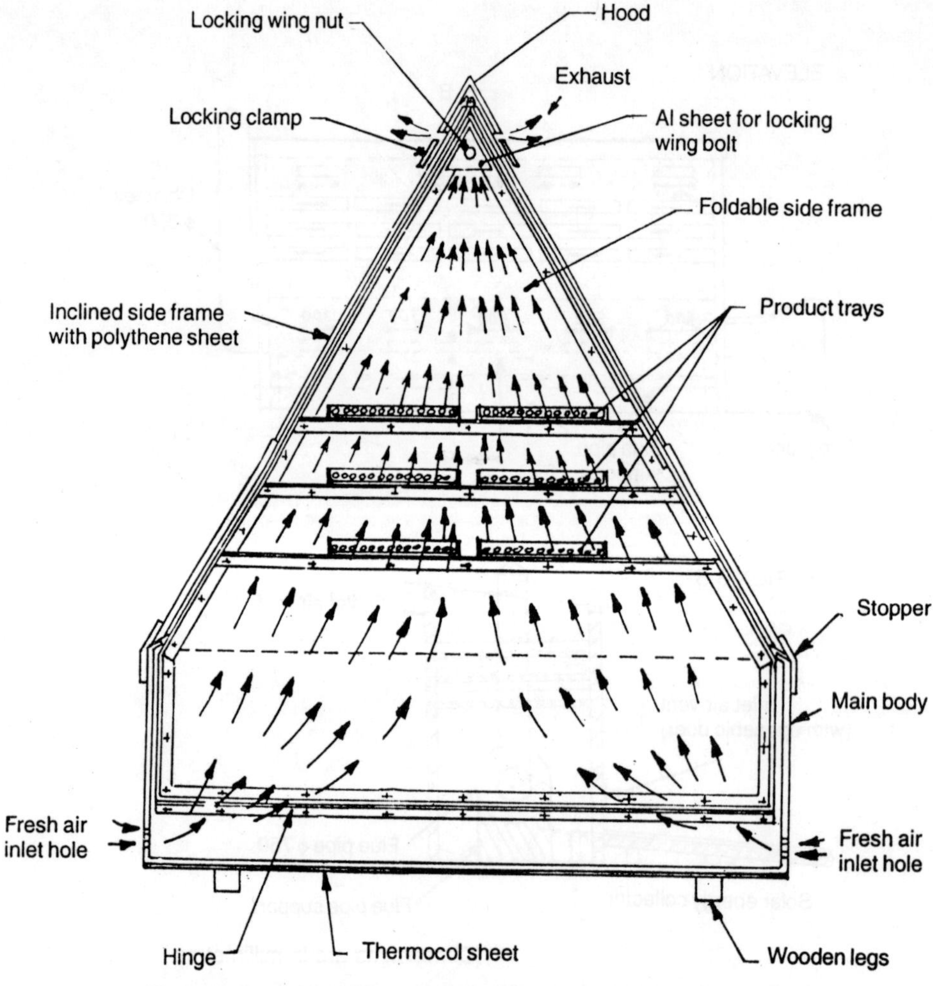

Fig. 4.8. Sectional view of the solar dryer for agricultural products.

The materials that can be used for the transparent cover are glass or plastics. Glass does not deteriorate with age, whereas plastic is affected by solar radiation. In many instances it may be wise to use plastic material if the risk of breakage of glass is high. It is also possible to use glass and plastic on double glazed collectors. The arrangement should, however, be such that the plastic material can withstand the heat and it is advisable to use plastic as the outer covering. This arrangement can also act as a protection for the inner glass cover.

The collectors can be made from materials such as wood, aluminium or galvanized iron. The main objective is that there should be minimum air leakage from the collector, no possibility of the insulation getting wet, easy to manufacture, and low cost. It is possible to use wood and galvanized iron sheets to make good air heaters. In fact, various designs have used these materials for small and large solar collectors. If wood is used it should be treated against attack by insects and painted as protection from wet conditions.

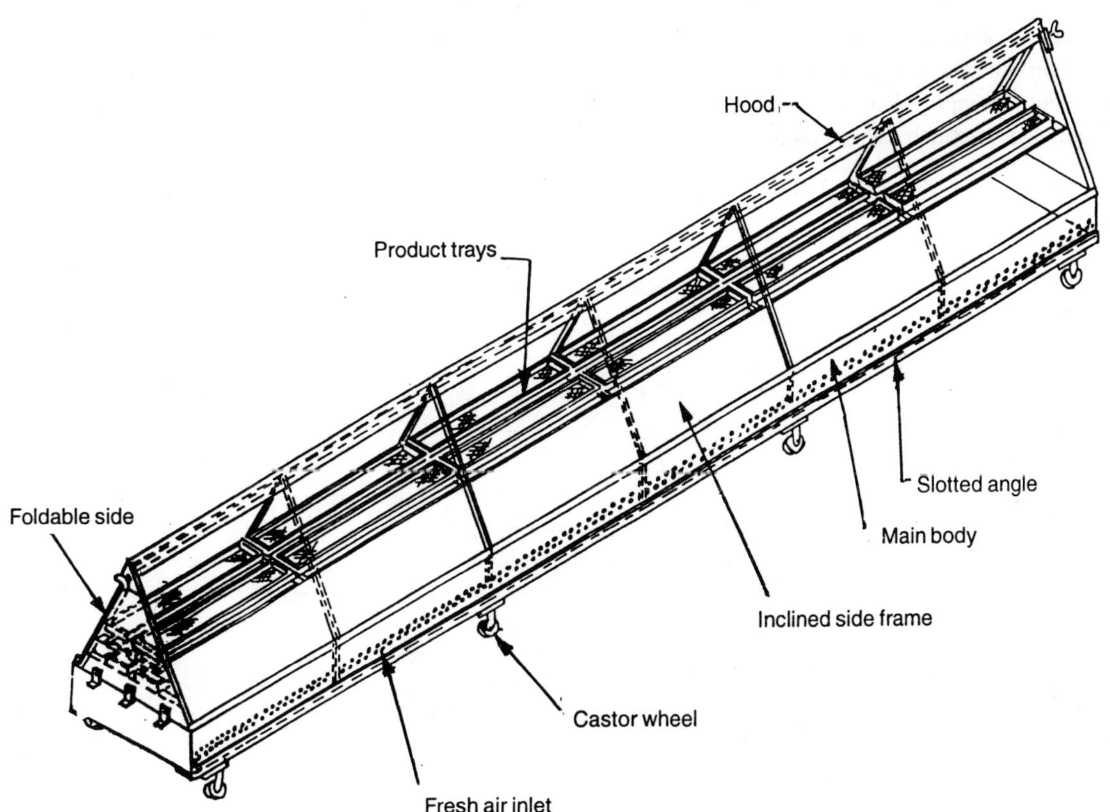

Fig. 4.9. Isometric view of the 100 kg. solar dryer.

Insulating materials are numerous and it should be noted that the use of a locally available thermal insulator does not necessarily imply cheaper equipment because the thickness needed is often greater compared to other materials such as foam or fibre glass wool. For example, if straw is used, a cavity of about 15 cm may be needed compared to 5 cm for fibre glass wool, and more building material is needed for the dryer using the straw.

The drying chamber may be made from plywood or sheet metal. Transparent materials such as glass or plastic are used for one or more of the walls in mixed mode dryers. The shelves on which the crops are placed are made from simple wire mesh.

Materials that are needed for solar crop dryers are simple and available in most developing countries. There may be cost restrictions because of import taxes that causes prices to be high for some simple materials. This problem can, however, be minimized by finding alternatives in some instances.

EFFECT OF DRYING ON THE QUALITY OF VEGETABLES

Dehydrated commodities can lose some of their nutrients during the dehydration process. Oxidation is a primary cause of loss, particularly in the case of ascorbic acid, but non-oxidative losses also occur. There is evidence that non-enzymatic browning reduces the value of the protein. It is observed that glucose reacts with acetic, citric or lactic acid in a model system to produce brown pigments and carbon dioxide. Oxygen accelerates the process. Heat is also considered as a main factor responsible for protein damage. Heat damage is due to a time-temperature relationship. There is evidence that heat damage is much more likely to occur when the initial moisture content is high. Based on the

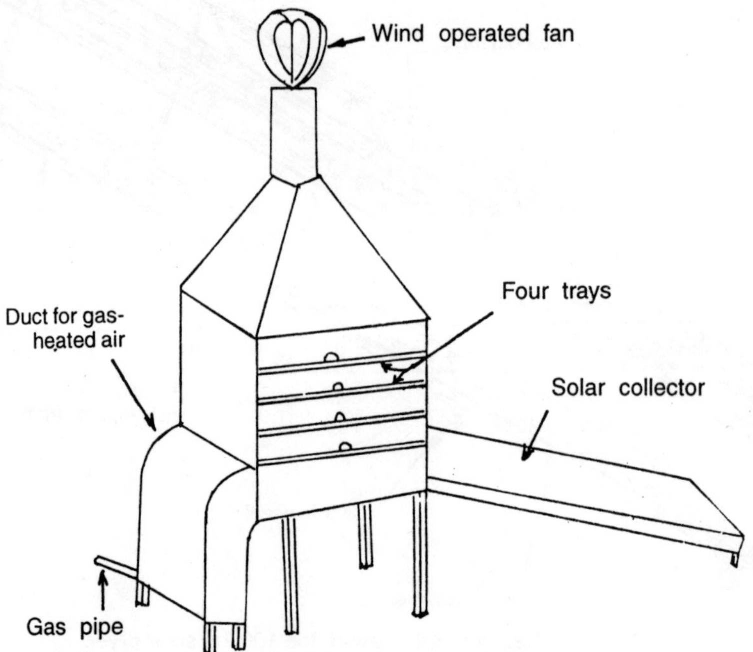

Fig. 4.10. Solar vegetable dryer with auxillary gas heater (capacity of 85 kg on four trays).

results of the Regional Research Laboratory of USA, the metal grid reduces losses during drying of riced potatoes, as stainless steel trays hasten drying and reduce the length of drying time.

It is very important when drying green vegetables to have a final product that meets with consumer acceptance. Good colour, and consequently carotene content, can be retained by blanching. Appropriate storage conditions can help to retain the desired qualities as well.

STORAGE PROBLEMS OF DRIED VEGETABLES

High moisture content will contribute to quality deterioration and indirectly to a decrease in quantity. Even under the most controlled conditions, the stored produce cannot be granted as mould-free. The infestation of moulds in the presence of moisture increases the respiration rate of moulds. In the process of respiration there is a liberation of heat and water, and therefore, the moisture content of produce increases which in turn increases the reaction until the produce spoils. The chemical analysis showed no serious changes. Moisture content, protein, ether extract, fibre, carbohydrate and ash content were almost the same. The fungal load did not exceed 100,000 spores/g in the examined samples. However, further tests will be needed productwise to be conclusive.

Specific technical problems

There are a number of technical problems related to solar drying technology in developing countries. Solutions of these problems are not simple and they constitute interesting and complex research areas. The following are specific technical problems that need resolution to enhance the diffusion of solar energy for drying.

(A) Lack of rigorous field data studies : Despite the use of solar energy for drying for many years, there is still a lack of rigorous field studies on actual process performance under a variety of operating conditions. Therefore, many installations are still over- or under-designed, which leads to a waste of resources or a mismatch with user's needs. An inadequate substantive database also makes rigorous comparison of techniques, new or existing, and their economic assessment. There is an urgent need, therefore, to closely monitor a number of existing units to provide this substantive database.

(B) Lack of understanding of fundamentals : The study and development of solar drying system depends on a collaboration between the architect, the thermotechnics engineer and the solar physicist. It is because of this multidisciplinary nature of solar technology, that, it has not received much attention from research and development groups in developing countries. Consequently the technology has not achieved the efficiency that is possible based on current knowledge in developing countries. Furthermore, considerable time and resources have been lost in pursuing designs and approaches that are at odds with the fundamentals of this technology. For solar drying technology to reach its full potential in developing countries, therefore, there is an urgent need to enhance the understanding of the technology's fundamentals among research and development personnel.

(C) Optimization : There have been few attempts to rationally optimize the existing designs to reduce capital costs and improve performance based on the factors mentioned above. It is evident from literature survey that considerable improvements could be achieved through simple optimization techniques, leading to enhanced viability of this particular solar energy application.

Solutions to these problems are not simple, especially while using this technology at medium range temperate applications.

ECONOMIC VIABILITY

In addition to technical considerations there are also economic aspects that need to be closely examined because of scarce fináncial and material resources in most developing countries. The economic viability of solar drying is difficult to assess, even at this stage of development mainly because of (a) lack of substantive data particularly in the area of industrial and feed lot applications and (b) absence of an agreed methodology that allows economic data to be compared among the various applications, under varying conditions, thus enabling rigorous economic comparison between solar and other renewable energy technologies or conventional energy sources.

There is an urgent need to more fully understand the social and economic factors that control determination of any solar energy technology, so as to ensure that the technology diffuses as rapidly as possible.

SOCIAL AND INSTITUTIONAL BARRIERS

Despite the importance of social and institutional implications of the technology, there have been few studies on this topic. The most important of inhibitions to diffusion of solar drying technology are the causes and consequences of poverty. Secondly, the most striking reason why some innovations are successful while others are not is the extent to which a new technology meets the intended user's need.

At the level of the societal group, there are several factors that tend to inhibit the spread of solar energy technology for drying. As soon as more social benefits in terms of health, convenience, leisure etc., are realized, solar energy technologies will gain in attractiveness. Social feasibility, however, will be highest among the poorer families.

At the national level strong government interest and backing is necessary if solar drying technology is to be disseminated rapidly. This backing should be manifested in a national infrastructure that coordinates activities, provides technical assistance, and allocates resources. In addition, there should be a close cooperation between implementing and research and development agencies. Government can also have an important influence on the financial viability of solar drying technology through tax incentives, subsidies etc.

STRATEGIES TO PROMOTE DIFFUSION OF SOLAR DRYING TECHNOLOGY

(1) **Monitoring of existing field units :** Though considerable information exists on the laboratory performance of solar drying systems, there is a dearth of substantive data from actual performance of field units over a long period. It is recommended that a small number of units that are representative of a variety of techniques be selected in several countries. These should be monitored closely for 2-3 years to evaluate their performance in relation to laboratory units. This would also help to identify persistent problem areas that require further research and development.

(2) **Application of standardized testing and evaluation methodology :** Some of the technical and economic information on solar drying collected from developed and underdeveloped

countries is difficult to compare and evaluate because of different reporting methds and evaluation techniques. Hence, there is an urgent need to evolve a standard methodology to evaluate the technical and economic performances of various solar units.

(3) **Technology transfer through collaboration :** Various solar systems must be made appropriate for local conditions. Since the conditions vary significantly between developed and developing countries, possibilities of technology transfer are quite limited. However, there is wide scope for technical collaboration between experts/institutions of developing and developed countries, especially for optimizing the process parameters. This could be accomplished by courses, workshops and direct links between appropriate institutes in developing and developed countries.

SOCIO-ECONOMIC STUDY

There is an urgent need to understand the social and economic factors that control the dissemination of solar drying technology. These factors include financial viability, availability of initial capital investment, technical back-up and infrastructure, and appropriateness of design to match available technical skills and social customs.

Real participation and self-help initiatives by users may slow down the diffusion process on a short-term basis, but will lead to successful implementation of solar energy program in the long-run. Adoption of solar drying units becomes economically more attractive in areas where the price of alternative fuel is high and the availability uncertain. Higher adoption rates could be reached through Energy Integrated Approach.

STUDY OF SOLAR ENERGY UTILIZATION IN INDIA

Solar energy can be used in two ways, directly and indirectly. The direct means include thermal and photovoltaic conversion, while the indirect means include the use of water power, the winds, biomass and the temperature differences in the ocean.

Systematic efforts to harness solar energy were initiated in India in the early 1950's with emphasis on low-grade thermal process such as solar energy water heating, distillation, space heating, drying and cooking. However, there was no national solar energy policy until 1973. Only after the oil shock in 1973, did this renewable energy sector begin to attract attention. In 1973, an expert panel on solar energy was set up in the Department of Science & Technology and space fields on solar thermal investigations were declared as priority areas. But the real fillip came in 1992 with the Prime Minister's decision to upgrade the department to a full-fledged ministry. With that, the attitude to the sector changed drastically. Earlier efforts to tap non-conventional energy sources were mainly through Government-funded small demonstration projects.

In the last few years, NPL, New Delhi and Central Buildings Research Institute, Roorkee have developed considerable know-how on flat plate collector technology. The work on silicon solar cells is being done in the Solid State Physics Laboratory, New Delhi and on Cds/Cu2s cell at IIT, New Delhi, Jadavpur University, Space Technology Centre, Trivandrum and the Central Electrochemical Research Institute, Karaikudi (Tamil Nadu).

The Central Salt and Marine Chemicals Research Institute, Bhavnagar and the Central Arid Zone Research Institute (CAZRI), Jodhpur have done considerable work on designing better and cheaper solar stills for producing drinking water from brackish or non-potable waters.

Solar dryers for agricultural products and medicinal plants have also been developed. Several institutions in the country, *viz.*, P.A.U., Ludhiana, Regional Research Laboratory, Jammu and CAZRI, Jodhpur have done pioneering work in this area.

Solar water heaters and solar pumps are being developed at the NPL, New Delhi and Central Mechanical Engineering Research Institute, Durgapur. Since last few years, G.N.D. University, Amritsar is also contributing in a significant way by encouraging research work in this vitally important field by starting 2 years M.Sc. (Energy Science) course for students.

In India, the present annual capacity for power generation from renewable sources is around 200 MW and is likely to be increased to 2000 MW by 1996-97. Now there is a realization that solar power is cheaper way to electrify lakhs of remote villages which may otherwise take years to get energized through conventional sources. Wind farms have now become viable with state electricity boards in Gujarat and Tamil Nadu allowing private investments in this field. The budget has also offered significant concessions to import equipment for minihydel projects and wind farms, though most of these equipments are now being made in the country as well.

A FEW MILESTONES

Low temperature solar thermal systems have been installed in all the states of the country under "Solar thermal extension programme". The status of various solar system installations as on 31.03.1992 is given in Table 4.1. But all these steps are not enough to promote pollution-free sources and more needs to be done. Cities should start tapping renewable energy source for their needs. After all, the realization has dawned that tapping renewable sources of energy, especially the solar option, is the only way out for the impeding doom caused by pollution. At BARC 10, 25 and 100 kg capacity foldable solar dryers have been designed, fabricated, installed and used by various farmers, traders, cooperatives, agricultural universities, Agro Industries Development Corporation of various states for drying of

Table 4.1. Status report on solar system installations

S.No.	*System*	*Number*	*Capacity (L/day)*	*Collector area (m^2)*
1.	Industrial/commercial water heating systems	4,829	91,38,000	1,86,172
2.	Domestic solar water heating system	10,127	13,47,700	26,905
3.	Solar cookers	2,27,483	—	—
4.	Solar dryers/solar air heater	150	80	80
5.	Solar timber kilns	71	—	—
6.	Solar stills (water distillation systems)	9,651	—	—
7.	Greenhouse	25	—	—
8.	Solar huts	250	—	—
9.	Passive bulidings	6	—	—
10.	Experimental solar thermal power plant	2	—	—
11.	Deep wind pump (gear type) as on 31.08.1992	151	—	—
12.	Street lighting systems	34,832	—	—
13.	Water pumping systems	1,171	—	—
14.	Small power plants	50	—	—

Source : Invention Intelligence, October 1994, page No. 502.

different agricultural commodities *viz.* fruits, vegetables, fruit juices, chillies, pepper, cardamom, cocobeans, ginger, turmeric and other spices, medicinal herbs, fish and fish products, home-made products like *papads, vadi* and extruded snack food items. Number of entrepreneurs and industries have shown keen interest in utilizing these solar dryers due to various advantages *viz* no cost in processing, drying takes place in clean and enclosed environment hence, no chances of product getting affected during drying due to insects, flies or rodents. Dryers are made out of very cheap and plentily available materials *viz.* aluminium sheets, aluminium wire mesh, angles, plastic, thermocol and wood materials. These dryers are lighter in weight and foldable hence could be stacked one above the other when not in use. These dryers have also been demanded by pharmaceutical and chemical industries for drying their products. More than 10 to 12 industries have shown interest in manufacturing these units. The dryer has been patented on BARC's name.

CONCLUSION

The paper has attempted to highlight the utilization of solar energy in agricultural sector by using solar energy as a heat source in operating crop and food dryers. There is information available in the literature that can be used to design workable solar crop and food dryers for various applications. But tests have to be carried out locally before the suitability of a particular dryer can be established, and further research and development work including manufacturing techniques have to be carried out in this area of utilizing solar energy for agricultural sector efficiently.

REFERENCES

Bassey, M.W., 1978 : A recommended formula for predicting total solar radiation in Sierra Leone. *11th Biennial WASA Conference 13-18 March, 1978*, Lome Togo.

Bassey, M.W., 1980a : Solar crop drying in Sierra Leone. *International Symposium on Solar Energy Utilization.* 10-20 August, 1980, London, Ontario, Canada.

Bassey, M.W., 1980b : Development of solar dryers for agricultural use in Sierra Leone. Interim Report IDRC Project, 3-P-78-0113.

Bongirwar, D.R., K.K.V. Nair and K.B. Patil, 1994 : Machinery/equipment for food processing in small-scale industry. *Industry. Indus. Prod. Finder*, June : 197-208.

BRI, 1979 : A survey of solar agricultural dryers. (Brace Research Institute), *Technical Report T.*, 99, Qubec, Canada.

Gupta Sneh, 1990 : Energy consumption in India. *Yojana*, 34.

Hendel, C.E., 1960 : Effect of drying and dehydration on food nutrients. In : *Nutritional evaluation of food processing*, Ist Ed., (Eds. R.F. Harris and H. Von Loesecke), pp. 148-157.

Lewis, V.H., W.B. Esselen, Jr. and C.R. Fellers, 1949 : Non-enzymatic browning of food stuff Production of carbon dioxide. *Industrial and Engineering Chemistry*, 41 : 2587-2591.

Nair, K.K.V. and D.R. Bongirwar, 1992 : Development of 100 kg capacity solar dryer for agricultural products. *Indian Chemical Engineering Congress* Manipal, India. December, 19-22, 28 p.

Nair, K.K.V. and D.R. Bongirwar, 1993 : Economic aspects of raisin production by solar dryer, *Chemical Engineering Congress 1993* Bombay, 266-267.

Noyes, R.W., 1982 : *The sun-our star.* Harvard University Press, Cambridge.

Satcunanathan, S., 1973 : A crop dryer utilizing a two pass solar heater. *International Congress : The sun in the service of mankind.* Paris, France, 2-6 July.

Selcuk, M.K. *et al.*, 1994 : Development and theoretical analysis and performance evaluation of shelf type solar dryers. *Solar Energy Journal, IG*, 81.

Sharma, V.K., 1990 : Strategies to enhance the viability of solar drying technology in developing countries. *Invention Intelligence*, 547-548.

Singh, D.P. and K.P. Prabhakar, 1994 : "Making solar energy a viable energy option", *Invention Intelligence*, 29 : 499-502.

Sinha, R.N., 1971 : Interrelationship of physical, chemical, biological variables in the deterioration of stored grains. In : *Grain storage : part of a system.* (Eds. R.N. Sinha and W.E. Muir), AVI Publishing Co. Connecticut, U.S.A.

Sootha G.D., 1993 : Solar Energy Application. *Yojana*, 23/24, 19 p.

Suresh, N., 1993 : Going for clean energy. Spectrum, *The Tribune*, 8th July.

Yaciuk, Gordon (Ed.), 1982 : Proceeding of workshop held at Edmonton. Alberta, 6-9 July, 1981, Ottawa, Ont. IDRC, p. 104.

Chapter 5

Water and Airborne Infections and their Control

K.S. Manja

Recent advances in microbiology emphasized the need for the control of pathogens in water and air. Since the source of pathogens entering the environment is infected persons, their sneezing, coughing and excreta, it is necessary to contain them to control infectious diseases. Human excreta and sewage are the major sources of water pollution. Effective treatment of drinking water and regular monitoring of its quality employing simple methods is the need of coming century. Multi-drug-resistant strains of highly pathogenic bacteria especially tubercle bacillus have emerged in recent years which may force World Health Organization to declare global emergency on this account. The paper discusses the efficient methods of decontamination of water and air.

INTRODUCTION

Waterborne diseases are the major causes of human suffering in the developing countries. Further, in developing countries 80% of the infectious diseases are waterborne and 50% of the deaths among the children are due to polluted drinking water. Even the developed nations like United States of America reports 20,000 cases of waterborne infections per year (Olson and Nagy, 1994) though the reported cases in the middle of the 20th century was much less. Increased reports of waterborne infections could be due to better surveillance and partly due to greater pressure on urbanization leading to disproportionately lower civic facilities. Lack of access to safe drinking water and lack of basic sanitation in developing countries contribute immensely to the outbreak of waterborne infections. Emerging problem faced by the sanitary microbiologists in providing safe drinking water is the resistance of microorganisms to disinfectants and sterilizing agents (Sobsey and Olson, 1983; Russel, 1991). Since the antibiotic resistance in bacteria is plasmid-mediated, it is possible that the drug-resistant bacteria in drinking water could transfer this property to other pathogenic bacteria in gut.

The most common waterborne infectious diseases are typhoid, cholera, shigellosis, viral-associated diarrhoea and infectious hepatitis (WHO, 1993). Infectious hepatitis outbreak of New Delhi in 1955-56 was one of the largest epidemics of waterborne infections due to faecal pollution of drinking water (Dennis, 1959). Such outbreaks of waterborne infectious hepatitis were being reported frequently in India (Manja *et al.*, 1982; Naik *et al.*, 1992).

Most of the Indian cities do not have safe sewage disposal system. Further, some of the untreated city sewage is even being used for spray irrigation. This can spread the enteric pathogens upto a kilometre downstream in the air. Majority of the human infections are transmitted either through air or water. Some of the airborne viruses like foot and mouth disease virus can be transported for hundreds of miles by wind (Donaldson *et al.*, 1982). Infected airborne particles can easily carry the pathogens of respiratory tract namely *Mycotuberculosis,* influenza virus, fungal spores, mycoplasma, rickettsia etc. (Collins, 1983). To cause effective infection in the respiratory tract, the aerosols should be less than 5 μm in diameter. Only the particles of these dimensions can reach alveoli since larger particles get trapped in the nose and upper respiratory tract (Hatch, 1961). The most common sources of airborne infections are the patients suffering from respiratory infection, hospital waste and infected animals suffering from zoonotic infections like anthrax, plague, brucellosis, etc. Cooling tower water which appears to be most inocuous could be the source of *Legionella* infections (Mattison, 1980) in closed environment. Even the aircraft environment get contaminated with respiratory pathogens (Masterton and Green, 1991). Aircrafts are now being used for the transport of critically ill patients who are suffering from communicable diseases.

Airborne epidemic of group - A meningococcal meningitis outbreak erupted among the pilgrims travelling to Mecca and Saudi Arabia (Moore *et al.*, 1989). In recent years a great emphasis was given to chemical pollution of air and water ignoring the microbial pollution of environment that can cause chain of infections among the susceptible population. Further, the microbes found in hospital air usually are drug-resistant to common antibiotics and the infections caused by such agents are difficult to treat (Cox, 1987). Even the research laboratory is no exception in harbouring the pathogenic microbes in air. Smallpox virus of microbiology laboratory in Birmingham University formed aerosol from the vial infecting an unvaccinated worker of the same building (DHSS, 1980). In this case, the person entered the common enclosure connected through the air-conditioned duct of the laboratory. Infected persons continuously discharge the pathogens into the environment and every effort should be made to contain these microorganisms.

SOURCES OF MICROBIAL CONTAMINATION OF WATER AND AIR

Over 97% of the available water is saline and only 3% can be exploited for drinking purpose. Even in this small percentage of water, the major portion is in glaciers and polar regions. Therefore, the entire global pollution depends on this very small percentage of water present in the underground and rain water stored in the reservoirs for drinking purpose. Rapid urbanization exerted tremendous pressure on this scanty resource. Continuous use of groundwater much faster than the recharge rate added further to the microbial pollution of groundwater. Therefore, the quality of groundwater that exists today is quite different from what it was in the early part of this century, Gibson *et al.* (1986)

reported that improper sewage through stone strata can cause groundwater pollution. Human activities on the watershed are the major contributors of pathogenic microorganisms of public health concern. Farm animals and wildlife also contribute human pathogens to the water source in variety of ways. In the tropical region people depend on rainwater storage reserviors for raw water which are exposed to a large number of enteric pathogens of man and animals (Canong, 1986). More than 67% of waterborne disease outbreaks took place due to poor quality of raw water (Sobsey and Olson, 1983) in the United States during mid-seventies. Contaminated source of water needs to be treated more effectively to disinfect the water since normal treatment process may not be effective due to high microbial load. Pojasek (1977) emphasised the need for source water protection. Microbial regrowth and biofilm formation in the pipeline also contribute microbes to drinking water and cause infection (Ireland *et al.*, 1983).

Patients suffering from respiratory infections continuously discharge pathogenic microbes into the air. Highly virulent and drug-resistant microorganisms could be isolated from the hospital environment. Though the hospital waste should be incinerated (Collins and Kennedy, 1992) to avoid microbial pollution of air, only very few hospitals do so. Microwave-based clinical waste decontamination units have also been proposed (Hoffman and Hanley, 1994). The spores of anthrax in the air can cause epidemic in a closed environment (Plotkins *et al.*, 1960). We are yet to evolve safe air quality in public places. In 1976 there was an outbreak of pneumonia with 15-20% mortality that occurred among the legionaires attending 58th convention in Philadelphia in an air-conditioned hotel, 34 deaths were reported out of 221 cases. It took 6 months for the Centre for Disease Control, USA to isolate and identify the bacteria and named it *Legionella pneumophila*. The causative agent harbouring the water of cooling tower went into the airconditioning duct of the hotel as aerosol and caused the outbreak (McDade *et al.*, 1977). Pathogenic microbes could be found in municipal wastes as well (Mose and Reinthaler, 1985). Dust in the air surrounding the incinerators could carry pathogenic bacteria (Peterson, 1971). Airborne microorganisms are generated during collection, sorting and land-filling of municipal waste (Crook *et al.*, 1987). One of the most harmful airborne microbes is the multi-drug-resistant *Myco. tuberculosis* from the chronic AIDS patients and WHO is on verge of declaring global emergency on this count (Spinney, 1996).

PATHOGENIC BACTERIA IN WATER

Enteric pathogens are the predominant group of microorganisms present in raw water. Since the germicidal efficacy of chlorine reduces with temperature, waterborne pathogens survive for longer time in winter. But bacterial regrowth in distribution system occurs as a thin film and was not known till recently mainly due to the difficulties in sampling techniques. Therefore, there is a possibility of microbial load increasing in the system without any leakage of pipes. The most common pathogenic bacteria that can cause waterborne epidemics are *Salmonella, V. cholerae, Campylobacter* and enteropathogenic *E. coli*. More recently *Legionella, Aeromonas, Yersinia enterocolitica* are also known to cause waterborne epidemics (Lee and West, 1991; Schubert, 1991; Weber *et al.*, 1981). Apart from these known enteric pathogens opportunistic pathogens like *Pseudomonas, Proteus, Citrobacter* are also encountered in drinking water. These are capable of causing infections in immunologically compromised individuals like convalescents, newborn babies, aged individuals and post-operative cases with immunosuppression. This has greater significance in water supplies to hospitals, nursery schools, old age homes where a large number of susceptible people live.

Chlorination of water kills majority of the bacterial pathogens, but at lower levels of chlorination it causes sublethal injury. The enteric pathogens like *Yersinia, Salmonella* and *Shigella* are more resistant to chlorine injury than others (Le Chevallier *et al.*, 1985). These injured bacteria, though difficult to recover by normal microbiological methods, are capable of producing waterborne infections. This situation underestimates the public health hazard of drinking water. Therefore, the presence of injured pathogens in drinking water is a potential health hazard. *Legionella* spp. are widespread in aquatic environments. They may get colonised on the biofilms formed on the internal walls of water distribution system. Dirt and micronutrients present in water provide sufficient nutrients to these bacteria. *Campylobacter* spp. which are widespread among animals, gain entry into the drinking water system through their excreta. Outbreaks of *Campylobacter* enteritis due to polluted drinking water has been reported recently (Blazer *et al.*, 1984). Improper treatment of drinking water was reported to be the reason for outbreaks of *Aeromonas* infection (Schubert, 1991). Therefore, the present treatment systems in some situations might permit the growth of bacteria in drinking water causing waterborne infection. Storage of treated water unusually for long periods could also permit the growth of psychrophiles. Further, it was reported that the pathogens might be present in absence of *E. coli* or coliforms in water (Ramteke *et al.*, 1992).

PATHOGENIC VIRUSES IN WATER

Effective control of waterborne bacterial diseases could be achieved in the developed countries mainly due to the use of chlorine in drinking water supplies. Enteric viruses are more resistant to chlorine than the bacteria. Therefore, waterborne viral infections made their appearance in recent years. Further, the drinking water is not routinely checked for the presence of viruses and no standard has been laid for viruses in drinking water. More than 100 pathogenic viruses that were excreted by man may find their way into drinking water. Among them enteroviruses, rotaviruses, ECHO viruses and hepatitis A viruses are the most important ones and often cause human infections. It was reported that these naturally occurring viruses are more resistant to chlorine than those cultured in the laboratory environment (Paymert, 1981; Shaffer *et al.*, 1980). Improper treatment of drinking water and heavy contamination of raw water can cause extensive waterborne disease outbreaks (Dennis, 1959; Hejkal *et al.*, 1982). Though a large number of disinfectants have been tried to kill the viruses in water but these chemicals in general have been found to be less effective in killing viruses than coliforms (Alverez and O'Brie, 1982; Gowda *et al.*, 1981). Therefore it is advisable to ascertain the source of raw water for possible viral contamination. If the viral contamination is anticipated then more effective disinfectants like oxidizing agents need to be used for drinking water.

PROTOZOA IN WATER

Generally protozoan infections are endemic and have long incubation periods. Therefore, outbreaks of parasitic infections are uncommon. They have frequently been isolated from the biofilms of distribution systems. *Giardia lamblia* and *Entamoeba histolytica* are the known pathogens reported to be present in grossly contaminated drinking water supply (Ramsay and March, 1990). Cysts of protozoa are more resistant to chlorine than bacteria or viruses. A large scale waterborne Cryptosporidium outbreak took place in Georgia when the river water was used after routine treatment. The filtration system was not efficient in removing the protozoa and chlorination failed to destroy the protozoa (Hayers *et al.*, 1989).

PATHOGENS IN AIR

Pathogens cannot grow or multiply in air because of the absence of nutrients but can be transmitted through air for long distances. Higher wind speeds that assist in increased dehydration of droplets resulting in breaking down of larger particles. It has been reported that the pathogens could be isolated from air up to 60 km from the site of epidemics (Hough-Jones and Wright, 1970). The major sources of airborne pathogens include infected patients through sneezing, coughing, talking, excretory products, and sewage. Sneezing, aerosolises the pathogens into air. Smaller droplets of 50 μm or less stay in air and travel long distances. Faecal discharge and other human excretory products are the most important sources of indirect air contamination. Flying birds and insects can transmit microorganisms from place to place and redistribute pathogens in air. The discharge of pathogens to air from infected animals can occur by breathing, sneezing or by splash from falling urine and faeces (Hough-Jones and Wright, 1970).

Sewage treatment plants through activated sludge and trickling filters can create aerosols and droplets containing pathogens. Air surrounding sewage treatment plant may contain > 10^5 bacteria per m^3. Presence of coliform, faecal coliforms and faecal streptococci are taken as the index of air quality surrounding the sewage treatment plant (Lembke and Kniseley, 1985). Microbial endotoxin upon 48 μg/g of airborne dust was detected in samples from swine and poultry rearing units (Thedell *et al.*, 1980) and could be a health hazard to the animal handlers (Cole, 1983). Clusters of bacteria survive longer in air than the individual cells (Handley and Webster, 1995). Pathogens in incinerator emissions, municipal waste and clinical wastes can enter the air causing respiratory infection (Collins and Kennedy, 1992). Cooling towers of airconditioning system can also be the source of respiratory infection due to *Legionella* (Yamamoto *et al.*, 1991).

DETECTION OF PATHOGENS IN WATER AND AIR

Intestinal microorganisms are taken as indicators of faecal pollution of drinking water. If they are present than it is presumed that more pathogenic organisms of intestinal origin might be there, hence such water samples are considered unsuitable for drinking. Coliform group of organisms are universally accepted indicators of faecal pollution of drinking water. They are gram negative, rod shaped bacteria which could grow at 35-37°C, producing acid and gas in lactose bile broth. Coliforms that could grow at 44°C are typical indicator bacteria of human origin. For the enumeration of coliforms in drinking water, multiple tube method or membrane filtration methods are used (WHO, 1993). Multiple tube method is based on statistical probability which employs inoculation of different volumes of test water into the liquid media and incubating at 35°C. If the coliforms are present in the inoculated tube, acid and gas is produced in the medium. The number of tubes showing acid and gas are matched with the precalculated statistical table (HMSO, 1969) giving the number of coliforms in 100 ml of water. Coliforms as an index of water quality has also been questioned (Dutka, 1973) since sme of the enteric pathogens were isolated from drinking water in absence of coliforms.

Coliphage is an alternate indicator of faecal pollution. These are the viruses infecting coliforms and their presence in drinking water indirectly forewarn the presence of coliforms. Coliphages are abundant in sewage and indicate the sewage contamination of water. They survive much longer than coliforms and can act as an indicator of faecal pollution. It has been evaluated in a field study and found to be reproducible (Wetsel *et al.*, 1982). In a more recent study (Martins *et al.*, 1989), the coliphage test showed least correlation with other microbiological tests. It is possible that the coliphage test is less sensitive indicator of faecal pollution when compared to other tests or it only indicates sewage pollution rather than faecal pollution from human source. Coliphage in water are enumerated by the number of plaque forming units or number of zones on a lawn of host *E. coli* strain plated on solid media. Plaque is formed due to the lysis of host strain by the infection of virulent coliphage present in the water. In this test (APHA, 1989) water sample is mixed with molten trypticase soy broth and host *E. coli* strain. Molten mixture at 45°C is poured on to a petridish and incubated for 6 hrs. Development of clear zone on the agar indicates the presence of coliphage.

Bachrach and Bachrach (1974) have employed radiometric method for the detection of coliform in water. More simpler gas chromatographic method of detection of *E. coli* has been proposed by Newman and O'Braien (1975). The method is based on the principle that *E. coli*, while utilising lactose, produces ethanol and acetic, which could be detected by gas chromatograph. Colorimetric method for the detection of *E. coli* has also been reported by Warren *et al.*, (1978). The test uses chromogenic galactoside which is degraded by the enzyme b-galactosidase produced by *E. coli*. More simpler qualitative screening method for faecal pollution has been reported and evaluated (Clark, 1968; IDRC, 1990). This presence/absence test employs lactose broth in 50 ml quantity and water sample of 100 ml quantity (Dutka, 1991). After overnight incubation, the content of bottle was observed for the production of acid and gas. The appearance of acid and gas is the indication of coliforms, *E. coli*, or *Streptococci*. As a screening test it was found to be more sensitive than the standard procedures.

An extremely simple and inexpensive field test for faecal pollution of drinking water was developed and found to be specially useful for large scale screening of urban as well as rural water supplies (Manja *et al.*, 1982). In this large study; it was established that the presence of faecal coliform in drinking water is always associated with the H_2S producing organisms. The test is based on the detection of H_2S producing organisms in water. They include *Citrobacter, Salmonella, Proteus, Arizona, Klebsiella* and H_2S producing intestinal anaerobes. The new test was successfully used to monitor water quality during hepatitis outbreak in Gwalior. The new method of H_2S strip test for drinking water was evaluated by International Development Research Centre in an eight country three continent study (IDRC, 1990). The study concluded that this test is the best and simplest method to test remote water supplies. In a study in Latin America (Castillo *et al.*, 1994) coliphage test, MPN test and H_2S strip test were evaluated using untreated and treated drinking water. It was shown that the test is 10% more sensitive than the MPN method since it detects *Clostridium* also providing greater margin of safety. Improved and simpler methods for water quality have been reported (Audiena *et al.*, 1995; Manja, 1995) which could be effectively used in the laboratory as well as in the field.

Detection technique for airborne microbes are primarily the same as the methods used to sample dust and other airborne particles. Gravity settling plate method is the simplest and oldest method though it is less sensitive (Willian *et al.*,1972). More effective way of sampling the air for microbes is to use filtration technique. Through suction, microbes can be captured and cultured on a suitable media. For capturing more fastidious microbes, liquid trap method could be used where the air is sucked through liquid media and microbes are retained in the media which grow on incubation (Edmonds, 1979). Impinging the microbes on semisolid culture media is an effective method of sampling air. Different devices like Anderson sampler, slit sampler, drum sampler, multiget monitor employ this principle (Al Dagal and Fung, 1990). Anderson sampler is the most commonly employed tool for air sampling. It has six stages with gradual reduction in particle size down to 0.2 to 0.5 μm. Centrifugal air sampler can quantify the microbes with respect to the volume of air sucked (Errington and Powell, 1969). Though there are many devices for sampling microbes in air but each one has its own advantages and disadvantages (Edmonds, 1979).

SAFE WATER AND AIR QUALITY

Boiling of water is the safest method of ensuring the absence of pathogenic bacteria. But this is the most expensive and energy intensive process. In case of emergency this is the only method of choice. Though the spores are resistant to boiling and it is unlikely that the bacterial spores can cause waterborne infection. Commercial chlorine tablets are available to disinfect drinking water for household purpose. There should be minimum contact time of 30 min. for the chlorine to kill the pathogens. A survival straw containing iodinated resin to inactive the pathogenic bacteria in drinking water is being marketed. When the water passes through this tube containing iodinated resin it slowly releases 2 ppm of iodine which instantaneously kills the pathogens. Though high intake of iodine is contraindicated for pregnant woman andpatients of thyroid disorder but it will be highly useful to troops stationed in remote areas where they do not have accessibility to treated water. Ultraviolet rays which have low penetration power have been used mostly for surface sterilization. Most of the countries use liquid chlorine for safe drinking water but in Norway 30% of water supplies are treated with UV light. Efficacy of UV light depends on environmental factors like turbidity, metal ions, organic pollutants and flow rate of water. Biggest disadvantage of UV light is the lack of residual effect. But by employing electrochlorinator this could be overcome. This system cracks chlorine from sodium chloride producing 0.5 to 1 ppm residual chlorine which can suppress microbial regrowth if any. Chlorination of water at 2-5 ppm with residual level of 0.2 ppm normally ensures safe drinking water in municipal water supplies. During waterborne epidemics, an extra safety measure has to be taken and even the super chlorination is advocated. Only way to ensure safe drinking water supply is to employ simply microbiological water testing kits at plant operator level more frequently so that the corrective measures are taken before any waterborne disease outbreak appears.

Waste disposal system of hospitals should be streamlined to avoid any possible leak into the environment. Ideally the laboratories and hospitals handling highly pathogenic microbes should work in negative pressure system to prevent dissemination of these microbes outside. There is an obvious and urgent need to improve air quality in public places to control airborne infection.

The control of pathogens in outdoor air is more complicated than indoor environments like operation theatres. At present there is no laid down standards for permissible level of pathogens in public places or in hospitals. However a working standard has been laid down for operation theatres (Thomas *et al.*, 1972). Ultraviolet rays are the single source that can reduce microbes in air of outdoor and indoor environments. Air filters up to 0.45 μm effectively reducing the microbial load in the air. Use of biocides in cooling towers can reduce the growth of *Legionella* which may aerosolise through water spray (Yamamoto *et al.*, 1991). Further, indoor climate can be monitored employing recently reported technique of fungal index estimation (Abe *et al.*, 1996). With an objective of achieving health for all in 2000 A.D. it is necessary to have safe water and air free from pathogens though it may be difficult to achieve.

FUTURE RESEARCH

Microbial ecology and microbial interaction in mixed culture are not completely understood. Majority of antimicrobials are tried to pure cultures. Therefore,it is not clear how they are effective in natural environment. There is a need to develop simple methods for estimating survival of pathogens in natural conditions. Since the validity of coliform estimation in drinking water is being questioned, we need to develop simple and more sensitive method of testing drinking water quality. It is necessary to decrease the disease transmission through water. This only can provide better protection of public health. There is a need for simple method of viral detection in drinking water which is operator based.

There is no definite standard for air quality in public places, hospital wards, or in food processing plants. Though the pathogens in air could cause infection in man, we do not have effective methods for pathogen detection in air. The indoor air could be made safe, employing filtered air or disinfectants but the public places still remain health hazard if the patients or tuberculosis or other respiratory infections are in large numbers. The problems are further compounded if they are suffering from the infections due to drug resistant microbes. These are the problems which need immediate attention to control infectious diseases and to attain the goal of "Health for all in 2000 A.D."

REFERENCES

Abe K., Y. Nagao, T. Nakada and S. Sakuma, 1996 : Assessment of indoor climate in an apartment by use of a fungal index. *Appl. Environ. Microbiol.*, 62 : 959-963.

Al Dagal and D.Y.C. Fung, .1990 : Aerobiology - A Review. *Food Science and Nutrition*, 29: 333-340.

Alverez, M.A. and R.T.O. Brien, 1982 : Mechanism of chlorine inactivation of poliovirus by Chlorine dioxide and iodine. *Appl. Environ. Microbiol.*, 44 : 1064-1071.

APHA, 1989 : *Standard Methods for the Examination of Water and Waste Water.* 17th Ed., American Public Health Association, Washington, D.C.

Audieana, A., I. Perales and J.J. Borrego, 1995 : Modification of Kanamycin Esculin Azide Agar to improve selectivity in the enumeration of faecal streptococci from water samples. *Appl. Environ. Microbiol.*, 161 : 4178-4183.

Bacharach, U. and Z. Bacharach, 1974 : Radiometric method for the detection of coliforms in water. *Appl. Microbiol.*, 28 : 169-171.

Blazer, M.J., D.N. Taylor and R.A. Feldman, 1984 : *Campylobacter infection in man and animals* (Ed. J.P. Butter), CRC Press, Florida, USA.

Canong, M.J., 1986 : *Waterborne pathogens of US Virgin Islands.* Tech. Rep. 25, Carribean Research Institute. Charlotte Amalie, St. Thomas V.I., USA.

Castillo, G., R. Daurte, Z. Rluiz, M.T. Marucic, B. Hanarato, R. Mareado, V. Coloma, V. Lorea, M.T. Martins and B.J. Dutka, 1994 : Evaluation of disinfected and untreated drinking water supplies in Chile by the H_2S paper strip test. *Water Res.*, 28 : 1765-1770.

Clark, J.A., 1968 : A presence/absence test providing sensitive and inexpensive detection of coliforms, faecal coliforms and faecal streptococci in municipal water supplies. *Can. J. Microbiol.*, 14 : 13-18.

Cole, E.C., 1983 : *An aerobiological analysis to determine respiratory disease potential for poultry workers exposed to high concentrations of biological aerosols.* Ph.D. Dissertation University of North Carolina, Chapel Hill, USA.

Collins, C.H., 1983 : *Laboratory acquired infections.* Butterworth and Co. Ltd., London.

Collins, C.H. and D.A. Kennedy, 1992 : The Microbiological hazards of municipal and clinical wastes. *J. Appl. Bacteriol.*, 77 : 1-6.

Cox, C.S., 1987 : *The Aerobiological pathway of microorganisms.* John Wiley & Sons, New York.

Crook, B., S. Higgins and Y. Lacey, 1987 : Airborne gram negative bacteria associated with the handling of domestic waste. In : *Advances in Aerobiology*, (Eds. G. Boehm and R.M. Leuschner), Birkhauser Verlag, Basel.

Dennis, J.M., 1959 : 1955-56 Infectious hepatitis epidemic in Delhi, India, *J. Am. Water Works Assoc.*, 51 : 1288-1298.

DHSS, 1980 : *Report of the investigation into causes of the 1978 Birmingham Smallpox occurrence.* (Chamm, R.A. Shooter), Her Mejestys Stationary Office, London, UK.

Donaldson, A.I., J. Gloster and J.D.L. Harvey, 1982 : Use of prediction models to forecast and analyse airborne spread during the foot and mouth disease outbreaks in Britain, Jursey and the Isle of Wright in 1981. *The Veterinary Record*, 110 : 53-57.

Dutka, B.J., 1973 : Coliforms are inadequate index of water quality. *J. Environ. Health*, 36 : 39-46.

Dutka, B.J., 1991 : Water, *Food Laboratory News*, 7 : 32-42.

Edmonds, R.L., 1979 : *Aerobiology, The ecological system approach.* Dowden, Hutchinson and Ross, Stroudsburg, P.A.

Errington, F.P. and E.O. Powell, 1969 : A cyclone separator aerosol sampling in field. *J. Hyg.*, 67 : 387-391.

Gibson, G.R. Jr., D.L. Johnstone, D.O. Clever and E.E. Geldriech, 1986 : Public health microbiology of lakes and reservoirs. In : *Lake and Reservoir Management*, Vol. 2, North Amer. Lake Manage. Soc., Washington, D.C.

Gowda, N.M.M., N.M. Trieff and G.J. Stantom, 1981 : Inactivation of poliovirus by Chloramine T., *Appl. Environ. Microbiol.*, 42 : 469-476.

Handley, B.A. and A.J.F. Webster, 1995 : Some factors affecting the airborne survival of bacteria outdoors. *J. Appl. Bacteriol.*, 79 : 368-378.

Hatch, T.F., 1961 : The distribution and deposition of inhaled particles in the respiratory tract. *Bacteriological Reviews*, 25 : 237-240.

Hayers, E.B., T.D. Matte, T.R.O. Brien, T.W. McKinlay, G.S. Logsdon, J.B. Rose, B.L.P. Unger, D.M. Ward, P.F. Pinskey, M.L. Cummings, M.A. Wilson, E.G. Long, E.S. Hurwitz and D.D. Jaurank, 1989 : Large scale outbreaks of cryptsporiosis due to contamination of filtered water supply. *New Eng. J. Med.*, 320 : 1372-1376.

Hejkal, T.W., B. Keswick, R.L. Labelle, C.P. Gebra, Y. Sanchez, D. Dressman, B. Hafkin and J.L. Melnik, 1982 : Viruses in community water supply associated with an outbreak gastroenteritis and infectious hepatitis. *J. Am. Water Works Assoc.*, 74 : 318-321.

HMSO, 1969 : *Bacteriological Examination of Water Supplies.* Report No. 71, Her Majesty's Stationary Office, London, UK.

Hoffman, P.N. and M.J. Hanley, 1994 : Assessment of microwave based clinical waste decontamination unit. *J. Appl. Bacteriol.*, 77 : 607-612.

Hough-Jones, M.E. and P.B. Wright, 1970 : Studies on the 1967-68 foot and mouth disease epidemic. *J. Hyg.*, 68 : 253-258.

IDRC, 1990 : *Use of simple inexpensive microbial water quality test.* Project No. MR 247C, International Development Research Centre, Ottawa, Canada.

Ireland, R.W., B.W. Cooper and L.D. Bowen, 1983 : A review of Sydney's experience with bacterial contamination in the reticulation system. *Fed. Convent Aust. Water Waste Water Assoc.*, 10 : 44-86.

Le Chevallier, M.W., A. Singh, D.A. Shieman and G.A. Mc Feters, 1985 : Changes in virulena of waterborne enteropathogens with chlorine injury. *Appl. Environ. Microbiol.*, 50 : 412-419.

Lee, J.V. and A.A. West, 1991 : Survival and growth of *Legionella* spp. in the environment. *J. Appl. Bacteril.*, 70 : 121S-129S.

Lembke, L.L. and R.N. Kniseley, 1985 : Airborne microorganisms in a municipal solid waste recovery system. *Can. J. Microbiol.*, 31 : 198-201.

Manja, K.S., M.S. Maurya and K.M. Rao, 1982 : A simple field test for the detection of faecal pollution in drinking water. *Bulletin of Wld. Hlth. Orgn.*, 60 : 797-801.

Manja, K.S., 1995 : *Microbes for better living.* (Eds. R. Sankaran and K.S. Manja), Conference Secretariat, DFRL, Mysore, India.

Martins, M.T., A.E.I. Sharawi, B.J. Dutka, V.H. Pellizari, G. Ribeiro and E.F. Matsumoto, 1989 : *Toxicity Assessment*, 4 : 329-338.

Masterton, R.G. and A.D. Green, 1991 : Dissemination of human pathogens by airline travel. *J. Appl. Bacteriol. Symposium Supplement*, 70 : 31S-38S.

Mattison, G.F., 1980 : Legionellosis : Environmental aspects. *Annals of the New York Academy of Sciences*, 353 : 67-70.

Mc Dade, J.E., C.C. Shepard, D.W. Fraser, T.R. Tsai, M.A. Radus and W.R. Dowdle, 1977 : Legionnaire's disease, isolation of bacterium and demonstration of its role in other respiratory disease. *New Eng. J. Med.*, 297 : 1197-1203.

Moore, P.S., M.W. Reeves, B. Schwartz, B.G. Gellin and C.V. Broome, 1989 : International spread of an epidemic group A *Neisseria meningitidis* strain. *Lancet*, 2 : 260-262.

Mose, J.R. and F. Reinthaler, 1985 : Microbial contamination of hospital waste and household refuse. *Zentralbatt fur Bakteriologie and Hygiene I. Abt. Orig. B.* 181 : 98-110.

Naik, S.R., R. Agarwal, P.N. Salunke and N.N. Mehrotra, 1992 : A large waterborne viral hepatitis epidemic in Kanpur, India. *Bull. Wld. Hlth. Orgn.*, 70 : 597-604.

Newman, J.S. and R.T. O'Brien, 1975 : Gas chromatographic presumptive test for coliform bacteria in water. *Appl. Microbiol.*, 30 : 584-588.

Olson, B.H. and L.A. Nagy, 1984 : *Advances in Applied Microbiology*, Vol. 30 (Ed. A.I. Laskin), Academic Press, London, UK.

Paymert, P., 1981 : Isolation of viruses from drinking water at Pont-Viau water treatment plant. *Can. J. Microbiol.*, 27 : 417-420.

Peterson, M.L., 1971 : *Pathogens Associated with Solid Waste Processing.* US EPA Publication No. SW 49, Washington DC, Government Printing Office.

Plotkin, S.A., P.S. Brachman, M. Utell, F.H. Bumford and M.M. Atchison, 1960 : An epidemic of inhalation anthrax. The first in the 20th Century. *Am. J. Med.*, 29 : 992-1001.

Pojasek, R.B., 1977 : *Drinking Water Quality Enhancement Through Source Protection*. Ann. Arbor. Science, Ann. Arbor. Michigan.

Ramsay, C.N. and J. March, 1990 : Giardiasis due to deliberate contamination of water supply. *Lancet*, 336 : 880-881.

Ramteke, P.W., J.W. Battacharjee, S.P. Pathak and N. Kalra, 1992 : Evaluation of coliform as indicators of water quality in India. *J. Appl. Bacteriol.*, 72 : 352-356.

Russel, A.D., 1991 : Mechanism of bacterial resistance to nonantibiotics, food additives and food and pharmaceutical preservatives. *J. Appl. Bacteriol.*, 71 : 191-201.

Schubert, R.H.W., 1991 : *Aeromonas* and their significance as potential pathogens in water. *J. Appl. Bacteriol.*, 70 : 131S-136S.

Shaffer, P.T.B., T. Metcalf and D.J. Sproul, 1980 : Chlorine resistance of polioviruses recovered from drinking water. *Appl. Environ. Microbiol.*, 40 : 1115-1121.

Sobsey, M.D. and B.H. Oslon, 1983 : *Assessment of Microbiology and Turbidity Standards for Drinking Water.* (Eds. P.S. Berger and Y. Argaman), EPA. Washington D.C., USA.

Spinney, L., 1996 : Global emergency as T.B. toll mounts. *New Scientist*, 2023 : 8.

Thedell, T.D., J.C. Mull, M.E. Gladish and M.J. Peach, 1980 : A brief report of Gram negative bacterial endotoxin levels in airborne and settled dust in animal confinement buildings. *Am. J. Ind. Med.*, 1 : 3-5.

Thomas, M.E., E. Piper and I.M. Maurer, 1972 : Contamination of an operating theatre by gram-negative bacteria. *J. Hygiene*, 70 : 63-73.

Warren, L.S., R.E. Bemoit and J.A. Jessee, 1978 : Rapid enumeration of faecal coliforms in water by colorimetric galactosidase assay. *Appl. Environ. Microbiol.*, 35 : 136-141.

Weber, G., G. Stanek, N. Massiezek and M.F. Klenner, 1981 : *Yersinia enterocolitica* in drinking water. *Zentralbl Bakteriol. Abt. I. Orig. Reiche. B.* 173 : 209-216.

Wetsel, R.S., P.E. O'Neil, J.F. Kitchens, 1982 : Evaluation of coliphage detection as a rapid detector of water quality. *Appl. Environ. Microbiol.*, 43 : 430-443.

William, J.S., N.M. Macknight and H.W. Wilson, 1972 : Hospital airborne bacteria as estimated by the Andersin Sampler versus the Gravity Culture Plate. *Am. J. Clin. Pathol.*, 58 : 558-561.

World Health Organization, 1993 : *Guidelines for Drinking Water Quality*, Vol. 1, 2nd Edition, WHO, Geneva.

Yamamoto, H., T. Ezaki, M. Ikeda and E. Yabbuchi, 1991 : Effects of biocidal treatments to inhibit the growth of *Legionella* and other microorganisms in cooling towers. *Microbiol. and Immunol.*, 35 : 795-803.

Chapter 6

Role of Agroforestry in Conservation of Degraded Environment

K.S. Bhatia

Water, soil and vegetation are the most vital natural resources for the survival of mankind. The role of agroforestry in conserving these resources can hardly be over emphasised. In view of indiscriminate widespread deforestation, excessive grazing, continued land degradation problems due to serious erosion, depletion of soil fertility, environmental deterioration and ecological disturbance resulting from heavy pressure of fast expanding population on the one hand, and acute shortage in the supply of food, fuel, fodder, fibre, wood etc. on the other, agroforestry as a conservation land use system, has tremendous scope as a practical solution to environmental stability. Role of various agroforestry systems in control of environmental pollution, improvement of soil fertility and control of soil erosion has been discussed. A brief account of some of the recent information on the effectiveness of agroforestry systems in conservation of degraded land has been given in this paper.

PROBLEM OF EROSION AND ENVIRONMENTAL DEGRADATION

In India, out of 329 m.ha. of total geographical area, about 150 m.ha. is subjected to water and wind erosion (Table 6.1). About 25 m.ha. area has been subjected to degradation due to the exploitative type of agriculture. It is estimated that about 16.35 tonnes of soil is lost annually from every ha of India's land amounting to 5334 m tonnes for the whole country. Approximately 29% of this soil goes to the sea, 10% gets deposited in dams reducing their storage capacity by 1-2% every year and 61% gets transported from one place to another mostly getting settled on river beds.

To increase food production, we have increased the area under cultivation by clearing away most of our forests at the rate of 1.5 million hectares, with the result that the effective area under forest trees has been reduced to a mere 13% of our total land area, as against the desirable 33%. With this we are faced with serious ecological, environmental and socio-economic crisis. It has been estimated that by the end of this century we have to produce about 250 million tonnes of food grains, over 2000 million tonnes of green and dry fodder, 350 million tonnes of fuel wood and about 60 million

m^3 of timber for our increasing human as well as livestock population of the country which would be 1000 million and 600 million respectively, besides amelioration of our polluted environment. Soil erosion bring about environmental deterioration ecological disturbance and reduce the productivity of the soil. It is estimated that about 2.5 m tonnes of nitrogen, 3.8 m tonnes of phosphorus and 2.6 m tonnes of potash are lost every year from our country. This loss is perhaps more than the production of fertilizers in India which was about 3.1, 1.0 and 0.7 m tonnes of N, P and K respectively in 1981-82.

ROLE OF TREES IN CONTROL OF EROSION

The role of trees in erosion control is one of the most widely acclaimed reasons for including trees on farmlands that are prone to erosion hazards. The trees reduce erosion by enriching and binding the weaker surface soil, providing protection against erosion with canopy and leaf litter and impeding the velocity and erosivability of surface runoff by stem, surface roots and litter (Wiersum, 1984). Thus trees play an important role in soil and water conservation in all. localities. Though they may not be a cure for all problems of erosion, they are, by far, the cheapest tool in the hands of soil conservationists for regulating stream flows, reducing peak of floods, preventing erosion of soil and sedimentation of reservoirs and river channels.

Table 6.1. Problems of soil erosion and land degradation in India

(Area in m.ha.)

1.	Total geographical area	329.0
2.	Area subject to water and wind erosion	150.0
3.	Area degraded through special problems	25.0
	(a) Waterlogged	6.0
	(b) Alkaline soil	2.5
	(c) Saline soil including coastal sandy areas	5.5
	(d) Ravines and gullies	3.9
	(e) Area subject to shifting cultivation	4.4
	(f) Ravines and torrents	2.7
4.	Total problem area	175.0
5.	Annual average loss of nutrients from land estimated at 2 and 3	5.4 to 8.4 m.t.
6.	Average loss of production of net developing ravines estimated at 3(d)	3 m.t.
7.	Average annual rate of encroachment to arable land by ravines	8 m.ha.
8.	Total flood prone area	40 m.ha.
	(a) Average area affected by floods	9 m.ha.
	(b) Average cropped area affected by floods	9 m.ha.
9.	Total drought prone area	260 m.ha.

Source : Fertilizer Statistics, 1987-88

MAINTENANCE OF SOIL FERTILITY

It has been estimated that considerable amount of soils are lost through erosion every year from about 80 m.ha. of the cultivated lands, carrying away about 8 m.t. of nutrients. The inclusion of compatible and desirable species of trees in agroforestry system can result in marked improvement in soil fertility due to increased organic matter content of soil through addition of leaf litter, efficient nutrient cycling, efficient sharing of nutrients among the components, additional nutrient economy because of different nutrient absorbing zones of the root systems of the component species and nutrient release or availability (Nair, 1983).

Table 6.2. Soil fertility as influenced by tree species after 12 years at 0-15 cm depth

Tree species	*O.C.*	*N*	*Exchangeable (m e%)*	
	(%)	*(%)*	*Ca*	*K*
No tree (control)	2.86	0.10	9.7	1.1
Eucalyptus globulus	5.41	0.60	16.4	1.2
Acacia mearnsii	5.71	0.65	13.8	1.7

Source : Venkataraman, C., *et al.* (1983) Ind. For. 109(6).

Table 6.3. Influence of tree species on fertility of soil after 5 years at 0-15 cm soil depth

Tree species	*pH*	*O.C.*	*Available (kg/ha)*	
		(%)	P_2O_5	K_2O
Control	8.2	0.49	13	436
Luecaena	7.8	0.78	10	408
A. nilotica	8.0	0.80	19	848
D. sissoo	8.0	0.95	18	616
Pongamia	7.7	0.94	15	891

Source : Chandrashekariah, A.M. (1986), Ph.D. Thesis, U.A.S., Dharwad.

AMELIORATION OF CLIMATE BY TREES

Trees are essential to life on Earth. They moderate temperature and affect pollution, noise, wind and water. As trees grow, they provide a home for wild life and products like timber, fuel, fodder, fibre and other minor forest products for our daily use. The daily transpiration from a single tree can produce an estimated cooling effect of more than a million British Thermal Units. This is equal to 10 room-sized air-conditioners operating 20 hours a day. Because of the "green house" effect of waste particles in polluted air, the air temperature may be 20°F higher in urban areas than it is in nearby rural areas.

Trees absorb polluted air, and emit air richer in oxygen and somewhat freer of pollutants. As per data of the United States, a growth of 1 ton of wood releases atleast 1.1 tons of oxygen and absorbs atleast 1.5 tons of carbon dioxide. According to various studies made, three-fourths of conversion of CO_2 back to oxygen takes place in the ocean, but trees play an important part on the land (Madan, 1971).

EROSION AND ENVIRONMENT

Soil erosion is the single most important cause of land degradation. When most severe form of erosion like ravines, landslides and slips occur, the denuded land goes almost out of cultivation and is only fit for some permanent vegetative cover of trees. Failure of communication such as road, rail, telephone, electricity etc. are very common due to landslides and riverbank erosion etc. Groundwater resources are deficient due to increased rate of runoff and decreased percolation and infiltration of soils. Desertification is on increase and renewable resources are getting depleted. Wind erosion and shifting sand is going on at an alarming rate because of the reduction in the vegetative cover. Erosion affects environment in two ways :

Water pollution by water erosion

Water erosion pollutes water and is very dangerous. The water may contain toxicants unable to sustain any form of aquatic fauna. The flowing streams carry huge quantities of chemical fertilizers, pesticides and other toxic elements. It has been reported that forest runoff contains less nitrate contents then agriculture runoff. Runoff water in streams and rivers transport sediment load and pollute pond, river and sea water by soil particles, thus making drinking water a scarce resource (Singh and Bhardwaj, 1986). Adequate supply of safe drinking water is essential for improving public health.

Air pollution by wind erosion

The abrasive action of wind results in detachment of soil particles and are carried miles away from original place. The dust storms are often unbearable and people in villages, towns and cities in arid regions have to face inconveniences due to prolonged dust inhalation. The stagnating water can pollute air by foul smelling gases. The common pollutants of air are injurious gases such as carbon monoxide, sulphur dioxide, etc., and dust of coal particles from chimneys, soil particles etc. The plants act as air filter for its purification by absorbing and arresting injurious gases and particles.

NEED FOR ECOLOGICAL BALANCE

The population in India, with a growth rate of about 2.3%, is expected to cross one billion at the end of this century leading to the increase in demand for basic necessities *viz.*, food, fuel, fibre, fodder and timber. India is sustaining 16 per cent of global population on 2.5% of world's geographical area. There seems to be tendency of growing ecological imbalance due to pressure of population on forest. There has been a large scale deforestation over the past 30 years. Moreover, production from these forests is far below our expectation due to poor fertility and productivity because of erosion. Biomass production is only 0.5 cum/ha/year as against the world average of 2.1 cum/ha/year. Forest cover has declined from 22% in 1952 to 14.10% in 1982 as against 33% envisaged by the National Forest Policy. According to the survey conducted by the National Remote Sensing Agency, the good forest with a density of more than 30% cover only 11% of total 329 m ha of land mass.

Depletion of ozone layer and increased build-up of carbon-dioxide in the atmosphere, green house effect and global warming are some of the serious consequences of environmental degradation caused by excessive deforestation and heavy pressure of population. Due to population pressure it is impossible to allocate agricultural lands for growing forests. Therefore, advantage of forests can be harnessed by growing trees on marginal lands or with agricultural crops through agroforestry practices. This would ensure moderation of climate and ecological balance would be restored.

ENVIRONMENTAL BENEFITS FROM AGROFORESTRY

Agroforestry systems aim at growing of woody perennials along with agricultural crops on the same nit of land either in some form of spatial mixture or temporal sequence. KIng and Chandler (1978) defined agroforestry as a sustainable landuse system that maintains or increases total yields by combining food (annual) crops with tree (perennial) crops and/or animals simultaneously or sequentially on the same unit of land, using management practices that suit the social and cultural characteristics of the local people and the economic and ecological conditions of the area. Bene *et al.* (1977) expressed the main benefits of agroforestry usually in two major forms : Productivity and subtainability (conservation). Vergara (1982) describes ecological benefits of agroforestry as under :

(i) Reduction of pressure on forest.

(ii) More trees available to protect areas from environmental deterioration.

(iii) More efficient recycling of nutrients by deep-rooted trees.

(iv) Reduction of surface runoff, nutrient and soil loss.

(v) Improvement of micro-climate.

(vi) Improvement of soil fertility by addition of organic matter through leaf fall.

To achieve these objectives of conservation of soil and ecosystem, agroforestry systems are classified (King, 1980; Huxley, 1984) as agri-silviculture, silvi-pasture, agri-horti, etc. The use of these systems in conservation of soil and water has been stressed by several workers (Vaishnava and Narwadkar, 1989).

AGROFORESTRY AND SOIL CONSERVATION

Agroforestry is an age-old practice, followed traditionally in different forms in different parts of India. In arid parts of Rajasthan and Gujarat, *Prosopis cineraria* are grown with cereals. *Zizyphus mauritiana* (Ber) is identified as a most promosing multi-purpose tree species (MPTs) for arid areas of Rajasthan and poor degraded land of Bundelkhand region. Tree species like *Acacia nilotica, Azadirachta indica, Dalbergia sissoo,* and *Tamarix articulata* are found growing along the boundary or within the cropped area. Recently *Eucalyptus tereticornis* is being grown very extensively mostly on field bunds in many parts of India. Very recently a great amount of attention has been paid to *Leucaena leucocephala* while it is being popularised on its reportedly good qualities, very little factual information is available about its interaction with agricultural crops (Mittal and Singh, 1983) and/or soil conservation impacts. The usefulness of *Populus ciliata* as promising species for soil conservation in hilly areas was reported by Mathur *et al.* (1982). Das (1980) reported that *Alnus nepalensis* trees were grown on hill slopes for fuel wood purpose which subsequently acted as the terraces for paddy fields. Itnal (1986) reported that in agri-silviculture study in black soils, the runoff loss of rain water was least with *Acacia auriculiformis, Albizzia lebbek* and *Acacia nilotica.*

On a marginal land in North-Western Himalayan region at Dehradun, growing of trees (*Eucalyptus, Leucaena*) and grass (*Chrysopogon fulvus*) along with cereal crops (wheat, maize) reduced the runoff and soil loss substantially, the reduction being more under *Eucalyptus* than under *Leucaena* (Narain *et al.*, 1988). The soil loss of 47 t/ha in cultiváted fallow decreased to 0.9 to 5.0 t/ha under maize + *Encalyptus* and maize + *Leucaena*, respectively. Likewise, the runoff 26 per cent of rainfall in cultivated fallow was reduced to 3.6 to 8.9 per cent under maize + Eucalyptus and maize + *Leucaena*, respectively (Table 6.4).

Dhruvanarayana *et al.* (1986) reported that agri-horticulture system proved better in control of runoff and soil loss as compared to agriculture alone (Table 6.5). It is clear from data that the runoff loss and soil loss was highest in case of agriculture, medium in agri-horti system and lowest in case of forests.

In silvi-pasture system of agroforestry, the soils are improved greatly through continuous incorporation of organic residues and litters which help in improving organic carbon, nitrogen and phosphorus content of soil (Table 6.6).

Table 6.4. Soil loss and runoff under agroforestry system at Dehradun

Land use	*Soil loss (t/ha)*	*Runoff loss (% of rainfall)*
Cultivated fallow	46.70	26.4
Maize	17.70	18.3
Maize + *Leucaena*	5.00	8.9
Maize + *Eucalyptus*	0.91	3.6
Chrysopogon fulvus	0.33	1.6
Grass + *Leucaena*	0.13	0.6
Grass + *Eucalyptus*	0.02	0.1
Leucaena	0.04	0.4
Eucalyptus	0.01	0.1

Source : Annual Report, CSWCRTI, Dehradun, 1988.

Table 6.5. Soil loss and runoff under agri-horti system compared with agriculture and forest

Type of system	*Soil loss (t/ha)*	*Runoff loss (% of rainfall)*
Agriculture (Jhumming)	40.95	6.45
Agri-horti with terracing	2.96	3.11
Forests (Bamboo)	0.21	0.28

Table 6.6. Physico-chemical properties of soil under silvi-pasture and open grassland after seven years

Silvi-pasture	*pH*	*EC (m mhos)*	*O.C. (%)*	*Available*		*Dry forage yield (t/ha)*
				N (kg/ha)	*P (kg/ha)*	
L. leucocephala	7.2	0.18	0.98	237	16.3	5.28
A. nilotica	7.5	0.22	0.71	216	15.6	4.47
A. lebbeck	7.6	0.28	0.68	208	15.0	6.63
A. procera	7.6	0.30	0.62	197	14.2	4.21
Open grassland	7.7	0.28	0.60	178	13.0	5.95

Productivity of forages from under-storey grasses was found to be highest with *Albizzia lebbeck* followed by *Leucaena leucocephala*. The latter tree species have greatly improved the soil productivity too (Hazra, 1990).

Alley cropping is essentially an agroforestry system in which food crops are grown in alleys formed by hedge rows of trees or shrubs (Kang *et al.*, 1984). The hedge rows are kept pruned during cropping to prevent shading and to reduce competition with food crops. Alley cropping affect the sustainability of crop yield and improve degraded soil by recycling of plant nutrients. Kang *et al.* (1984) observed twice the amount of soil organic matter in the soil due to periodic addition of Leucaena prunings (Table 6.7) and these helped in recycling large quantities of other plant nutrients.

Table 6.7. Recycling of plant nutrients with alley cropping of *Leucaena* with maize and cowpea

Treatments (kg N/ha)	*Leucaena prunings*	*pH*	*O.C. (%)*	*Exchangeable K*	*Exchangeable Ca*	*Exchangeable Mg*	*Bray P (ppm)*
0	Removed	6.0	0.65	0.19	2.90	0.35	27.0
0	Retained	6.0	1.07	0.28	3.45	0.50	36.2
80	Retained	5.8	1.19	0.26	2.80	0.45	25.6
LSD 0.50		0.2	0.14	0.05	0.55	0.11	5.3

Source : Kang *et al.* (1984).

NUTRIENT RECYCLING

One of the main benefits of perennial tree components in agroforestry system is the contribution of nutrients by nitrogen fixation, leaf fall, turnover of fine root biomass and recycling of nutrients in different soil layers. Unlike fertilizer nitrogen, the nutrients added by a tree gradually and slowly available *in situ* and are less likely to be lost through volatilization and leaching.

Studies at CRIDA (Hyderabad) with 760 mm rainfall have shown that 8 year old *Leucaena leucocephala* trees add about 43 kg N/ha as leaf-fall, seeds, petioles, podwells etc. (Table 6.8).

Table 6.8. Cumulative annual dry matter and nitrogen addition by *Leucaena leucocephala*

Component	*Dry matter (kg/ha)*		*Total nitrogen*	
	One-year old trees	*Eight-year old trees*	*One-year old trees*	*Eight-year old trees*
Podwell	82	200	1.05	3.17
Petiole	241	533	2.46	6.81
Leaflet	84	1115	1.37	18.80
Seeds	313	297	14.45	13.70
Total	720	2145	19.33	42.48

Source : Venkateshwarlu (unpublished).

PROSPECTS OF AGROFORESTRY IN SALT-AFFECTED SOILS

Salt-affected soils which are lying barren for decades occupy vast expanse (8 m.ha) in India. Work done on afforestation of salt-affected soil has been reviewed by Yadav (1980). In a study on the effect of *Prosopis juliflora* (mesquite) and *Leptochloa fusca* (Karnal grass) agroforestry system on soil properties 22 months after planting, Singh *et al.* (1988) found considerable improvement in soil as evinced by reduction in the values of pH and EC and increase in the contents of organic carbon and available nitrogen (Table 6.9).

Table 6.9. Effect of Mesquite-karnal grass agroforestry system on soil properties 22 months after planting

Soil depth (cm)	*pH*		*ECe*		*O.C.*		*Available N (kg/ha)*	
	BP	*AP*	*BP*	*AP*	*BP*	*AP*	*BP*	*AP*
0-7.5	10.0	9.2	2.1	0.6	0.3	0.5	101	151
7.5-15	10.1	9.5	2.0	0.5	0.2	0.3	96	112
15-30	10.1	9.7	1.3	0.8	0.2	0.3	78	189
30-45	10.2	10.0	1.3	0.9	0.2	0.2	63	60

BP = Before planting; AP = After planting.

SHELTERBELTS AND WINDBREAKS

Use of shelterbelts of trees, shrubs etc. for the protection of crops from erosive winds is a special form of agroforestry system. Shelterbelts when planted across and on the margins of agricultural fields effectively protect the crops and control sand drift. It has been reported that soil loss was 184 kg/ha in *Cassia siamea* type of shelterbelt, while it was 547 kg/ha in bare soil (without shelterbelt) during June (Table 6.10).

Table 6.10. Effect of different types of shelterbelts on soil erosion

Type of shelterbelts	*Total amount of soil loss (kg/ha) from 20th April to 26th June*		
	*2H**	*5H*	*10H*
Prosopis juliflora	93	609	351
Cassia siamea	92	277	184
Acacia torilis	106	494	300
Bare soil (no shelterbelts)	263	831	547

H* refers to the height of shelterbelts.
Source : Shankaranarayan *et al.* (1987).

SCOPE FOR FURTHER WORK

Agroforestry is regarded as a sound landuse system which emphasises "productivity" and "sustainability" at the same time. Agricultural research in the world has mostly been production or crop oriented. Only since the last decade, some attention has been given to sustainable agriculture production through agroforestry system. We can conclude that the role of agroforestry in meeting either present or future requirements of fuelwood, food, fodder and small timber and for environmental protection has been very well recognised in our country. It is now necessary to develop location-specific, need-oriented systems alongwith necessary support systems so that farmers can get the required seedling and other inputs easily and market the produce at competitive prices.

REFERENCES

Bene, J.G., H.B. Beall and A. Cote, 1977 : *Trees, Food and People*, IDRC, Ottawa.

Das, D.C., 1980 : Some aspects of shifting cultivation related to soil and water conservation. *Indian J. Soil Conserv.*, 8 : 53-59.

Dhruvanarayana, V.V., Ram Babu and C. Venkataramanan, 1986 : Soil erosion under different agro-climatic conditions. *Indian J. Soil Conserv.*, 14 : 39-50.

Hazra, C.R., 1990 : Forages from agroforestry based production systems in Bundelkhand region. *Internl. Symp. on Water Erosion, Sedimentation and Resource conservation* Oct. 9-13, 1990, 437-445.

Huxley, P.A., 1984 : Education for agroforestry : United Nations University NRTS - 23/UNDP Publication. *Social, Economic and Institutional aspects of agroforestry*, Tokyo, Japan.

Itnal, C.J., 1986 : Agroforestry in different climatic/edaphic zones in India - some examples of indigenous practices in Karnataka. ICAR-ICRAF training cum workshop on *Agroforestry*, held at CRIDA, Sep. 13, 1986, Hyderabad, India.

Kang, B.T., G.F. Wilson and L.T. Lawson, 1984 : Alley cropping - A stable alternative to shifting cultivation. *IITA*, Ibadan, Nigeria, 1-22.

King, K.F.S., 1980 : Multiple use research, Key note address, IUFRO/MBA conference. *Research on multiple use of forest resource*, May 1980, Flagstaff, Arizona.

King, K.F.S. and M.T. Chandler, 1978 : *The Wastelands*, ICRAF, Nairobi.

Madan, U.S., 1971 : Role of forestry in soil conservation. *Participant Jour.*, 5 : 19-21.

Mathur, H.N., R.P. Singh and K.C. Sharma, 1982 : *Populus ciliata* - A promising tree species for soil conservation in hilly area. *Ind. For.*, 108 : 599-604.

Mittal, S.P. and P. Singh, 1983 : Study on intercropping of field crops with fodder crop of "Su babul" (*Leucaena leucocephala*) under rainfed conditions. *Ann. Rep. Central Soil and Water Cons. Res. Ins.* Dehradun, India.

Nair, P.K.R., 1983 : Tree integration on farmlands for sustained productivity of small holdings. In : *Environmentally Sound Agriculture* (Ed. William Lockeretz). Praeger Publisher, New York, USA.

Narain, P., R. Singh and P. Joshi, 1988 : Studies on nutrient and water budget under different landuse systems. *Annual Report*, pp. 56 *CSWCRTI*, Dehradun, India.

Sheng, T.C., 1986 : *Watershed Conservation - A collection of papers for developing countries*. The Chinese Soil and Water Conservation Society, Taipei, Taiwan, Republic of China and Colorado State University Fort Collins, Colorado, USA, pp. 85-89.

Singh, R. and S.P. Bhardwaj, 1986 : Technology for checking soil erosion. *Indian Fmg.*, 36 : 24-28.

Singh Gurbachan, I.P. Abrol and S.S. Cheema, 1988 : Effect of spacing and lopping on a mesquite (*Prosopis juliflora*) Karnal grass (*Leptochloa fusca*) agroforestry system on an alkaline soil. *Expl. Agric.*, 25 : 401-408.

Vaishnava, V.G. and P.R. Narwadkar, 1989 : Role of agroforestry in soil conservation. *Proceedings of Eleventh International Congress on Agricultural Engineering*, Dublin, 4-8 Sept., 1989.

Vergara, N.T., 1982 : New direction in agroforestry : the potential of tropical legume trees. Working Group on AF Environment and Policy Institute, East West Centre, Honolulu, Hawaii, 13-22.

Wiersum, K.F., 1984 : Surface erosion under various tropical agroforestry system. In : *Symposium on the effect of forest land use on erosion and slope stability*. (Eds. C.L. O'Laughlin and A.J. Price), 232-240. Envir. Policy Inst. East-West Centre, Honolulu, Hawaii.

Yadav, J.S.P., 1980 : Salt affected soils and their afforestation. *Indian Forester*, 106 : 259-272.

Chapter 7 Mining and Mineral Processing Wastes in Indian Base-Metal Mines and their Environmental Management

K.L. Rai

The paper deals with the most serious, and perhaps also the most difficult, environmental problem related to the mining and mineral-processing wastes in the Indian base-metallic mines. The unfortunate fact, that this problem has already assumed highly unpalatable and alarming proportions in most of the base-metal mining districts of the country, has been highlighted. Although lot of concern about safeguarding the mining environment has been aroused and voiced in the recent years, much remains to be done for realistic appraisal and effective control of the environmental fallouts of the current practices of waste disposal and management in our base-metal mines, in general. Two case-studies—one of the Malanjkhand Copper Project and the other of Rangpo mining district, have been carried out in this connection.

INTRODUCTION

The mining and mineral-processing wastes generated during the exploitation of base-metallic ores ad their treatment in various processing plants, in general, pose the most difficult, and often the most expensive environmental problem to be rectified in any base-metal mining venture. These wastes, unless properly handled and disposed of, often pose serious threats to ecology and environment, being the primary source of heavy metal pollution of the land-surface and groundwater-resources in and around a mining project. Proper management of all types of wastes generated in a base-metal mining project, therefore, is essential in the best interest of sustainable mining of the resources in question.

Proper management planning concerned with the above-stated wastes comprises of firm programmes ensuring that :

(a) they are disposed of in such a manner that they will not endanger the public or worker safety,

(b) shall not upset the ecological balance and the environment, in general,

(c) will remain stable and environmentally benign in perpetuity with little or no maintenance, and
(d) depending upon the potentialities, can be effectively and economically reclaimed to a useful state, if needed, at any stage in future.

PRESENT STATUS IN INDIA

In India, the concept of environmental management of mining and processing wastes in base-metal mining areas as integral part of comprehensive mine-planning has been introduced rather late during the mid-eighties, essentially after the promulgation of the Environment (Protection) Act of 1986 and, somewhat stringently, after the Environment Impact Assessment Notification of 1994 under it. Submission of EMP with requisite budgetary provisions for this aspect as integral part of the comprehensive mine plan is now mandatory as per the above referred notification dated 27th January, 1994 issued by the Ministry of Environment and Forest, Govt. of India.

Table 7.1. Solid wastes in some base-metal mines of India

Deposit/Project	*Source*	*Soil wastes*	
		Rate of generation per annum	*Method of disposal*
Indian Copper Complex of HCL, Mosabani, Bihar	Mines	7200 t	80% of it is used in underground slopes as support-fill
	Mill-tailings	1.4 Mt	50% of deslimed tailings are conveyed underground to backfill stoped-out areas
	Smelter and converter (Malanjkhand)	42,000 t	Dumped along the river-banks
Khetri Copper Complex of HCL, Khetri, Rajasthan	Mines (Khetri, Kolihan and Chandmari)	1.1 Mt	Dumped in low-lying areas or on the hill-slopes or along the river-banks
	Mill-tailings	1.3 Mt	Disposed of in tailings dam
	Smelter slag	60,000 t	Reprocessed after crushing and milling in the concentration circuit
Malanjkhand Copper Project of HCL	Mines	4.26 Mt	Dumped on banging wall side
	Mill-trailings	1.90 Mt	Disposed of in tailings dam
Rampura Agucha	Mines	1,400,000 M^3	Spoil dumps
Rango (Zn-Pb-Cu), Sikkim	Mines	Over 100,000 t	Dumped along the river-banks

In mining and refining sector, the environmental regulations are assuming more and more stringent shape and will affect all stages of the production processes from the disposal of mine-spoils and mill-wastes to that of the refinery-slimes and sludges. Most of the mining companies have now set up environmental divisions of departments which are charged with the responsibilities to regularly monitor the dumping of mine-wastes and tailings and also carry out reclamation work and assessment of environmental impacts from time to time.

In almost all major base-metal mining districts of the country, where mining and mineral processing of ores has been going on for the last several decades, the lack of proper and scientific environmental management of the mining, milling and processing wastes until a few years ago has cumulatively led to serious environmental problems which now defy the best-intentioned efforts and mitigation measures. The present status of generation of various types of solid wastes and liquid effluents in some of the Indian base-metal mines and the methods of their disposal are summarised in Tables 7.1 and 7.2 respectively. The data illustrates the dimensions of the problems concerned with the mining and milling wastes. According to an estimate, the total waste rock to be handled per annum is of the order of about 9 million tonnes (Rathore *et al.*, 1994).

Table 7.2. Liquid effluents in some base-metal mines of India

Source	*Discharged quantity of effluent (m^3 per month)*	*pH value/Quality/Remarks*
A. Indian Copper Complex of HCL, Bihar		
(i) From the mines		
Mosabani	22,000	6.5 to 6.9
Rakha	80,000	4.0 to 5.0
Badia	2,500	5.0 to 6.5
(ii) From the concentrator plant		
	Not estimated	Effluents of tailings water, having high percentage of suspended solids, are discharged in tailings pond
(iii) From the Smelter, Refinery and Sulphuric acid plants		
Nickel sulphate plant	20,000	A modern integrated neutralisation system for plant units being planned
B. Khertri Copper Complex, Rajasthan		
(i) From the fertiliser plant	1,20,000	Auto-controlled neutralisation system installed
(ii) From other plants (including concentrator, smelter and refinery)	1,80,000	Acid effluents environmentally hazardous due to increased solubility and availability of toxic metal iron
C. Malanjkhand Copper Project, M.P.		
Concentration plant	3,75,000	Discharged into the tailings pond

In most of the mining areas, the waste dumps and liquid effluents from mining and processing plants have been responsible for general environmental degradation, often assuming hazardous proportions. The solid wastes, unless stabilised, are often prone to slumping, sliding and blowing effects of strong winds. In course of time, part of them find passage to drainage channels and reservoirs, silting and choking them thereby contributing to increased flood-levels, which in turn, lead to pollution and spoiling of fertile agricultural land, surface water resources, flora and fauna of the downstream terrain. Long-term efforts of such environmental pollution on the health, efficiency and longevity of local habitants in such areas, remain a matter of bitter disputes — sometimes to the extent of legal battles between the companies concerned and the environmentalists or the N.G.O.'s or directly the public itself.

In the recent years, increased governmental control with more and more stringent stipulations and legal requirements for maintaining ecological balance, regular environmental monitoring, pollution control and restoration of land for mining, disposal of mining and processing wastes is being planned and programmed progressively in a phased manner under scientific and result-oriented waste management technologies. There is increased realisation of the need for stabilisation of waste-dumps and pollution control of all attributes of the environment. This is best illustrated through the details of selected case-studies.

SELECTED CASE-STUDIES

A. Malanjkhand Copper Project, Balaghat District, M.P.—Project and its environmental setting

The copper deposit of Malanjkhand (22°5"-22°1' 48.5", 80°42'33"-80°43'7") located about 90 km North-East of district town of Balaghat in Madhya Pradesh, is estimated to contain approximately 800 million tonnes of mineable copper ore at 0.8% cut-off with 0.004%, molybdenum, 0.2 gm/t gold and 6 gm/t of silver. It is under active exploitation at present by Hindustan Copper Ltd. in the form of a large mechanised open cast mine upto a depth of 200 m from the surface producing 2 Mt/year of copper ore at a total excavation of 13.5 Mt of mineralised rock (Subhedar, 1986). With the success of ongoing trial-runs for chemical and bacterial heap-leaching of the disseminated ore, specifically from the hanging wall mineralisation, it has been planned to enhance the total production of copper ore to 3 Mt/year in near future.

The location of the mining project (Fig. 7.1) in Central Indian highlands of Bihar Plateau (ground level : 580 MRL) that constitutes a prominent water-shed for the Banjar river, a tributary of Narmada river flowing northwards, and Sone river, tributary of Godavari river flowing southwards and its proximity to the famous Kanha National Park, a reserve forest of rich biodiversity, has invoked serious consideration regarding the impact of the mining project on the surrounding environment. Unfortunately, this angle does not appear to have received much attention during the mine planning stages of the project.

At present rated capacity of 2 Mt/year of production of ore, the mining-, milling- and processing-wastes, both solid and liquid, being generated in the opencast mine and its milling/processing plants (Table 7.1) are fraught with the dangers of adverse environmental impacts in the long run. These aspects need to be tackled effectively through long-term comprehensive mining and environmental management planning. Fig. 7.2 depicts the existing plan for environmental management in the leasehold area of the project.

(a) Mining and milling practices

The open-pit mine is located in the central part of the arcuate Malanjkhand Hill. When fully developed as per the present mining plans, the pit dimensions will be 2,200 m (length) × 640 m (width) × 276 m (depth) covering 23 planned benches. The cumulative overall waste/ore ratio, which had been approximately 7 : 1 upto the 10th bench level, is destined to be ultimately 4 : 6 : 1.

Difficult mining and milling conditions are ascribed principally to very hard and abrasive nature of the ore which comprises 92% quartz and a complex jointing and shearing system of its host-rocks. Generation of oversize boulders during blasting due to the jointing system in the rock creates additional problems of sizing the rock and its transportation. The current haul distance through a fleet of haul-trucks average 0.85-1 km for transporting the ore to the primary crusher and 2.2-3 km for disposing

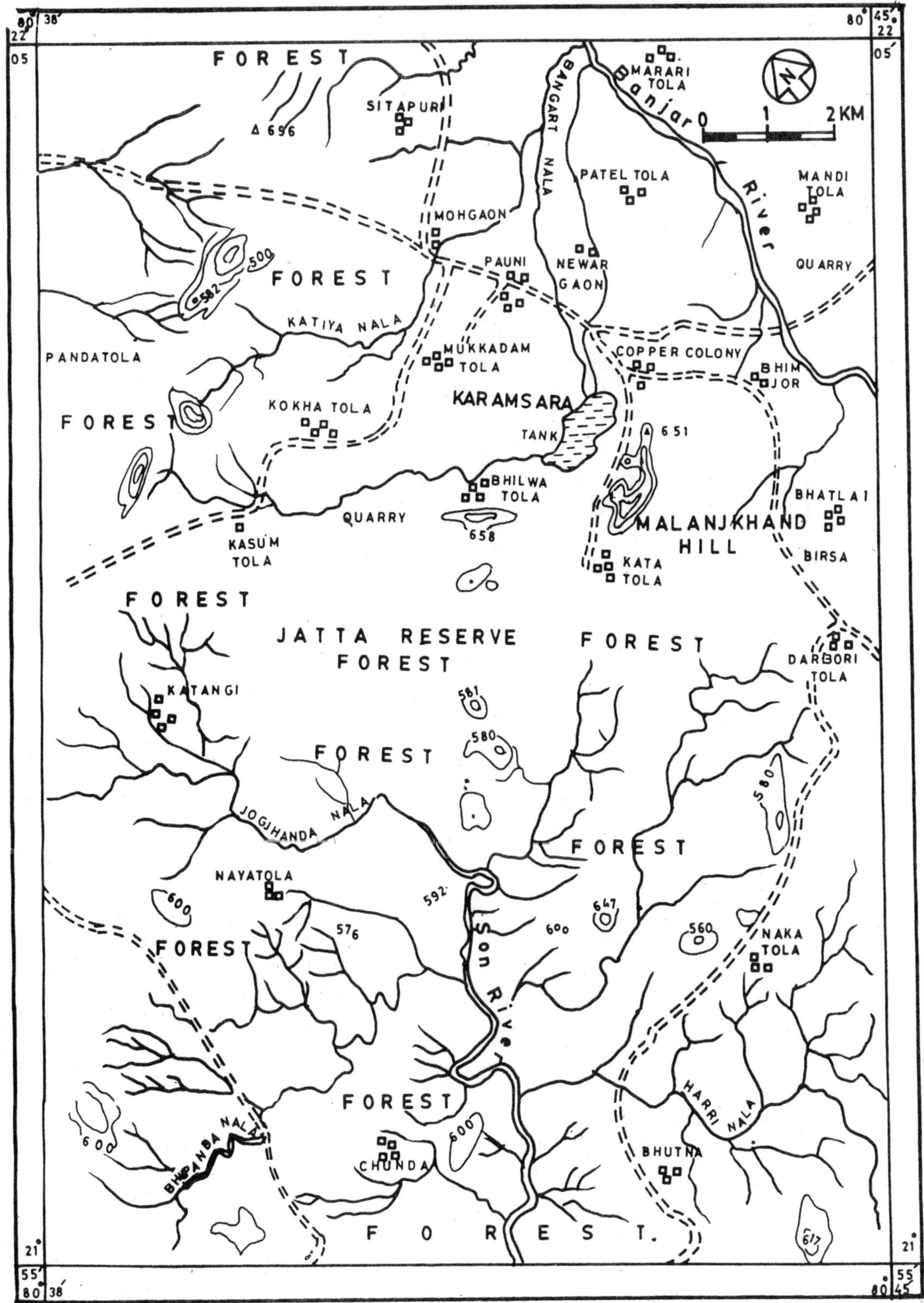

Fig. 7.1. Physiographic map of Malanjkhand region.

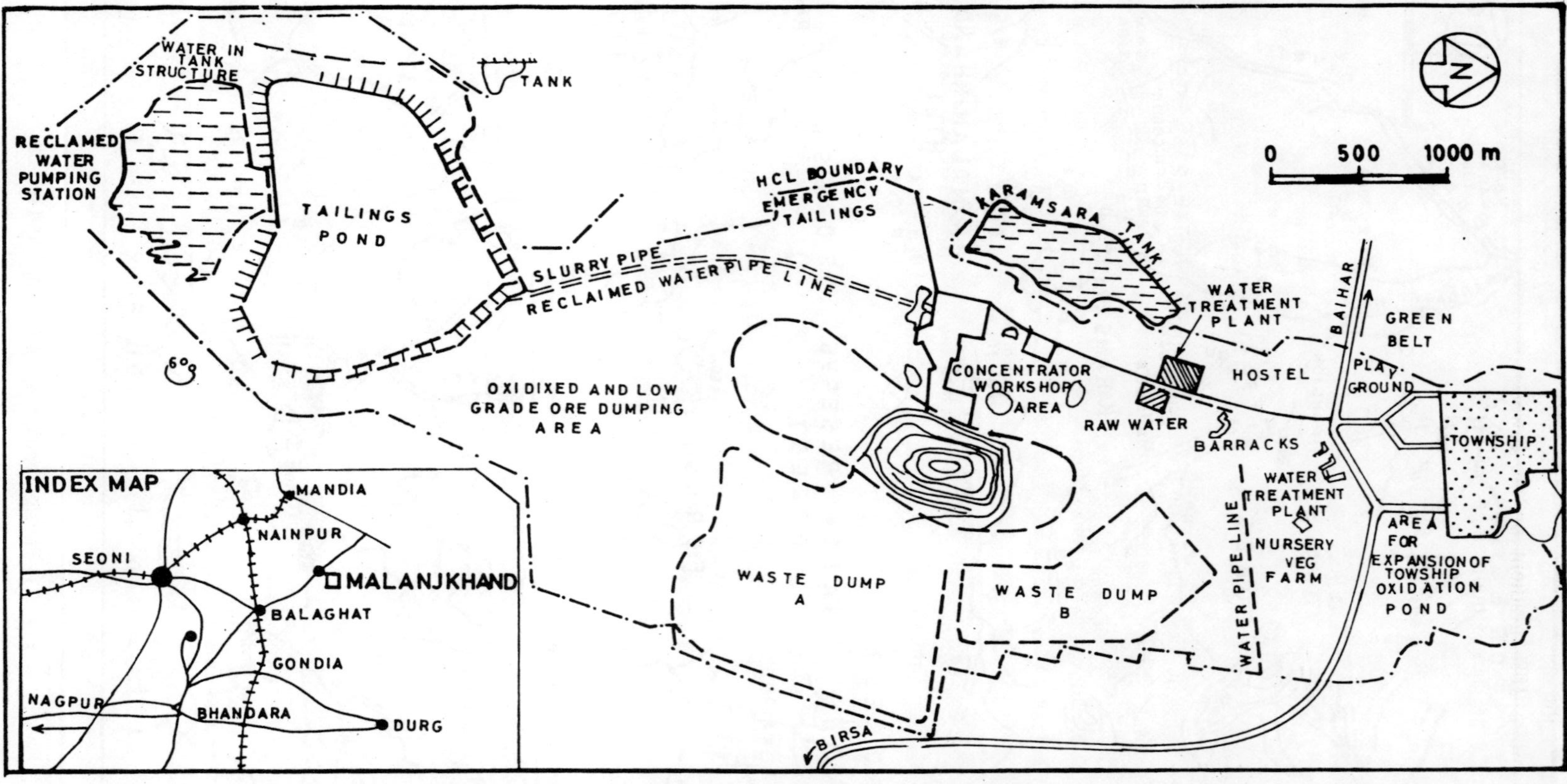

Fig. 7.2. Environmental management plan of Malanjkhand project (after M.C.P.).

of the mine waste at the spoil-dumps. The ultimate pit design envisages haul roads with an optimum width of 22 m and a 10% gradient on the lowest five levels.

Separate dumps are being maintained for (a) the waste rock/spoils, (b) oxidised ore, and (c) the lean or marginal ore in addition to a stockpile of the ore adjacent to the primary crusher. Three 20 m lifts have been planned in the areas of spoil dumps over the life of the mine to give total eventual height of 60 m. The upper surfaces of the dumps are graded to 1% gradient to facilitate drainage outwards.

(b) Solid wastes

Stripping and open casting of the arcuate Malanjkhand hill has, over the years, resulted in creation of laige soil-dumps, which now form prominent landmarks in otherwise peneplained country and are the source of aesthetic degradation of the area in their present form. For reasons of short-term gains, these dumps were allowed to be made on the hanging-wall side of the mineralised quartz reef. Vast potentialities of the deposit, discovered at depth in the layer years, however, have exposed the fallacy of thse gains. Consequently, these dumps are now proving to be the biggest liability of the mine, not only as impediments to detailed exploration in depth, but also as the source of heavy metal pollution of land and water resources of the area. Data representing the average concentration of various metallic elements in the host-granitoids of the Malanjkhand vein, that constitute the main bulk of the solid mine-wastes, is given in Table 7.3. Accordingly, fairly high contents of Cu, Zn, Mo, Sn, W and F in the waste-rock should receive the attention of planners.

In the original/approved mining plans/DPR, there is no provision to stabilise the dumps by vegetation and convert the same into agricultural land or pastures. There is no provision either to reclaim the

Table 7.3. Minor and trace element contents in the host and country-rocks of Malanjkhand copper deposit

	Average in normal graphites (ppm)	*Pipardhar (grey) granodiorite (Country rock) (ppm)*	*Malanjkhand (pink) granite (Host-rock) (ppm)*
Cu	12	89	906
Zn	50	55	146
Pb	20	12	19
Mo	1.5	19	42
Sn	3	14	30
W	1.5	13	43
Ni	3.4	31	27
Co	2.3	26	16
Cr	14.2	127	63
Rb	192	93	109
Sr	248	499	300

open-pit land by backfilling or its use as a lake for recreational and other purposes. In all probability, the entire agricultural land of about 900 hectares shall be lost/damaged for ever.

Beside the solid waste referred to above, the opencast mining and milling operations generate substantial amount of fugitive dust which is generally wind-blown to far-off places owing to the location of the mine and the concentrator plant in situations vulnerable always to high winds. The surrounding villages, particularly those located along the directions of the strong winds are therefore prone to airborne dust pollution. Except for the provision of dust extractors at the crushing plant and other strategic points of the mill and vehicle-drawn water sprays on the mine roadways, no other major dust-control programme has been implemented in the mine and concentrator plant possibly because of the fact that the drift of surface dust is essentially seasonal and not problematic at present.

(c) Liquid effluents

(i) Mine water : The mine water from the opencast mine is discharged, without interim treatment, at the rate of about 1000 m^3/day in rainy season, through open drains into the neighbouring Karamsara tank located near Karamsara village, about 0.5 km east of the mine. The environmental vulnerability of this practice is a matter of dispute, as this water is toxicated with heavy metals, mainly Cu, Zn, Mo, Ni and sulphur.

(ii) Process water from the benefication plant : The benefication plant consumes a major quantity of the water during the process of grinding as well as disposal of the tailings in the form of a slurry. This slurry and other liquid effluents are pumped through three sets of tailings pumps operating in series, delivering approximately 15000 m^3 of liquid effluents daily through rubber-lined, 250 mm diameter, pipelines into a tailings pond located about 4 km south of the concentrator plant (Fig. 7.2).

At the tailings disposal area, a starter dam of earth-fill construction, 18 m high at present has been built. This will ultimately be raised to a height of 47 m using the tailings themselves. The tailings area is equipped with an efficient decant system. The tailings are settled out and the decant water is recycled into the benefication plant. Although the effluent water from various sections of the plant is not allowed to drain to the surrounding area and is pumped to waste water treatment plant before being let out for irrigation and other purposes, a part of it gets discharged alongwith the mine water into the nearby Karamsara tank.

Other sources of pollution of the surface water resources include the effluents and leachates from the heap-leaching operations of the oxidised and marginal/low-grade ores and the leach liquor run-off from the spoil-heaps consequent on surface precipitation. The latter is usually an acid sulphate solution due to high pyritic content of the mine-spoils and unless the spoil-heaps are stabilised and revegetated, the effective control on the incidence of this highly acidic run-off from large spoil-dumps during and after the monsoon rains, is going to be only a mirage.

The probability of pollution of the groundwater from the subsurface seepage of mine-water during its flow along the open drains and also from other polluted surface waters like the leach liquor run-off from the spoil-heaps, can not be ruled out in view of the high incidence of joints, faults, and shear-fractures throughout the mine area. Pollution of groundwater due to long-term contaminated seepage of acid sulphate solutions and toxic elements from the base metals of tailings pond and chemico-bacterial leach heaps is another possibility which needs critical study and systematic investigation. A preliminary study on the metal toxicants in the groundwater of the dug-wells from Malanjkhand area, accomplished by Babu and Trivedi (1987) has revealed that Mo, Ni, Co, Pb and Sn are within the permissible limits for drinking purposes, whereas Zn and W are abnormally high and therefore, need special attention. This is corroborated by the apprehensions of Das (1987) that the quarry water may be intoxicated with heavy metals as the ore-body contains traces of Zn (0.1%), Mo (0.1 to 0.001%), Ni, Co, Ag, Ga, Sn, V, Au beside copper, sulphur etc. The practice of discharging this water coupled with a part of the effluents from the benefication plant into the Karamsara tank will, in his opinion, "pollute the latter beyond permissible limit". Keeping this possibility in view, the practice under reference needs earliest possible review and rectification as the water from the Karamsara tank is used extensively by the villagers and also by the State Irrigation Department for irrigating neighbouring lands during major part of the year.

A general consciousness about the environmental vulnerability of mining and milling operations in Malanjkhand Project has grown considerably in the recent years. As a result of it, the mining company is already seized with the problem of environmental protection as reflected well in the Project's environmental management plan. It is hoped that the abatement measures undertaken in this plan shall produce visible and quantum improvement in the mining environment of the project in near future.

B. Polymetallic Base-Metal Mine at Rangpo in Southern Sikkim

The Zn-Pb-Cu polymetallic ore-body of Bhotang Hill near Rangpo (27°10' : 88°30') in southern Sikkin represents the only base-metal deposit of the Lesser Himalayan belt under active exploitation at present. It is, however, a small mine with ROM at around 60 to 80 tpd. Its operation is being carried out since 1966 by the Sikkim Mining Corporation amidst numerous geo-environmental and techno-economic constraints, which are typical of the difficult Himalayan conditions (Rai and Rao, 1986; Rai, 1993).

Established ore reserves of the deposit are Zinc : 68.0 M.T., Lead : 37.3 M.T. and Copper : 0.64 M.T. over the proved strike length of 500 m in Bhotang Hill with varying width of the ore body from 1 to 12 m. The tenor of the ROM grade ore is estimated as 2, 52, 1.2 and 1.1 for zinc, lead and copper respectively. Other minor metals reported in the ores include As, Bi, Sb, Ni, Co, Ag, Au, Hg, Cr and Sn. The silver is extractable at the rate of 750-1000 gm per ton of ore.

(a) Mineralisation and host-rock characteristics

Base-metal mineralisation in the deposit is localised principally along a compressed belt of black carbonaceous phyllite and sericite-chlorite schist belonging to Upper Proterozoic Daling Group metasediments. The complex mineralogical and chemical composition of its ores have posed serious problems of ore beneficiation and extractive metallurgy particularly because of very high bismuth content of the lead which concentrates at an average of 0.9% against the desirable level of 0.05% in the present metal market. Similarly, the processed zinc concentrate has high iron-content at 10-15% against a maximum desirable level of 6%. The lead concentrate is simply unsaleable and, therefore, has to be stacked at the concentrator plant in the form of open dumps.

The mineralised host-rocks, in general, possess strong potentiality to release various heavy metals and toxicants during their chemical weathering. The average concentration of certain important heavy metals in various host-rocks of the deposit are given in Table 7.4. As may be observed from the given data, the main host-rocks of mineralisation, namely the chlorite schist and metabasites, possess fairly high concentration of metal toxicants like Zn, Cu, Pb, Ni, Co, As and Bi in them.

Environmental impact related to mining and milling wastes

Indiscriminate disposal of the overburden rock from an old open cast mine and mine spoils from the

Table 7.4. Mean concentration of heavy metals in the host-rocks of polymetallic sulphide mineralisation, Rangpo, Sikkim

Elements	*Host metabasites*	*Chlorite schist*	*Silicate/talc schist*
n	8	5	6
Cu	553	480	130
Zn	3040	1042	98
Pb	348	571	122
Ni	219	176	28
Co	118	204	33
As	35	122	5
Cr	20	15	20
Bi	29	24	25

In addition to above-stated toxicants, anomalous concentration of radioactive minerals has been reported from certain zones of the ore-body.

present underground mine on the steep hill-slopes, without follow-up measures for slope stabilisation, has led to frequent landslides that have often upset the entire progress of mining and milling of ores for months together. Subsidence of the backfilled parts of old quarry are also not uncommon. Large-scale dispersal of mining and milling wastes and finer tailings due to the torrential rains, facilitated by the gravity, have over the years led to progressive large-scale silting of the river bed of the Tista river. This, in turn, contributes to dispersal of the hazardous mine pollutants over wider areas in downstream direction, particularly after the floods, adversely affecting the quality of the agricultural land in the narrow river valley. The heavy metal toxicants, fixed in the top layers of the soil affect its fertility and reportedly are having adverse impacts on the yield of the local crops, mainly maize and rice. Environmental impact of the ongoing practice of disposing of the toxic liquid effluents from the mine and the concentrator plant directly into the river without any treatment has, in all probability, its own toll affecting the local faunal and aquatic life and flora over the years. This remains yet to be assessed through systematic periodic assessment.

In addition to the liquid effluents stated above, other sources of pollution of the surface and groundwater resources of the area apparently include the following :

* leachates released from the surficial weathering of mine-debris used as backfilling material in the subsidence areas;
* leachates released from the heaps of non-saleable lead-concentrates stacked in open near the concentrator plant;
* leachates/seepage of toxicants from the concentration pond as well as tailings slurry near the main adit of the mine.

A recent study of the effluent characteristics from the Rangpo mining area by Gupta and Prasad (1994) has revealed the unpotable nature of the river water in the downstream areas of the Rangpo mine. The relevant data obtained by them is presented in the Tables 7.5 and 7.6.

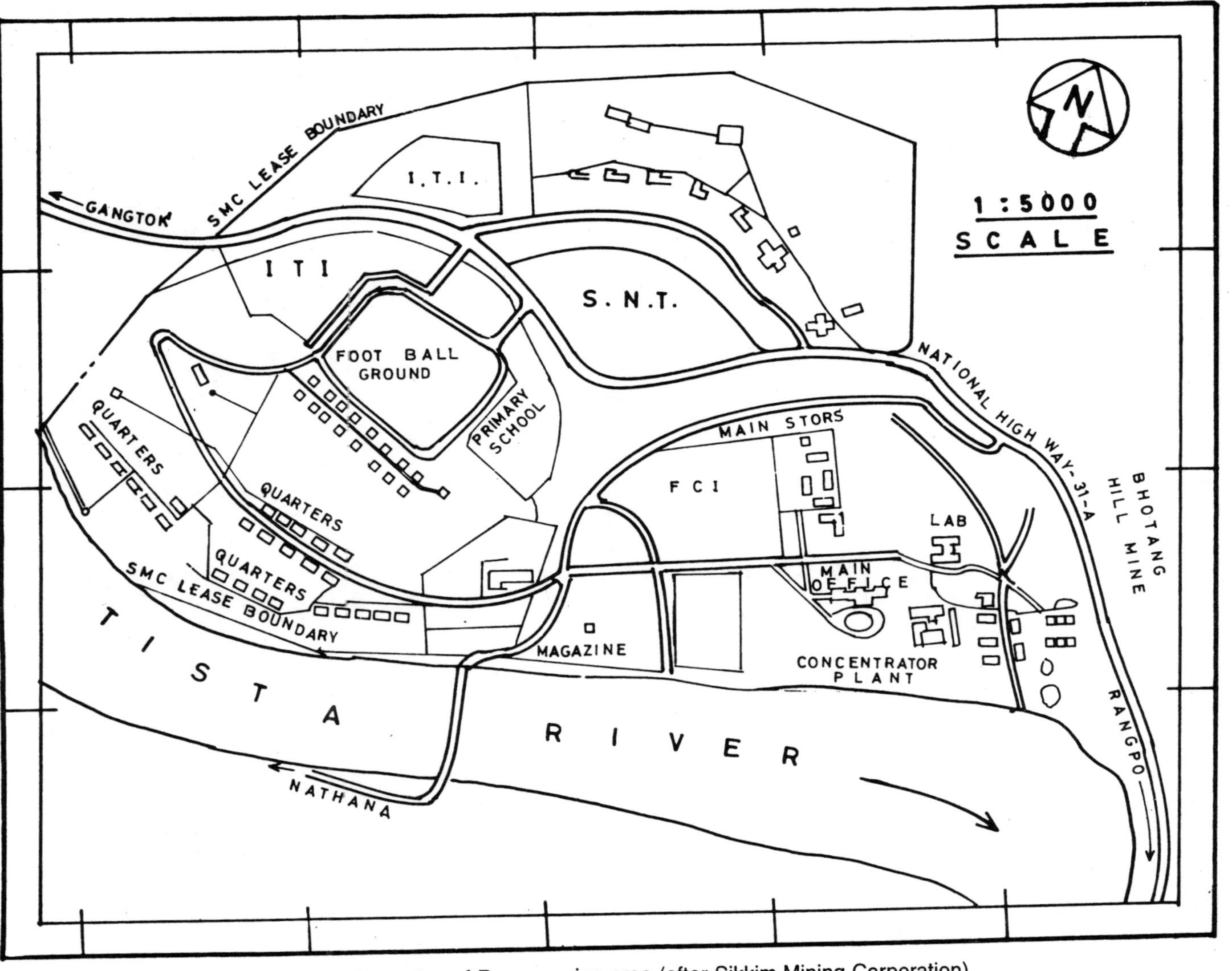

Fig. 7.3. Surface plan of Rangpo mine area (after Sikkim Mining Corporation).

Table 7.5. Chemical quality of percolation water in underground mine along fault and pumped out mine-water before discharging into Tista river (trace elements in ppm)

Sl.No.	*Type of water*	*pH*	*TDS Mg/L*	*THD*	*Cu*	*Pb*	*Zn*	*Fe*
1.	Percolation water	6.0	816	660	0.24	33	29	4.81
2.	Pumped out mine water	5.0	3616	440	1.80	277	35	5.42

Table 7.6. River Tista water quality with respect to tailing effluents (trace elements in ppm)

Sl.No.	*Name of water*	*Ph*	*TDS Mg/L*	*THD*	*Cu*	*Pb*	*Zn*	*Fe*
1.	Upstream near Common's bridge	6.8	34	20	—	—	0.3	0.8
2.	Downstream near Rangpo bridge	7.0	45	30	—	—	0.7	0.4

TDS - Total dissolved solid; THD - Total hardness

REFERENCES

Babu, S.K. and R.K. Trivedi, 1987 : Environmental aspects around Malanjkhand, Balaghat district, M.P. - a bane or boon. *Proc. National Convention of Env. Engrs. on "Impact of Environmental Protection on Future Development of India, Nainital*, pp. 253-262.

Das, J., 1987 : Current practices of environmental protection efforts in Indian Mines - a few case studies. *Publ. Indian Bureau of Mines, Nagpur*, pp. 235-268.

Das, J., 1989 : Surface mining - environmental aspects (Non-coal sector). *Ind. Mining & Engg. Jour.*, Special Number, pp. 541-547.

Gupta, P.K. and M. Prasad, 1994 : Best management aspects of tailing disposal for underground base metal mines in India. *Proc. Internat. Conf. Environmin., New Delhi*, pp. 259-266.

Rai, K.L., 1993 : Geological and geoenvironmental constraints in small-scale mining of Himalayan Mineral deposits - selected case studies. In : *Small-Scale Mining : A Global Overview*. (Ed. A.K. Ghose), Oxford & IBH Publ. Co., New Delhi, pp. 51-64.

Rai, K.L. and A.M.S. Rao, 1986 : Problems and prospects of small-scale metal mining in the metallogenic domains of the Himalayas. In : *Strategies for Exploration of Mineral Deposits in Developing Countries* (Ed. A.K. Ghose), Oxford and IBH Publ. Co., New Delhi, pp. 235-247.

Rai, K.L. and A.S. Venkatesh, 1990 : Malanjkhand Copper deposit - a petrological and geochemical appraisal. *Geol. Surv. India, Special Pub.*, 28 : 263-584.

Rai, K.L. and A.S. Venkatesh, 1993 : Geological setting and nature of copper-molybdenum mineralization in the intra-continental acid magmatic regime of malanjkhand, Central India. *Resources Geology*. Spl. Issue No. 15 : 169-181.

Rathore, S.S., S.C. Jain and R. Nath, 1994 : Impact of waste dumps on environment and its remedial measures - a case study. *Proc. Internat. Conf. on "Environmental issues in minerals and Energy Industry*, New Delhi, pp. 52-55.

Subhedar, S.S., 1986 : Resource potential of Malanjkhand copper deposit. An economic overview towards self-reliance of copper in India. *Proc. Nat. Sem. on Precambrians of Madhya Pradesh*, Jabalpur.

Chapter 8

Survey of Acid Rains in the Venice Region (Italy)

Francesco Zilio-Grandi* and Ing. L. Szpyrkowicz**

In the recent years, air quality and in particular rainwater composition have become matter of increasing concern all over the world. A vast research program has been undertaken in the Veneto region. A network of eight sampling stations ("background" stations), remote from motor traffic and industrial sources of air pollutants, and one "zero" station, was realized in 1988 in accordance with the EMEP statements. It gives meterological and physical data on air quality and reduces the wet only samples of meterological precipitations, the chemical composition of which is measured in the laboratory.

An automatic analyzer based on colorimetric methods is used for ammonium, sulphate, nitrate, phosphate and chloride determination and an Atomic Absorption Spectrometer for Ca, Mg, K and Na analysis. The analytical methods used and their detection limits are described. The methods used are based on those recommended by U.S. Standard Methods for water samples. The method for sulphate determination needed the most substantial modification, due to cation interferences.

Data on rain quality of the region in the period February 1989 - June 1995 are reported together with their substantial elaborations. The relationships between single ion concentrations and the iron budget enable rain water composition to be interpreted and the validity of the analytical methods to be checked. The average rain composition approximates a sulphate/nitrate mixture with an average pH of 5.24 and an SO_4^{2-}/NO_3^- mole ratio of 0.92.

INTRODUCTION

The progressive changes in the physical-chemical characteristics of atmospheric depositions, caused principally by the increase of industrialization, lead to alterations, sometimes serious, in different ecosystems. The hazards that acidic rain create for biological life and for surface water quality caused a rise in public interest and awareness. The main effects of the acidification of meteorological precipitations reported in the literature are : damage to forests (Lauterwasser, 1988; Sah and Melwes, 1989; Schultz, 1988), changes in aquatic plant communities, loss of game fish populations, changes in sediment metal concentration, etc. (Schofield, 1982). Lipfert *et al.* (1989) reported that acid aerosols also have negative effects on human and animal health, consisting lung function changes. Olem and Berthouex, 1989 reported that there is risk that acid deposition may alter cistern and tap drinking water quality especially by exceeding the limit for Pb. The study of such problems begins with sampling and analysing the physico-chemical characteristics of the precipitations (Morselli *et al.*, 1986).

The direct measuring of air and rain constituents is a fundamental tool for evaluating the air quality which, as has been shown elsewhere (Millan and Gangoiti, 1988), cannot be substituted with mathematical modelling of transport phenomenon, due to the multiplicity of factors which influence the scavenging process (non-existence of a universal washout coefficient, occurrence of chemical reactions in the atmosphere, different characteristics of background air mass). Numerous studies have already been undertaken on rainwater composition (Smith, 1987; Barrie *et al.*, 1987).

Anyway, standard sampling procedure and standard protocols in the risk of air quality assessment are needed for further comparison of the data from different sites (Voldner and Alvo, 1989). That is one of the reasons for creating sampling and monitoring networks.

National and citizens acid rain monitoring networks already operate in numerous countries. For example in the US, acid deposition has been measured since 1980, when the Energy Security Act, Public Law, 96-295 was enacted and the National Acid Precipitation Assessment Program was initiated. A major goal of the NAPAP has been to assess the sensitivity of the nation's surface waters to acidification by acidic depositions. Maps showing the acid neutralization capacity of surface waters have been created for all regions. Apart from this national network, a number of regional and local networks operate all over the US (Voldner and Alvo, 1989; Bolze and Beyera, 1989). E.M.E.P. (1977), developed a cooperative programme for monitoring and evaluation of long-range transmission of air pollutants in Europe and published a manual for sampling and chemical analysis EMEP/CHEM-3/77, O.E.C.D. (1979) developed a programme on long-range transport of air pollutants. The Economic Commission for Europe of United Nations signed at Geneva in November 1979 a Convention on long-range Transboundary Air Pollution.

Analysis of meteorological precipitations has already been undertaken in Italy (Bettinelli *et al.*, 1984; Cavallaro *et al.*,1984, 1986; Morselli *et al.*, 1986; Brocco *et al.*, 1986; Del Turco *et al.*, 1986, 1987; Greghi *et al.*, 1989; Dell'Atti *et al.*, 1988; Mosello *et al.*, 1988; Brovelli and Bassanino, 1988; Bassanino *et al.*, 1989; Cossu *et al.*, 1989; Desideri *et al.*, 1989; Basile and Celano, 1990; Bellandi *et al.*, 1995). Since 1975 the meteorological service of the Italian Air Force has furnished data on meteorological precipitation quality; the results have appeared in the literature (Ciattaglia, 1975 and 1981). In April 1972 a consortium of various research institutes was formed to monitor atmospheric depositions in Northern Italy; the results relating to October 1982 - December 1984 are reported in *Gruppo di Studio Nord Italia*, 1987. A report on the "Activity of the RIDEP (Rete Italiana per lo studio delle deposizioni atmosferiche) network in the period 1988-1992" is reported by Mosello (1993).

The Veneto is a région in the north of Italy; the whole region covers just over 18,000 sq. kms. and it has a population of about 4.5 million. The main town is Venice, but there are also other interesting towns (Padua, Vicenza, Verona) which are mainly characterized by the presence of important monuments and works of art and also by important industrial areas. The ecosystem is varied, particularly by the presence of a lagoon. In this situation the possible effects of acid rain are well known (Galloway and Likens, 1976 and 1978; Barrie *et al.*, 1987; and Smith, 1987).

La Giunta Regionale del Veneto, conscious of the importance of identifying the sources of rain pollution and the possible contribution of transboundary sources, decided to organize a regional network according to the directives of the Minister of Health (Republica Ital. 1982, Suppl. G.U.) and to European Environmental Monitoring Program - E.M.E.P. - (Broocco *et al.*, 1983).

The Veneto Regional Authorities passed specific acts (Giunta Regionale del Veneto, 1986-87) implementing the national directives established by the Health Minister (Repubblica Italiana, 1982, Suppl. G.U.). A total of nine monitoring-sampling stations are installed in the region with the aim of controlling air pollution levels (PHA, 1985) (Fig. 8.1).

	Station	*Site*	*Province*
*	0	Monte Cherz	BELLUNO
*	1	Cesiomaggiore	BELLUNO
*	2	Monte Cesen	TREVISO
*	3	Castelfranco	TREVISO
*	4	Caorie	VENEZIA
*	6	Campodoro	PADOVA
*	7	ERBE'	VERONA
*	8	Badia Polesine	ROVIGO

Fig. 8.1. Map of the Veneto region and sampling station locations.

In this way, apart from giving basic material for evaluation of the phenomenon in time, it was proposed to verify the existence of site differences inside the region influencing the air quality and to explore the possibility of influencing the local pollution sources.

In some previous papers, sampling and analysis of rain of the Veneto regional network were described and preliminary results for rain quality according to the E.M.E.P. research project have been reported (Cosma *et al.*, 1988A & B; Zitio-Grandi and Szpyrkowicz, 1991A & B) and afterwards data for the periods 1989-1991 and 1991-1993 have been evaluated using statistical methods (Zilo-Grandi and Szpyrkowicz, 1994A and B). Mantovan *et al.*, 1992 and 1995 characterized the rainwater sampled by Veneto regional network using a multivariate analysis.

Zilio-Grandi and Szpyrkowicz (1996) in a "Project on the Venice Lagoon System" monitored the data from a sampling station of atmospheric precipitation located in Giare near Mira (Venice) during the period April 1992 - March 1994 with the aim to give some more information about the influence which meteorological depositions have on the lagoon water. The measured parameters and the analytical methods were chosen in accordance with EMEP directives.

The paper presents the results of a long-term investigations regarding the physical-chemical characteristics of the precipitations and the air pollution in the Veneto region. The data discussed concern the measurements of air pollutants and of the wet deposition for nine stations in the region of Veneto carried out during the period from February 1989 to June 1995.

EXPERIMENTAL

Network description

In 1986, Veneto region complied with the requirement of Italian Law, n. 289 of 27 April 1982 to participate in an European Environmental Monitoring Program (E.M.E.P.) on air quality and on pollutants subjected to transboundary transport, and established in the north of the region a station for atmospheric control called "zero" station, where the location and particular site characteristics assure an air quality which can be considered as unpolluted, and which allows quality comparison with the other national or European stations.

Then, a change of program agreed with the Minister of Health made increased control possible in the region with the installation of a monitoring station, in this case of "background" one, with more parameters compared with those of "zero" station, sited in areas far from probable local industrial

Fig. 8.1. Map of the Veneto region and sampling station locations.

and/or urban influences on rainfall. In the years 1988-1989, taking into account the need to increase knowledge of the state of air quality in the region, the administration decided to install another seven stations with the same philosophy of sampling, analysis and location as the first two stations (in accordance with the E.M.E.P. projects).

The "background" sampling stations are located in areas with the following characteristics :

— away from highways and atleast 20 km from industrial sources of air pollutants,

— atleast 20 km away from intensively inhabitated areas (with more than 50,000 inhabitants),

— atleast 20 km away from industrial sources of air pollution,,
— atleast 5 km away from centres with more than 5,000 inhabitants,
— atleast 100-200 m away from single houses,
— at a distance of atleast 4-5 times the height of the nearest impediment.

The aim was to give a general picture of the pollution of rainwater in the region.

Sampling procedure

Given the objective of the research, the number of sites to be monitored and the technical and economical resources, it was decided to undertake continuous on-site analysis of all the rain events for pH and conductivity (easy to automatize) and to collect daily samples of a "wet-only" type for the analysis of a composite sample (a "wet-only" sample is obtained by the exposure of the sampler only in the case of a rain event).

In each station meteorological parameters (temperature, relative humidity, direction and speed of the wind, atmospheric pressure, precipitation volume) are measured, together with the pH and conductivity of the rain. A computerized system memorizes and manages the measured data.

The stations are equipped with :

- a system for rainwater collection, including a sampler capable of automatic operation for 12 days, a thermostat and a rain gauge with a cover operated automatically by a wet sensor.
- a system for particulate sampling, including a pump for air flow-rate regulation and 8 filter holders with cellulose filters.
- an air sampling system provided by a constant flow-rate sampling pump and 8 scrubbers.

The rain samples are collected on a daily basis; if more than one rain events occur on the same day, they are collected together and are considered as a single event sample. Air scrubber liquids and air filters are changed daily. The particulate and gas sampling systems also collect samples on a daily basis. The rain samples, the filters and the scrubbers are collected weekly and sent to the laboratory.

Analytical methods and apparatus

Storage and analysis of the samples were done as per the procedures recommended in "Methodologies and Quality Controls for the Study of the Chemistry of Atmospheric Depositions in Italy" published by the Ministry of the Environment and the National Research Council N.R.C. - Italian Institute of Hydrobiology (Mosello *et al.*, 1990). The fresh un-filtered samples were measured for the following characteristics :

* volume,
* pH, (ORION pH meter, mod. 420),
* conductivity at 20°C, (CRISON conductimeter, Mod. micro CM 2200,
* alkalinity (HCO_3^-), of the samples with pH > 5.6, by a direct titration with HCl N/100 and mixed green bromocresol-methyl red indicator).

The samples were filtered on a cellulose acetate filter and kept in a refrigerator at 4°C for subsequent analyses. Each single event rain sample was analyzed for alkalinity, anion content (chlorides, sulphates, phosphates, nitrates and nitrites) and cations (sodium, potassium, magnesium, calcium, ammonium); anions and NH_4^+ were determined within 2-3 days, while other cations were analyzed usually every 1-2 months.

Anions and ammonium were determined by an automatic Bran & Luebbe, Mod. TRAACS 800 apparatus, based on colorimetric analysis. The TRAACS 800 is a computer-controlled continuous-flow wet analytical chemical system.

The system is composed of a sampler and an analytical console, having two detection channels, enabling upto two different chemical analyses to be performed at the same time. The reagent changer module provides for reagent storage and sequencing. The changer allows automatic switch-over from one set of reagents to another on a manifold. The random access sampler can hold upto 120 sample cups in blocks of ten. Motion instructions for the probe and tray are initiated by commands from the central processing unit, which also transmits the commands to set the parameters and the conditions for the analysis (number of samples, the ratio of sampling and washing periods) and recalls the specific calibration curve. Portions of single samples are separated in the hydraulic system by air bubbles. The colorimeter assembly in the analytical console can accommodate two individual sample detector channels. The light source is a tungsten halogen lamp operating at 8.0 volts-dc. A lens system collimates light from the lamp through the interference filter, then focusses the lamp filament image onto the flowcell. Light not absorbed by the sample in the flowcell is transmitted to the dual silicon photodiode detector.

The analytical methods used for the single ion determination are the following (Brocco *et al.*, 1983; Technicon Traacs 800, 1986) :

— **Nitrates :** The reaction of the reduction of nitrates to nitrite by an alkaline solution of hydrazine sulphate is carried out using a copper catalyst; nitrites are then treated with naphthyl-ethylenediamine and sulphanilamide under acidic conditions to form a soluble dye which is measured colorimetrically at 520 nm. The detection range is 0.1-10 mg l^{-l} NO_3^-.

— **Chlorides :** The method consists in the liberation of thiocyanate from mercuric thiocyanate with the formation of unionized but soluble mercuric chloride. In the presence of ferric ion, the liberated thiocyanate forms a brown coloured ferric thiocyanate, the concentration of which is proportional to the original chloride concentration. The detection range is 0.1-10 mgL^{-1} Cl^-.

— **Ammonium :** The determination is based on the Berthelot reaction in which the formation of a blue coloured compound occurs when the solution of an ammonium salt is added to sodium phenoxide, followed by the addition of sodium hypochlorite. A solution of EDTA is added to the sample to eliminate the precipitation of calcium and magnesium hydroxide. Sodium nitroprusside is added to intensify the blue colour. The detection range is 0.1-10 mg L^{-1} NH_4^+

— **Sulphates :** At an acidic pH the sulphate ion reacts with barium, substracting it from a barium methylthymol blue complex and forming a barium sulphate. The loss of blue colour which occurs at an alkaline pH is correlated to the sulphate concentration and measured at 610 nm. The bivalent cations of Ca and Mg which interfere are eliminated by pretreating the sample on a cation exchange resin column. The detection range is 0.20-10 mg L^{-1} SO_4^-.

The above methods, initially used for the analysis of surface and waste waters, were optimized and adapted for the range of microconcentration, which characterizes meteorological precipitations. Alkali (Na^+, K^+) and alkaline earth (Mg^{2+}, Ca^{2+}) metal determination is conducted by atomic absorption spectrometry (Perkin Elmer Model 4000), using an air-acetylene flame oxidizing at 589, 766, 285 and 423 nm, respectively. HCO_3^- is determined by titration with HCl (N/100) with a methyl red bromocresol green mixed indicator.

DATA QUALITY CONTROL

The control of data quality is done using the procedures recommended by the Italian Ministry of the Environment. Ionic balance calculation and comparison of calculated and measured conductivity (see Morselli *et al.*, 1986; Mosello *et al.*, 1990; Mosello R., 1993; Mosello *et al.*, 1995).

Ionic balance determination

As all the principal ionic components of the rain were measured, the difference between the global concentration of the cations and that of the anions, expressed in μeq l^{-1}, must be near to zero. If ID is the difference between the concentration of cations and the concentration of anions and IS is their addition, the per cent difference (PD) is defined on the basis of the values of PD and IS : PD = ID * 100/0.5 * IS.

The analyzed rain sample is classified in three categories, as follows :

Table 8.1. Categorization of rain sample

Ionic concentration (μeq/l)	*Categories*		
	1	2	3
IS < 50	PD < = 60	PD > 60	
50 < = IS < 100	PD < = 30	30 < PD < = 60	PD > 60
100 < = IS < 500	PD < = 15	15< PD < = 30	PD > 30
IS > 500	PD < = 10	10 < PD < = 20	PD > 20

Each analysis whose PD does not fall in category 1 is repeated.

Comparison between calculated and measured conductivity

The conductivity measured for the rain samples is compared with the calculated value, obtained as the sum of the products of the concentrations of single ions (in μeq l^{-1}) and the corresponding equivalent ionic conductivity at infinite dilution (S.cm²/eq).

If CM (μS/cm at 20°C) is the measured conductivity and CE the calculated one, the difference % CD is given by : CD = 100 * (CE - CM)/CM.

On the basis of the values of CD and CM the sample is classified into three categories, as follows:

Table 8.2. Categorization of samples on the basis of CD and CM values

Conductivity (mS/cm)	*Category*		
	1	2	2
CM < = 50	CD < = 30	CD > 60	
CM > 30	CD < = 20	20 < CD < = 40	CD > 40

Each analysis whose CD does not fall in category 1 is repeated.

RESULTS

Rainfall

Volume and frequency of precipitation

Table 8.3a & b reports, for each station, the volume (in mm) of rain collected and analyzed since 1989, the rainfall in mm/year, taking into account the months when the station was active, and the mm for event. The rainfall data recorded during the years at each station, reported in Figs. 8.2 and 8.3, clearly shown the variability in time. The station of Monte Cesen with 1004 mm/year appears to be the most rainy. The Southern stations are less rainy, both as mm/year and as number of rain events. Comparison of the rainfall recorded at the stations and the rainfall relative to the analysed events for the period 1990-1994 showed that, on average, 90% of the total rain was analysed.

Table 8.3. Volume of rain collected in the stations in the period 1989-1995

YEAR	0	1	2	3	4	5	6	7	8
1989	573	944	1031	787		398	538	627	658
1990	578	608	936	554	182	429	409	207	508
1991	555	697	1114	695	522	590	466	573	478
1992	672	1027	813	728	750	699	417	498	339
1993	688	965	900	547	622	539	370	383	417
1994	533	873	953	686	819	787	340	475	390
1995	229	532	697	523	466	656	561	288	459
total volume (mm)	3829	5644	6444	4520	3362	4099	3101	3051	3250
month of activity	78	77	77	77	57	72	77	65	77
mm/anno	589	880	1004	704	708	683	483	563	506
number of events	541	486	450	409	275	334	316	361	381
mm/event	7.1	11.6	14.3	11.0	12.2	12.3	9.8	8.5	8.5

Data from partially active stations are in grey: for 1995 only the first six month have been active.

Acidity

The analysis of the pH data assumes a particular importance as a synthetic index of the quality of atmospheric depositions. It can be seen that generally the rain in the region is not highly acidic, but in some cases pH is lower than 5.6 (a pH value of uncontaminated rain in saturation equilibrium with atmospheric CO_2 at 25°C). A significant percentage of cases with a pH higher than 5.6 for each station (from 6% to 73%) is observed. A significant decrease of pH can nearly always be related to high rain volumes, in accordance with Moselli *et al.*, 1986 and Greghi *et al.*, 1989.

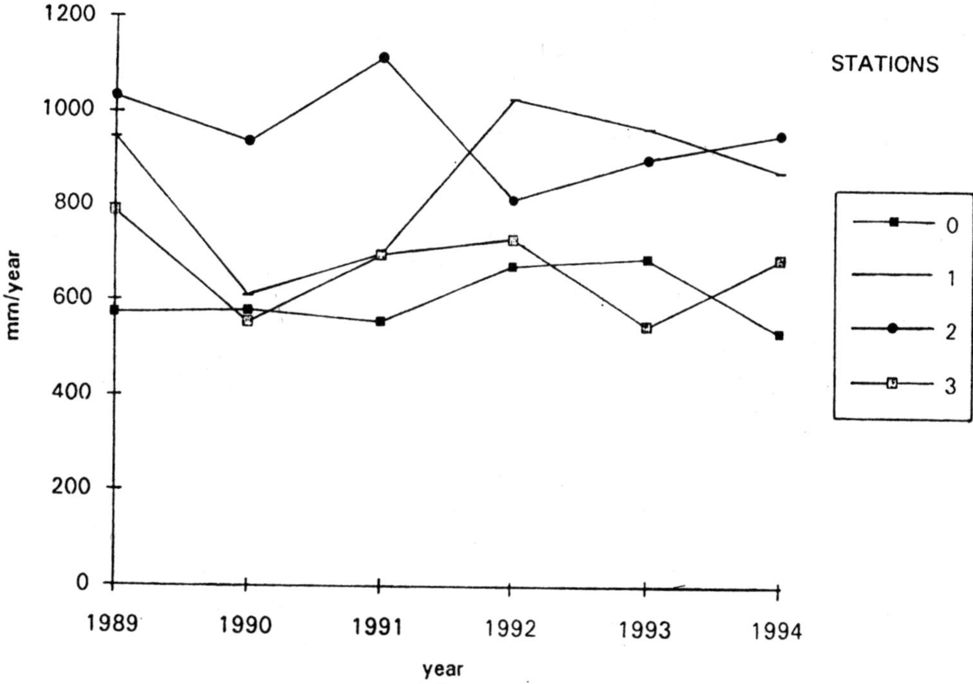

Fig. 8.2. Volume of rain in mm/year at stations 0-3.

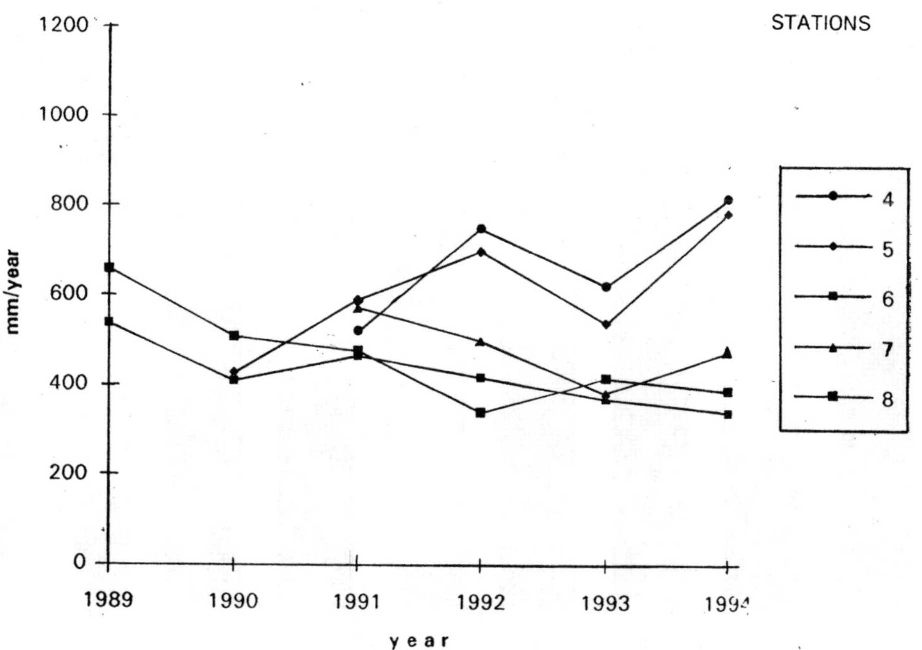

Fig. 8.3. Volume of rain in mm/year at stations 4-8.

Rains collected at the stations of Monte Cherz (73%), Monte Cesen (64%), Badia Polesine (49%) and Carrata (48%) proved more acidic while the rains of the stations of Castelfranco (41%), Campodoro (36%), Cesiomaggiore (23%) and Erbe (18%) were less acidic. At the station of Caorle only 6% of acid rain was found.

Table 8.4. Ranges of pH of nine stations

pH	Station number								
	0	1	2	3	4	5	6	7	8
number of events	541	486	450	409	275	334	316	361	381
min	3.95	4.17	4.05	4.45	4.76	4.27	4.48	4.41	4.50
max	7.39	8.25	8.02	8.09	8.00	8.18	8.12	8.19	8.20
pH<5.6	73	23	64	41	6	36	48	18	49
5.6<pH<6.5	23	39	25	41	42	42	35	47	34
pH>6.5	4	38	11	18	52	22	17	34	17

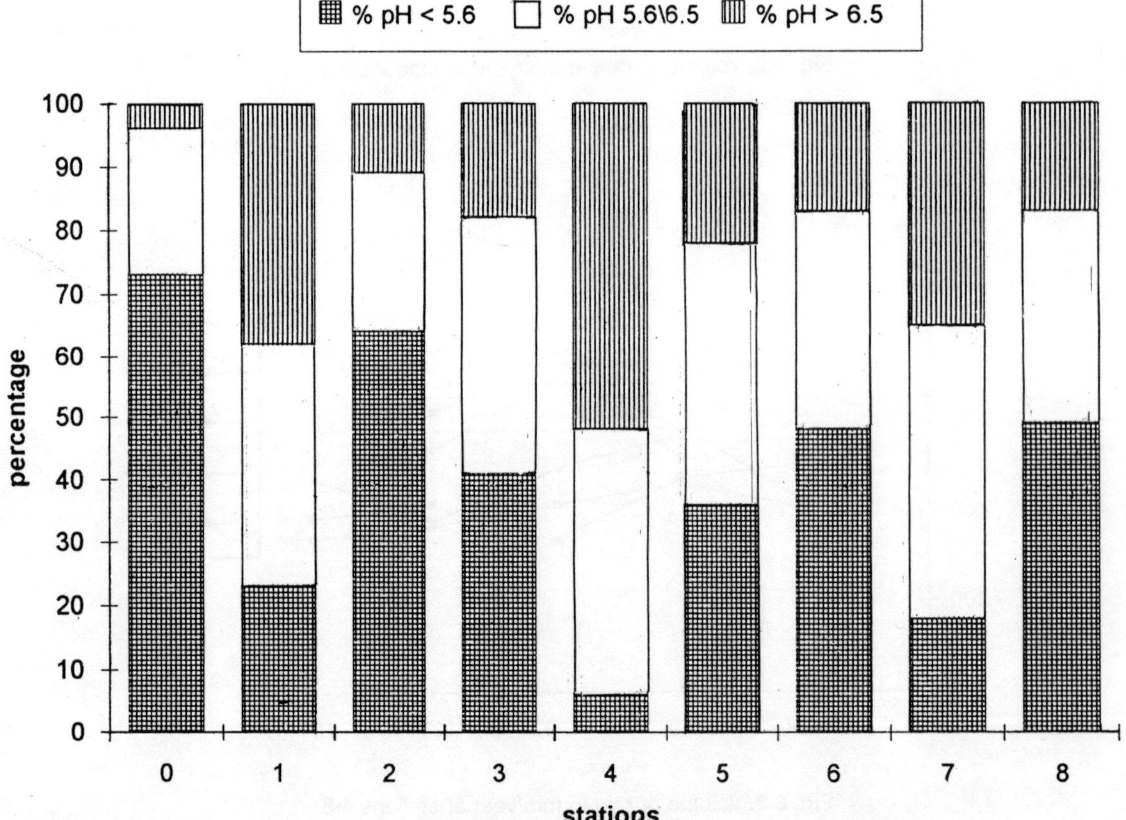

Fig. 8.4. Classification of pH.

Ionic strength and conductivity

Concentration of the ions have shown considerable values among the sampling stations throughout the period of the study. Site-specific variations are observed particularly for station No. 4 and 7 for all the ions; their concentrations are noticeably higher in comparison to the other stations. As can be seen, sulphates and nitrates are always the prevalent anions and ammonium and calcium the main cations.

Table 8.5 reports the rain volume-weighed mean values of the total ionic strength (sum of single ions, in μeq/l^{-1}) found on a single station in different years and for the entire period (1989-1995) as well as the corresponding value of conductivity, in μS/cm.

Observing mean values of ionic strength relative to the considered period, the rains of the "zero" station of Monte Cherz (160 μeq/l) was the least polluted, followed by Monte Cesen (250 μeq/l), Cesiomaggiore (319 μeq/l), Campodoro (332 μeq/l) and Castelfranco (389 μeq/l). The stations of Badia Polesine (479 μeq/l), Carrara (502 μeq/l), Caorle (551 μeq/l) and Erbe (552 μeq/l) were the most polluted. The mean values of conductivity, always relative to the entire period, confirm what is indicated by the data of ionic strength.

Figs. 8.5 and 8.6 report the annual mean values, weighed on the rain volume, of the ionic strengts measured on each station : stations 3, 4,6 and 7 showed the most variability. The trend of ionic strength of single event in function of time (in days), from 1989 to 1995, is shown in Figs. 8.7 and 8.8, which represent the data of linear regression ionic strength - time.

Table 8.5. Ionic strength, μeq/L, and conductivity, μS/cm

YEAR	STATIONS								
YEAR	0	1	2	3	4	5	6	7	8
1989	171	376	296	402		295	552	439	515
1990	201	282	220	319	565	346	591	421	496
1991	153	332	276	430	498	316	397	412	481
1992	156	269	258	334	521	360	487	682	521
1993	144	339	247	518	725	348	620	681	509
1994	152	315	223	384	555	317	585	674	469
1995	116	312	211	334	411	330	360	577	355
8995	160	319	250	389	551	332	502	552	479
COND	N.D.	19.5	17.5	24.4	30.5	21.0	32.0	32.3	30.6

In grey the data with partial activity of the station: for 1995 the data are relative only to the first six months.

Table 8.6 reports a statistical evaluation, obtained through determination of the value of students "t" test (Spiegel, 1994), from the difference between the initial and final values of ionic strength in μeq l^{-1} extrapolated by the regression lines for the period 23/02/1989-29/06/1995 (2318 days). These evaluations give only an indication on the trend of the phenomenon in the period considered, being influenced by several factors which include :

— shifts in meteorological behaviour,

— changes in pollutant sources,

— changes in sampling and analytical practices.

Hence, extreme caution is advised in attempting to assess cause/effect relationship, or anticipate future behaviour on the basis of limited record of data (Map3S/Raine Research Community, 1982).

Table 8.6. Linear regression between the ionic strength of single events, μeq/L

STATION	INITIAL VALUE	FINAL VALUE	% DIFF.	EVENTS	"t"	EVALUATION
5 - Campodoro	451	488	8	334	-1.24	INSIGNIFICANT
7 - Erbé	741	781	5	361	-0.94	NOT SIGNIFICANT
0 Monte Cherz	219	213	-3	326	0.43	NOT SIGNIFICANT
4 - Caorle	777	726	-6	275	1.20	NOT SIGNIFICANT
1 - Cesiomaggiore	496	417	-16	486	3.74	SIGNIFICANT
2 - Monte Cesen	354	299	-16	450	3.29	SIGNIFICANT
3 - Castelfranco	640	530	-17	409	3.18	SIGNIFICANT
6 - Carrara S.S.	922	535	-42	316	8.15	SIGNIFICANT
8 - Badia Polesine	987	472	-52	381	11.21	SIGNIFICANT

Calculation of the "t" of Student value $t = \frac{|\chi_1 - \chi_2|}{(2\sigma^2)^{0.5}} \cdot n^{0.5}$

χ_1=ionic strength extrapolated to the beginning of the period;

χ_2=ionic strength axtrapolated to the end of the period

σ = standard deviation of the linear regression; n = number of measurements

Critical values of "t" di Student for a freedom degree more than 120:
t = 1.96 at a probability level of 95%
t = 2.58 at a probability level of 99%.

The difference between the initial and final values extrapolated from the regression slopes is evaluated as:
SIGNIFICANT, at level of 99%, per "t" > 2.58
INDICATIVE, at level del 95%, with "t" in the range 1.98 - 2.58
INSIGNIFICANT at level of 99%, with "t" < 1.96.

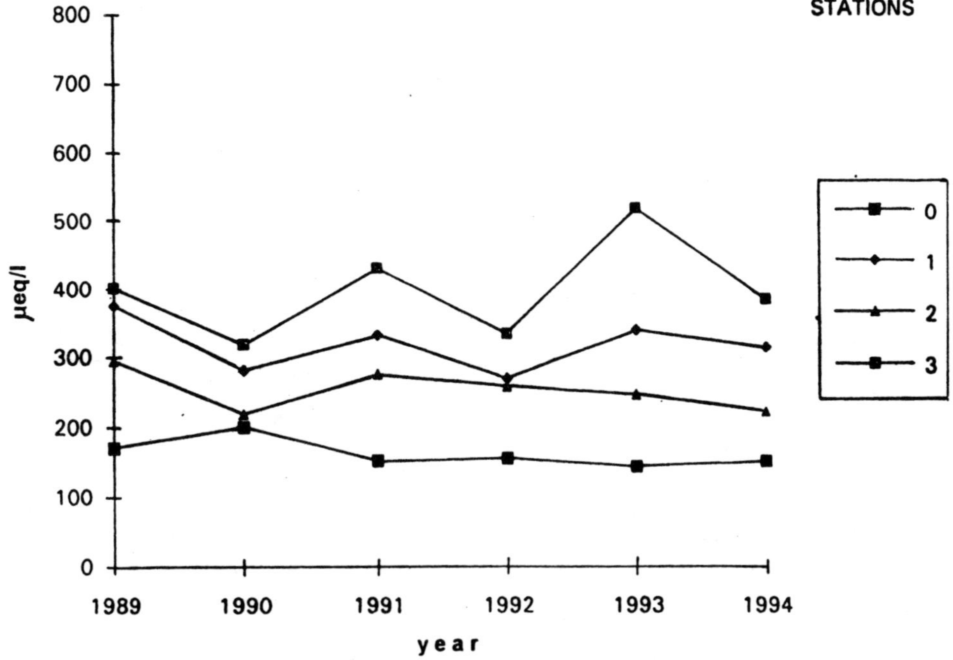

Fig. 8.5. Ionic strength in μeq/l at stations 0 to 3.

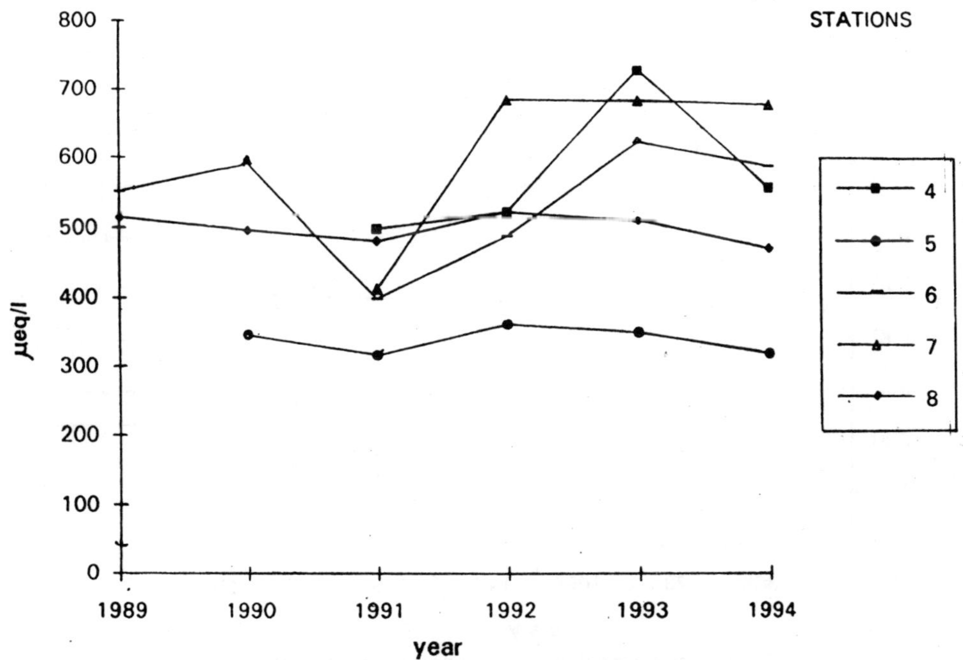

Fig. 8.6. Ionic strength in μeq/l at stations 4 to 8.

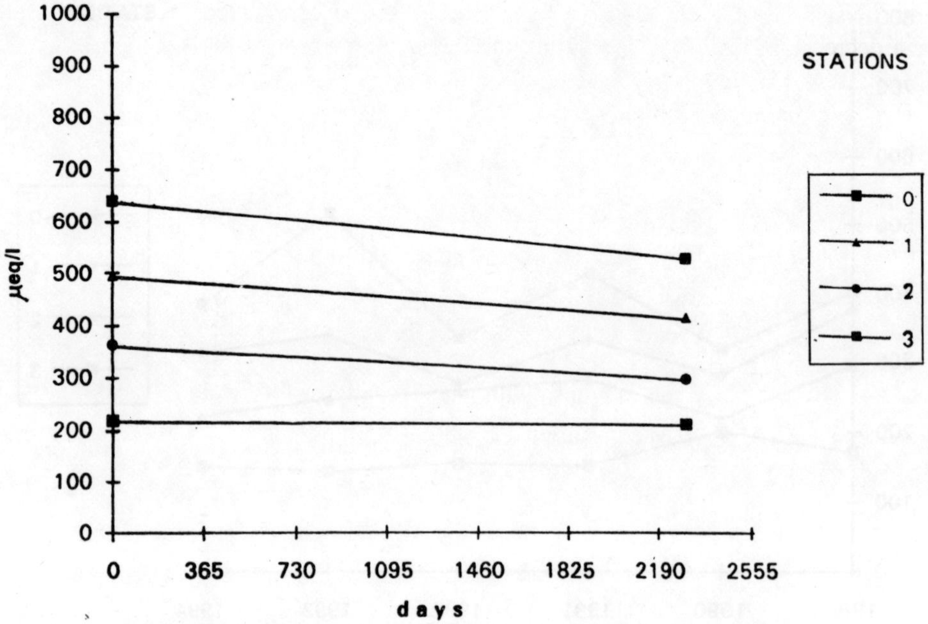

Fig. 8.7. Linear regression ionic strength - time of stations 0 to 3.

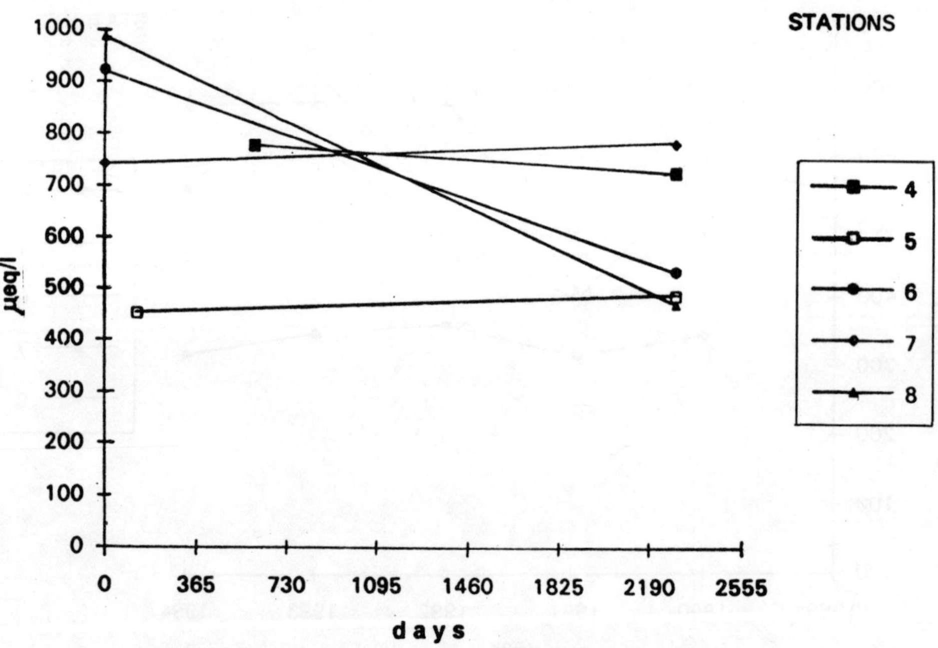

Fig. 8.8. Linear regression ionic strength - time at stations 4 to 8.

Whereas the stations of Campodoro, Erba, Monte Cherz and Caorle record a statistically insignificant increase or a decrease in ionic strength during the time of low entity (3-8%), the decrease is significant for the other stations and particularly substantial at Carrara and Badia Polestine (42-52%) (Figs. 8.9 and 8.10). For better understanding of the trend of total ionic strength in function of time, the regression of single ions against time has been plotted, for every station. As for the total ions, the per cent variation between the initial and final value of ionic concentrations is determined on the basis of extrapolated values from the regression lines for the entire period 1989-1995. The results obtained (Table 8.7), outline the following time trend during the study period :

1. SO_4^{-2} decreases noticeably on all the stations;
2. NO_3^- increases at Caorle (4) and Campodoro (5) and decreases at Carrara (6) and Badia Polesine (8);
3. NH_4^+ decreases on all the stations, with the exception of Caorle;
4. Cl^- decreases except at Monte Cesen (2) and at Campodoro (6), while Na^+ decreases on all the stations;
5. HCO_3^-, Ca^{2+}, K^+ and Mg^{2+} have a very variable trend;
6. On the stations of Carrara (6) and Badia Polestine (8), where the decrease of the total ionic strength was considerable (42-52%), a generalized decrease of all the single ions was observed.

Table 8.7. % variation of ionic strength, μeq/L, of total and single ions calculated from the data obtained from the regression slopes in function of time (in days)

STATION	IONS	Cl	NO_3	SO_4	HCO_3	NH_4	Ca	Na	K	Mg
0 - Monte Cherz	-3	-19	2	-16		6	45	-45	-17	92
1 - Cesiomaggiore	-16	-23	5	-32	10	-2	-4	-67	-9	-34
2 - Monte Cesen	-16	18	6	-40	3	-28	3	-32	47	-45
3 - Castelfranco	-17	-12	-8	-38	104	-26	7	-34	22	-32
4 - Caorle	-6	-25	44	-18	-4	32	3	-52	-8	-2
5 - Campodoro	8	29	53	-24	66	5	28	-8	75	8
6 - Carrara S.S.	-42	-17	-20	-52	-34	-33	-44	-60	-18	-63
7 - Erbé	5	-9	-7	-22	88	-7	44	-52	10	-15
8 - Badia Polesine	-52	-40	-34	-56	-80	-39	-59	-69	-32	-74

Table 8.8. Average concentrations of single ions weighted on the volume, μeq/L : period 1989-1995

ST	Cl	NO_3	SO_4	HCO_3	H	NH_4	Ca	Na	K	Mg	ANIONS	CATIONS	IONS
0	8	27	37	2	13	26	32	6	1	6	75	85	160
1	24	42	62	29	4	54	58	19	8	20	157	162	319
2	23	36	53	10	9	43	38	17	8	12	123	127	250
3	28	47	93	25	5	78	61	22	8	22	193	196	389
4	65	48	71	90	1	44	116	55	9	52	273	277	551
5	25	45	74	21	4	71	52	18	7	15	164	167	332
6	35	62	136	18	6	95	83	29	8	29	253	249	502
7	30	63	103	77	2	94	122	23	10	28	273	279	552
8	31	59	135	14	6	102	71	27	9	26	239	240	479

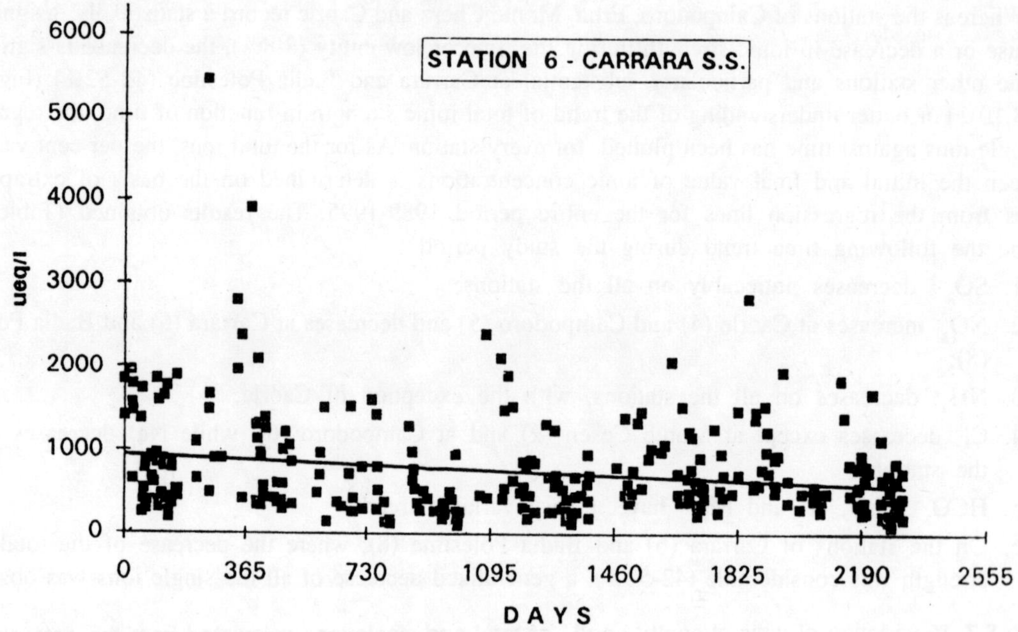

Fig. 8.9. Linear regression between ionic strength of single event, μeq/l and time (in days) at station 6.

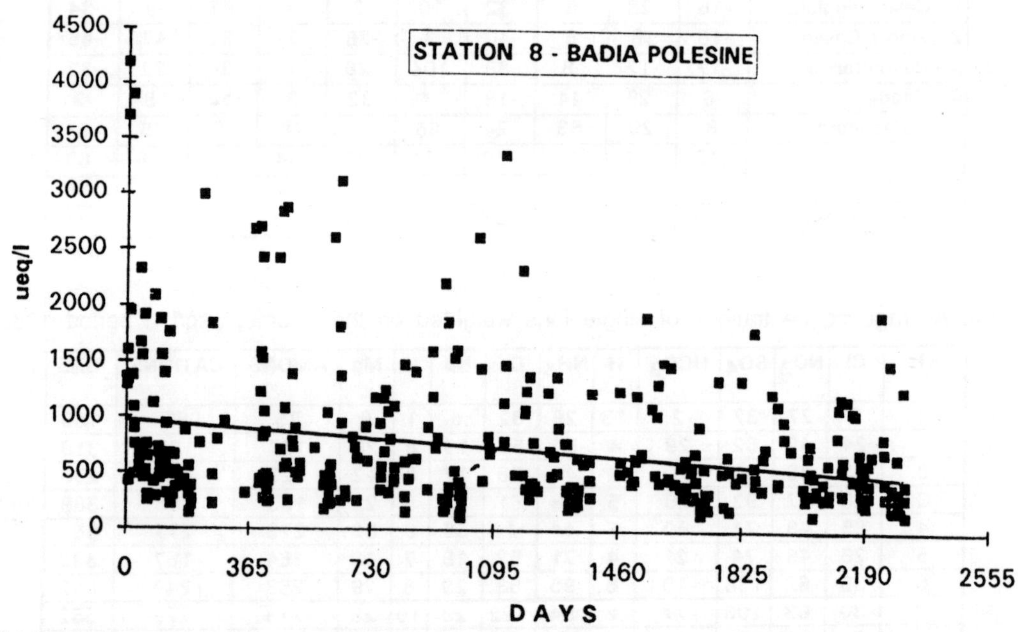

Fig. 8.10. Linear regression between ionic strength of single event μeq/l, and time (in days) at station 8.

Table 8.8 and 8.9 and Figs. 8.11 and 8.12 report the composition of rain on all the stations for the period considered expressed as mean volume-weighed concentration of single ions and as per cent contribution of single ions to the total ionic strength.

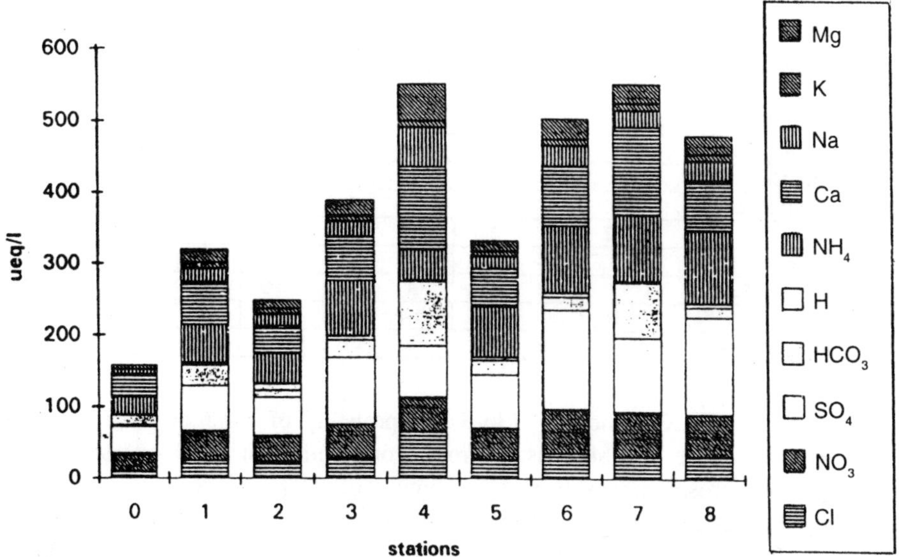

Fig. 8.11. Volume-weighed average concentration of single ion.

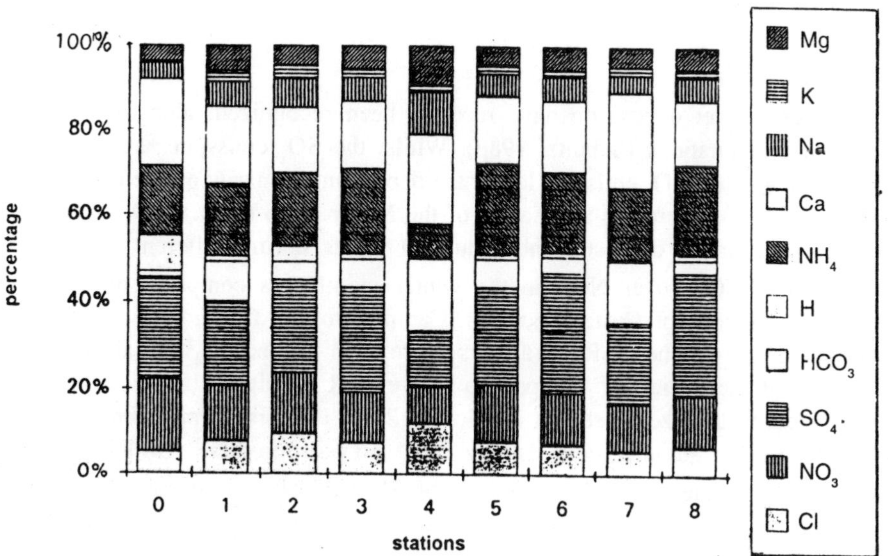

Fig. 8.12. Per cent contribution of single ion to total ionic strength.

On seven stations, including station "zero", the contribution of the main anthropogenic emissions, SO_4^{2-} and NO_3^-, is substantial (30-40%) and prevails over due to other emissions, probably of local origin, such as ammonia and particulars (Ca^{2+}, Mg^{2+}, K^+). On the stations of Caorle (n. 4) and Erbe

Table 8.9. Single ions % contribution to the total ionic strength : period 1989-1995

STATION	Cl	NO_3	SO_4	HCO_3	H	NH_4	Ca	Na	K	Mg
0 - Monte Cherz	5	17	23	1	8.1	16	20	4	1	4
1 - Cesiomaggiore	8	13	19	9	1.3	17	18	6	3	6
2 - Monte Cesen	9	14	21	4	3.6	17	15	7	3	5
3 - Castelfranco	8	12	24	6	1.3	20	16	6	2	6
4 - Caorle	12	9	13	16	0.2	8	21	10	2	9
5 - Campodoro	8	14	22	6	1.2	21	16	5	2	5
6 - Carrara S.S.	7	12	27	4	1.2	19	17	6	2	6
7 - Erbé	5	11	19	14	0.4	17	22	4	2	5
8 - Badia Polesine	6	12	28	3	1.2	21	15	6	2	5

(n. 7), on the other hand, HCO_3^- and Ca^{2+} (36-37%), probably of common origin, prevail over SO_4^{2+} + NO_3^- (22-30%) : for these two stations a strong contribution of local emissions is evident. The contribution of marine aerosol ranges from a maximum of 24% at Caorle to a minimum of 9% at Monte Cherz. The highest percentage contribution of H^+ (8.1-3.6%) is determined on the stations of Monte Cherz and Monte Cesen; the rain acidity in these stations can be considered also of medium degree (minimum pH = 3.95-4.05). The literature data indicate that the precipitation of sulphate content is much higher in the neighbourhood of major ammonia emission sources, it is believed that the presence of ammonia increases the washout efficiency for SO_2 (Pierson *et al.*, 1987). Analysing the data of Veneto region on precipitations, this relationship is also confirmed, particularly for the seventh station (Erba - in the south of the Region).

Molar ratio [(SO_4^{2-})/(NO_3^-)]

Nitrogen oxides in wet deposition have recently been recognized, along with SO_x, as the main precursors of acid precipitations (Schultz, 1988). Whilst the SO_x emissions are directly related to the sulphur content of fuels, the NO_x emission levels are dependent primarily on the combustion temperature (Norton and Levin, 1989). Therefore a variation of the Molar Ratio (M.R.) [SO_4^{2-})/(NO_3^-)] is a function of the different operation of domestic and industrial plants during different periods of the year.

The prevalence of SO_4^{2-} over NO_3^- in the winter months, as compared to the summer months, is significant on the stations of Castelfranco (3), Campodoro (5), Carrara (6) and Badia Polesine (8). On these stations moreover the M.R. is always more than 1; i.e. SO_4^{2-} is always more than NO_3^- with the exception of the station of Campodoro where M.R. is higher than 1 only in winter. On the other stations, including the "zero" station of Monte Cherz, there is no prevalence of SO_4^{2-} over NO_3^- in winter with respect to summer, and usually NO_3^- prevails on SO_4^{2-} (M.R. < 1) (Table 8.10).

In conclusion, the influence of the emissions of SO_x and NO_x from the domestic heating system appears to happen mainly on the stations in the south of the region. As far as the weighed molar ratio [(SO_4^{2-})/(NO_3^-)] is concerned, it was observed that its value varies noticeably during the year and assumes different numbers in each of the four seasons (Schultz, 1988). In the case of the present

Table 8.10. Molar ratio M.R. (SO_4^{2-}) / (NO_3^-) for the period 1989-1995

STATION		quarter				
		I	II	III	IV	TOTAL
0	data n.	35	200	202	101	538
Monte Cherz	M.R.	0.56	0.72	0.72	0.67	0.70
1	data n.	63	201	115	107	486
Cesiomaggiore	M.R.	0.70	0.78	0.72	0.92	0.78
2	data n.	64	169	119	98	450
Monte Cesen	M.R.	0.68	0.72	0.73	0.97	0.77
3	data n.	66	151	87	105	409
Castelfranco	M.R.	1.24	1.04	0.89	1.06	1.04
4	data n.	49	85	57	84	275
Caorle	M.R.	0.71	0.76	0.68	0.92	0.78
5	data n.	58	120	76	80	334
Campodoro	M.R.	1.10	0.85	0.84	0.94	0.91
6	data n.	63	126	58	69	316
Carrara S.S.	M.R.	1.45	1.12	1.01	1.40	1.23
7	data n.	59	134	80	88	361
Erbè	M.R.	0.92	0.87	0.85	0.99	0.90
8	data n.	71	140	72	98	381
Badia Polesine	M.R.	1.52	1.20	1.09	1.41	1.29

study, this ratio also varied in time and in function of the location of the stations. It assumed the following mean values in the period 1989-95 : 0.70, 0.78, 0.77, 1.04, 0.78, 0.91, 1.23, 0.90 and 1.29 respectively for the stations from 0 to 8. The mean volume-weighted M.R. for the region was 0.87 [*c.f.* Pierson *et al.*, 1987, who reported a $[(SO_4^{2-})/(NO_3^-)]$ molar ratio equal to 1.8 for the Allegheny Mountain (corresponding to an equivalent ratio in the rain of about 3.7), with a volume-weighted mean pH of 3.5]. The same authors more generally reported that the $[SO_4^{2-})/(NO_3^-)]$ ratio in the rain were in the range characteristic of summer rains in the north-west of the United States which show 2.3-4 equivalents of SO_4^{2-} per equivalent of NO_3^-.

Depositions

The atmospheric contribution or total deposition, expressed in μeq/mq (product of the rain volume-weighted mean concentration in μeq l^{-1} by the volume of rain in mm), represents the quantity of ions which deposits on the ground with the rain. Table 8.11 shows the depositions in μeq/mq measured yearly, the total depositions for the period 1989-1995 and the mean depositions of the period, reported for the year, on all the stations, taking into account the different periods of operation of the stations. The mean deposition, reported for the year, on the "zero" station of Monte Cherz is 94 μeq/mq/year, noticeably less than those of the other stations. The annual mean depositions of the "Background" stations (1-8) are between 243 μeq/mq/year for Badia Polesine and 390 meq/mq/year for Caorle.

In Fig. 8.13 the annual total depositions (meq/mq/year) for all the stations are represented, together with the ionic strength. There is evidently a smaller difference between the total depositions of the different stations by comparison with the difference between the corresponding values of ionic strength, correlated to the different volume of rain that fell on the different stations.

In Figs. 8.14 and 8.15 the annual total depositions at single station are reported. While the annual depositions at the "zero" station of Monte Cherz remain constant in time, the depositions of the "background" stations show significant variability. To evaluate the trend of depositions in time, the linear regressions between the depositions of single rainy events, in meq/mq, and the time in days from 23.2.1989 to 29.6.1995 (2318 days) have been calculated. The results are reported in Table 8.12 and depicted in Figs. 8.16 and 8.17.

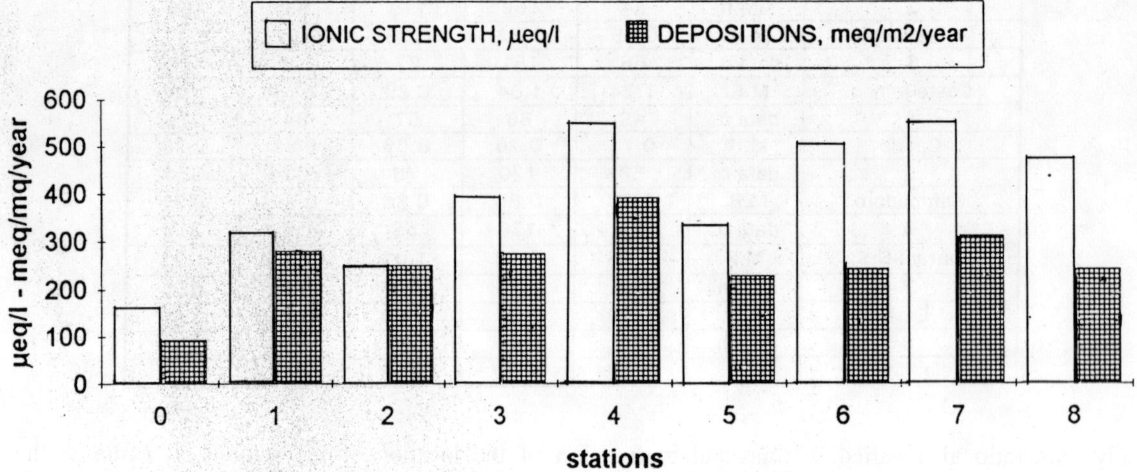

Fig. 8.13. Ionic strength, meq/l and depositions (meq/mg/year).

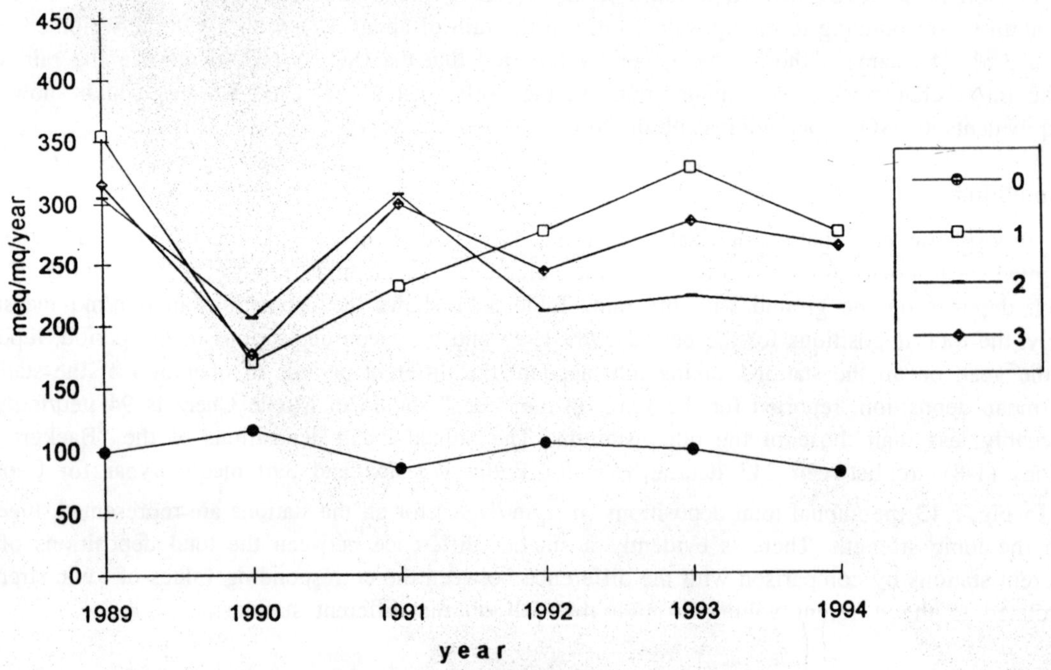

Fig. 8.14. Total depositions at stations 0 to 3.

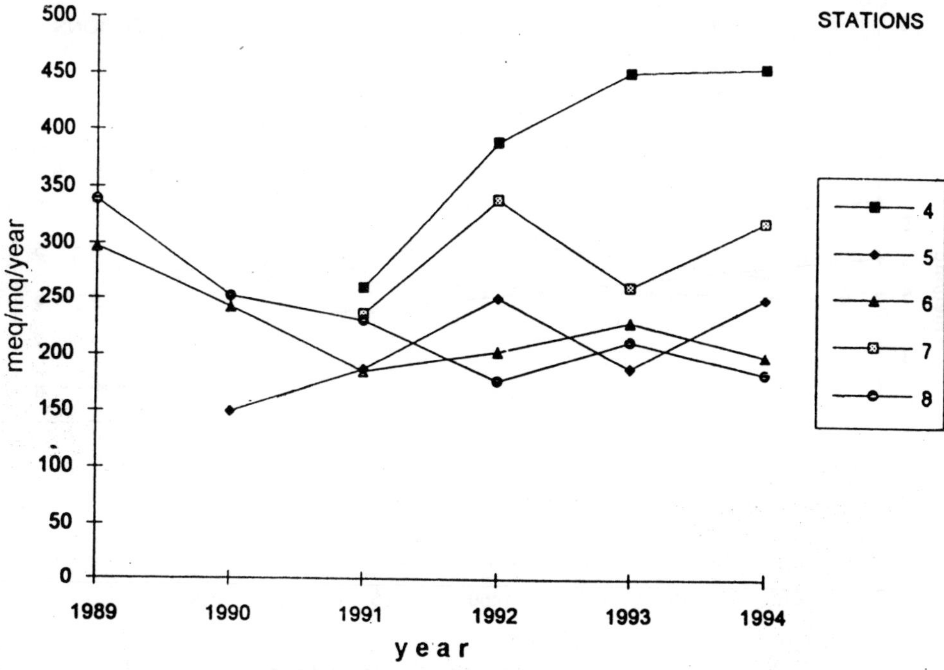

Fig. 8.15. Total depositions at stations 4 to 8.

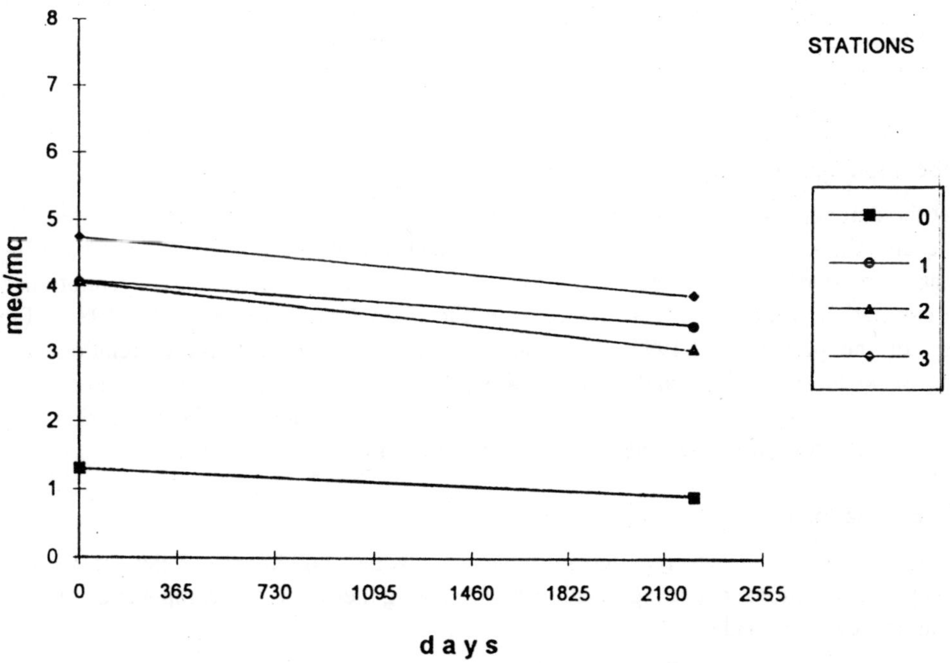

Fig. 8.16. Regression deposition - time at stations 0 to 3.

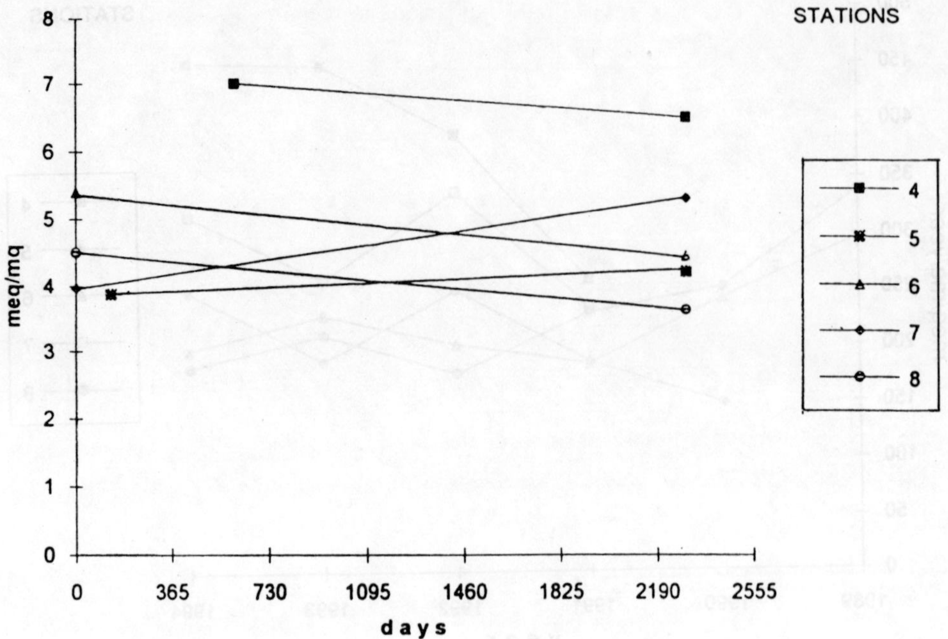

Fig. 8.17. Regression deposition - time at stations 4 to 8.

Taking into consideration the whole period 1989-1995, while the stations of Compodoro and Erba show a significant increase (9-34%), the remaining stations show a similarly significant decrease, particularly the stations of Monte Cesen and Monte Cherz (25-32%).

Rain composition correlations

To interpret the composition of the precipitations, the correlations between the single solutes, the solutes and volume and the conductivity and between some constituents of rain have been calculated. The slopes of the regression lines obtained for the rainwater volumes and single ion concentrations indicate a negative correlation, which is in accordance with Cavallaro (1984) and Mosello (1986). Table 8.13 summarizes for the period 1989-1995, the values of the most representative correlation coefficients "r" existing between single ions and Table 8.14 shows the values of the angular coefficients "A", the constants "B" and the correlation coefficients "r" of the equations of the regression lines (Y = B + AX) existing between the different components of rain.

Single ion correlations

From the analysis of the data reported in Table 8.13, which highlights correlations with r > 0.80, the following considerations regarding relationships existing between the components of rain in the different stations can be made :

Table 8.11. Depositions in meq/m^2

YEAR	STATIONS								
ANNO	-0	1	2	3	4	5	6	7	8
1989	99	355	305	316		117	297	275	339
1990	116	171	206	177	103	149	242	87	252
1991	85	231	307	299	260	187	185	236	230
1992	105	276	210	243	390	251	203	340	177
1993	99	327	222	284	451	188	229	261	212
1994	81	275	212	263	455	250	199	320	183
1995	27	166	147	174	192	216	202	166	163
1989 - 1995	612	1802	1613	1757	1851	1359	1558	1685	1556
DEPOS/YEAR	94	280	251	275	390	227	243	311	242

In grey the data with partial activity of the station.
For 1995 depositions, as all the other data are relative only to the first six months.

Table 8.12. Linear regression between depositions of single events, meq/mq, and time (in days)

STATION	INITIAL VALUE	FINAL VALUE	% Differ.	EVENT S	"t"	EVALUATION
7 - Erbé	3,97	5,31	34	361	-3,65	SIGNIFICATIVE
5 - Campodoro	3,88	4,23	9	334	-1,37	NOT SIGNIFICATIVE
4 - Caorle	7,01	6,50	-7	275	1,02	NOT SIGNIFICATIVE
1 - Cesiomaggiore	4,06	3,39	-16	486	2,91	SIGNIFICATIVE
3 - Castelfranco	4,75	3,87	-18	409	3,00	SIGNIFICATIVE
6 - Carrara S.S.	5,41	4,46	-18	316	2,76	SIGNIFICATIVE
8 Badia Polesine	4,50	3,64	-19	381	3,89	SIGNIFICATIVE
2 - Monte Cesen	4,10	3,06	-25	450	4,04	SIGNIFICATIVE
0 - Monte Cherz	1,31	0,89	-32	326	4,66	SIGNIFICATIVE

Calculation of the "t" of Student value $$t = \frac{|\chi_1 - \chi_2|}{(2\sigma^2)^{0.5}} \cdot n^{0.5}$$

χ_1=ionic strength extrapolated to the beginning of the period;

χ_2=ionic strength extrapolated to the end of the period

σ = standard deviation of the linear regression; n = number of measurements

Critical Values of "t" of Student for Freedom degree < 120:
t = 1.96 at probability level of 95%
t = 2.58 at probability level of 99%

The difference between inizial and final values extrapolated from the regression lines is evaluated as: SIGNIFICATIVE, at level of 99%, for "t" > 2.58
INDICATIVE, at level of 95%, for "t" between 1.98 and 2.58.
NOT SIGNIFICATIVE, at level of 99%, for "t" < 1.96.

* NO_3^- : Nitrate was very well correlated with NH_4^+ on all the stations, except Erba. The ratio between the concentrations expressed in meq l^{-1} shows that the predominance of NH_4^+ over NO_3^- is considerable in the stations of Castelfranco, Campodoro, Carrara, Erba and Badia Polesine (r = 1.49-1.66); it is limited on the stations of Cesiomaggiore and Monte Cesen (r = 1.19-1.28), while in the stations of Monte Cherz and Caorle NO_3^- prevails on NH_4^+ (r = 0.96-0.92). Consequently on seven stations out of nine the influence of local emissions caused by agriculture is significant, while on the other two stations, the influence of emissions from combustion prevails.
  NO_3^- is very well correlated with SO_4^{2-} on practically all the stations : however, the poor correlation of the two ions with H^+ (also considering only the acid events) prevent a sure attribution of these two ions to industrial activity, though it is probably the major source. The trend of the Molar Ratio $[(SO_4^{2-})/(NO_3^-)]$ nevertheless shows that the stations in the south of the region are affected by the emissions from the combustion plants during the winter period.
* SO_4^{2-} : Sulphate together with NO_3^- is very well correlated with NH_4^+ on all the stations and with Ca^{2+} and Mg^{2+} particularly on the stations of Cesiomaggiore, Carrara and Badia Polesine; the last two cations were probably influenced by local dust in some way.
* Cl^- : Chloride is well correlated with Na^+ on all the stations (a little less at Carrara and Badia Polesine) which confirms the common marine origin of the two ions; the correlation with Mg^{2+} and K^+ is less good.
* HCO_3^- : Biocarbonate is well correlated with Ca^{2+} on the stations of Erba, Castelfranco and Caorle, where the presence of specific sources of pollution can be supposed. It is very poorly correlated with Mg^{2+}.
* NH_4^+ : Ammonia together with NO_3^- and SO_4^{2-} is fairly well correlated with Ca^{2+}, K^+ and Mg^{2+} on the stations of Campodoro, Carrata and Badia Polesine. This multiple correlation can be attributed to dust raised from the soils (closeness of roads or agricultural cultivations).
* MM of rain : The linear correlation between the volume of rain and the concentration of single ions is always poor (r = 0.20-0.40).

K^+, Mg^{2+} and Cl^- are in any case more correlated with the volume of rain on the "zero" station of Monte Cherz. On the "background" stations on the contrary, ion concentrations which are relatively better correlated with the volume of the rain are, in order of NO_3^-, SO_4^{2-}, NH_4^+, K^+ and Mg^{2+}.

Table 8.13. Values of more representative correlation coefficients "r" between single ions present in the rain : period 1989-1995

STAZ	Cl- Na	Cl- Mg	NO3 SO4	NO3 NH4	NO3- Ca	NO3- Mg	SO4- NH4	SO4- Ca	SO4- Mg	HCO3 -Ca	HCO3 -Mg	Ca- Mg	n.data
0	0.76	0.54	0.75	0.78	0.64	0.59	0.83	0.70	0.57			0.74	326
1	0.77	0.52	0.77	0.82	0.70	0.68	0.71	0.59	0.75	0.62	0.58	0.81	486
2	0.82	0.60	0.78	0.80	0.54	0.70	0.82	0.65	0.72	0.75	0.45	0.79	450
3	0.88	0.36	0.75	0.82	0.46	0.53	0.87	0.82	0.67	0.87	0.63	0.81	409
4	0.91	0.55	0.82	0.77	0.58	0.76	0.75	0.84	0.77	0.83	0.48	0.69	275
5	0.87	0.67	0.71	0.86	0.46	0.70	0.87	0.64	0.83	0.79	0.36	0.78	334
6	0.70	0.56	0.79	0.86	0.71	0.76	0.85	0.82	0.85	0.64	0.47	0.88	316
7	0.82	0.62	0.69	0.68	0.36	0.49	0.82	0.36	0.58	0.90	0.53	0.66	361
8	0.71	0.63	0.75	0.83	0.64	0.68	0.82	0.84	0.81	0.63	0.59	0.90	381

Anions-Cations correlations

For all the stations, except for the station of Monte Cherz, the regression lines between the sums of the anions and cations (Table 8.14) show that the considered ions represent practically the total amount of soluble ionic species, the correlation coefficient is very high (r = 0.99). These correlations confirm the positive results obtained with the data quality control (calculation of the ionic balance and of conductivity) carried out on the analyzed samples. The correlation is less high for the "zero" station of Monte Cherz (coefficient of X = 0.86) and shows a major percentage difference between the sum of anions and cations. This difference can be attributed, atleast partially, to the lack of determination of the alkalinity (HCO_3^-) for a part of the samples. Its contribution is not negligible in the neutral and alkaline samples (with pH $\geq$ 5.6) which comprise 27% of the total.

Table 8.14. Values of angular coefficients "A", of the constants "B" and correlation coefficients "r" of the equations of the regression lines (Y = B + AX), relative to the components of the rain : period 1989-1995

STAZ	CATION ANION			(NH_4+H) (SO_4+NO_3)			Na - Mg			(Na+Mg) - Cl			NH_4, %			n.
	B	A	r	B	A	r	B	A	r	B	A	r	MIN	MAX	MED	data
0	30.2	0.86	0.92	9.2	0.40	0.83	4.3	0.29	0.47	3.8	1.34	0.72	4	255	36	326
1	5.1	1.00	0.99	13.9	0.45	0.80	7.7	0.56	0.55	19.6	1.02	0.74	9	190	54	426
2	0.5	1.03	0.99	7.4	0.50	0.86	8.8	0.75	0.59	5.9	1.14	0.81	6	157	46	450
3	0.3	1.02	0.99	6.1	0.59	0.91	17.8	0.36	0.40	23.5	1.04	0.73	10	167	58	409
4	-1.5	1.00	0.99	10.3	0-31	0.80	15.8	0.75	0.48	54.3	1.04	0.89	2	120	36	275
5	-0.8	1.02	0.99	7.1	0.61	0.93	6.4	0.75	0.59	10.3	1.10	0.87	10	300	63	334
6	-9.6	1.02	0.99	17.8	0.44	0.89	2.7	0.85	0.56	12.6	1.23	0.76	4	158	49	316
7	10.9	1.00	0.99	10.8	0.56	0.83	13.4	0.49	0.64	18.5	1.40	0.73	2	165	61	361
8	-1.9	1.01	0.99	21.5	0.49	0.87	13.2	0.59	0.67	14.5	1.40	0.73	10	108	54	381

Marine aerosol contribution

The existing relationship between the principal ions which define the marine aerosol are :

$Cl^- = 1.17\ Na^+$

$Na^+ = 4.34\ Mg^{2+}$

$Na^+ + Mg^{2+} = 1.05\ Cl^-$

The ratio between the volume-weighted mean concentrations of Cl^- and Na^+ ranges from 1.18 for the stations of Caorle, Carrara and Badia Polesine, to 1.24-1.35 for the other stations; here Cl^- is in fair excess of Na^+, the origin of which may be supposed to be exclusively marine. The correlation coefficient "r" is particularly high (0.91) for the station of Caorle, which clearly shows the same origin for the two ions, but it also manifests a consistently good correlation (r = 0.70-0.88) on the other stations. The regression between Na^+, Mg^{2+} and Cl^- (Table 8.14) also highlights the remarkable excess of Mg^{2+} with respect to Na^+ or the poor relationship existing between the two ions, just indicating their different origin. The contribution of marine aerosol, calculated from the data of Na^+, is at its maximum on the station of Caorle (23.8%); on the other stations, it ranges between 9.2% at Monte Cherz and 16.0% at Monte Cesen.

Neutralization factor for NH_4^+

Assuming that SO_4^{2-} and NO_3^- are present in the rain as salts of NH_4^+ (as has been seen, Cl^- is mainly of marine origin and is associated with Na^+ and partially with Mg^{2+}), it is possible to calculate if there is any defect or an excess of NH_4^+ with respect to the sum $[SO_4^{2-})/(NO_3^-)]$, obtaining the so-called "neutralization factor" :

$$fN[(NH_4^+)]\% = [NH_4^+)]/(SO_4^{2-}) + (NO_3^-)]$$

where the ionic concentrations are expressed in meq l^{-1} (Brocco *et al.*, 1986).

Whereas the neutralization factors of single samples, for all the stations, range between 2% and 300% (Table 8.14), the mean values of the period vary between 36% at Monte Cherz and Caorle to 63% at Campodoro. This means that a significant part of $[SO_4^{2-} + NO_3^-)]$ is often not neutralized by NH_4^+ and is consequently associated with other cations present in the rain and particularly with H^+, for acid samples, and Ca^{2+} and Mg^{2+}. Therefore, while 64% of acidity not bound to NH_4^+ on the station of Monte Cherz will result in binding to H^+ in the acid samples (73% of the total), on the contrary, on the station of Caorle the same percentage of SO_4^{2-} and NO_3^- not bound to NH_4^+ will be associated with Mg^{2+} and Ca^{2+}, which are present in considerable quantity and well correlated to each other.

Correlation $[NH_4^+ + H^+]$ vs $[SO_4^{2-} + NO_3^-]$

In accordance with Possanzini *et al.*, (1984) and Cavallaro *et al.*, (1984), the plots of $[NH_4^+ + H^+)$ vs $[SO_4^{2-} + NO_3^-]$ were dotted. The equations obtained for each station are reported in Table 8.14, together with the correlation coefficient. The plot slopes ranging from 0.31 to 0.61 and the intercepts different from 0 indicate that not all the sulphates and nitrates were present in the rain in the acid form. As can be seen, high values of correlation coefficients (always > 0.80) were obtained, which according to Morselli *et al.* (1986), is a constant part of the sum of $[(SO_4^{2-}) + (NO_3^-)]$ which is present in the rainwater in an acid form. As has been already said, SO_4^{2-}, NO_3^-, NH_4^+, and H^+ represent a substantial part of the total ionic strength in the rain, from 30% at the station of Caorle to 64% at the station of Monte Cherz. To verify the link between the total $[(SO_4^{2} + NO_3^-)]$ and $[(SO_4^{2-})/(NO_3^-)]$ in an acid form present in the rain, the regression lines $[NH_4^+ + H^+]$ vs $[SO_4^{2-} + NO_3^-]$ have been calculated. For all the stations this correlation (Table 8.14) confirms what was reported above, i.e. a constant percentage of the total $[SO_4^{2-} + NO_3^-)]$ is present in the rain in an acid form; this percentage (defined by the angular coefficient "A") ranges between 31% for Caorle and 61% for Campodoro. If only the acid samples are considered (1502 events from the total of 3553 for all the stations), these percentages vary between 40% for the station of Monte Cherz and 68% for the station of Erbe.

Multiway data analysis

Data consisting of 1174 observations over the period February 1989-December 1991, have been elaborated with statistical methods by correlating the various measured parameters, rain volume, conductivity and ionic strength. Principal component analysis (PCA) was used in an attempt to describe the structure of relations between the solutes in wet deposition (Fscoufier, 1985). Some extensions of PCA (Interstructure-Compromise-Infrastructure : ICI - the main references of ICI are Coppi and Bolasco (1988) and Lavit (1988), were considered in order to evaluate differences between the various relations between pollutants in each of the stations. Descriptive statistical analysis shows that, besides the station site, precipitation volume is a relevant classification criterion for the rain sampled in the region network. Interstructure analysis highlighted the differences among the several occasions (station per volume), in terms of covariance structure among pollutants. Differences were particularly noticeable for low volumes. The representation of the interstructure showed the mountain stations in the north of the region to be less polluted, as opposed to the more polluted rural stations, in the south. The remaining stations were intermediate. This fact reflects the territorial distribution of industrial and population settlements in the region. The analysis of the compromise made it possible to verify that the solutes, most important for the definition of a common structure were the same ones, which defined the first principal component of single structures. Infrastructure analysis showed a strong first common factor composed principally of all solutes (PO_4^{3-} and H^+ excluded) in all the situations defined by classification criteria. It was found that conductivity is fairly close to the first common factor, as might be expected. It is therefore important to consider its variations in different conditions described by classification criteria. The conclusions of the multiway data analysis did not show any practical differences with the previous results obtained by more approximate methods.

AIR POLLUTION

Particulates-Sulphates-Sulphur dioxide

Table 8.15 summarizes, for all the stations, the data concerning the concentration of particulates, sulphates and sulphur dioxide in air (minimum, maximum and mean values) relative to the period 1989-1995. In the "background" stations (1-8) the annual mean concentrations vary between 34 and 48 μg m^{-3} for particulates, between 3 and 7 μg m^{-3} for sulphates and between 9 and 12 μg m^{-3} for SO_2, which are clearly within the level set by the Italian standards (Republica Italiana, D.P.C.M. 20.3.1983 and D.P.R. n. 203, 20.5.1988).

On the "zero" station of Monte Cherz, the annual mean concentrations are 13 μg m^{-3} for particulates, 3 μg m^{-3} for sulphates and 1 μg/m^{-3} for SO_2, clearly below the concentrations measured in the "background" station. The statistical elaborations of annual data carried out according to the Italian law comparing the measured data with the air quality standards, shows, in any case, that the limits are always largely respected.

Table 8.15. Minimum, maximum and mean values of daily concentrations of particulates, sulphates and sulphur dioxide measured at the network stations, together with the number of valid data : period 1989-1995

STAZ	PARTICULATES μg/m^3				SULPHATES, μg/m^3				SO_2 μg/m^3			
	n.data	min. conc.	max conc.	mean conc.	n.data	min. conc.	max conc.	mean conc.	n.data	min. conc.	max conc.	mean conc.
0	2179	0.2	198	13	2182	0.3	48	3	2128	0.2	40	1
1	1515	3	177	36	1515	1	22	4	1812	1	79	9
2	1361	4	169	34	1361	1	27	4	1611	1	98	10
3	1578	3	307	47	1578	1	35	5	1385	1	93	11
4	1268	3	182	37	1268	1	31	4	1336	1	94	11
5	1405	3	298	44	1405	1	44	5	1644	1	119	11
6	1328	3	156	40	1328	1	28	5	1356	1	95	12
7	1218	3	188	48	1218	1	47	7	1363	1	73	10
8	1416	3	375	45	1416	1	40	6	1360	1	134	12

Regression of air pollution data against time

The linear regression between the data of concentration (in mg m^{-3}) of single daily determinations of particulates, sulphates and sulphur dioxide in air and time in days are calculated, from 1989 to 1995. The differences between the initial and final values are extrapolated from the regression lines. To standardize the evaluation, the data of concentration corresponding to the days when these data were available contemporarily have been used. The results are reported in Table 8.16. The results of the statistical elaboration point out as follows :

Particulates in air

A 12% statistically significant increase of the concentration in air was recorded at the station of Erbe. At the other stations, on the contrary, a generalized statistically representative decrease of 4-51% was found on all the stations, with the exception of Castelfranco. While for the "zero" station of Monte Cherz, the concentration of particulates in air varies from 16.2 to 11.6 mg m^{-3}, for the "background" stations the concentrations decrease from 40-64 to 28-50 mg m^{-3}.

Sulphates in air

Sulphate concentrations increased on the stations of Monte Cesen, Castelfranco, Caorle, Carrara, Erbe and Badia Polesine, always in a statistically significant manner. On the contrary, they decreased on the remaining three stations of Monte Cherz, Cesiomaggiore and Campodoro, in an equally significant way. As already seen, the concentrations of SO_4^{2-} in air are quite low, around 3-7 mg m^{-3}.

Sulphur dioxide in air

SO_2 in air clearly decreased on all the stations, with concentration between the beginning and the end of the period varying from 24% at Erbe to 78% at Monte Casen. While the concentration of SO_2 in air for the "zero" station of Monte Cherz varies from 1.35 to 0.78 mg m^{-3}, for the "background" stations it changes from 11-20 mg m^{-3} to 4-9 mg m^{-3}. As already stated, these evaluations give only an indication on the trend of the phenomenon in the considered period, being the phenomenon itself, influenced by several factors which include :

— shifts in meteorological behaviour,
— changes in pollutant sources, and
— changes in sampling and analytical practices.

Thus, extreme caution is advised in attempting to assess cause-effect relationship, or anticipate future behaviour on the basis of limited record of data (Map3S/Raine Research Community, 1982).

Correlations between the atmospheric pollutants : SO_4^{2-} and SO_2

As already seen, the mean annual and quarterly concentrations of SO_4^{2-} and SO_2 in air are very low in all the stations, particularly in the "zero" station, and range around 5 to 10 µg m^{-3}. This presence of SO_4^{2-} in air is due mainly to the oxidation of SO_2 caused by photochemical activity in the atmosphere. The Molar Ratio, M.R. = $SO_4^{2-}/(SO_4^{2-} + SO_2)$ and the correlation coefficient "r" existing between the 2 series of data, for every quarter and for the entire period of determinations, was calculated in order to define and estimate the fraction of SO_2 converted to SO_4^{2-}.

Table 8.16. Atmospheric pollution from 01.01.1989 (day 1) to 30.06.1995 (day 2372)

Linear regression between PARTICULATES in air, µg/mc and time in days.
Calculation and evaluation of the value of "t" of Student

STATION	INITIAL VALUE	FINAL VALUE	DIFF. %	EVENTS	"t"	EVALUATION
7 - Erbé	44.3	49.5	12	1089	-4.02	SIGNIFICANT
3 - Castelfranco	47.7	45.9	-4	1248	1.59	INSIGNIFICANT
6 - Carrara S.S.	45.7	36.0	-21	1165	9.01	SIGNIFICANT
2 - Monte Cesen	39.9	27.9	-30	1295	12.38	SIGNIFICANT
0 - Monte Cherz	16.2	11.6	-31	2055	9.71	SIGNIFICANT
8 - Badia Polesine	55.1	38.1	-31	1205	14.63	SIGNIFICANT
4 - Caorle	45.8	29.5	-35	1143	15.62	SIGNIFICANT
1 - Cesiomaggiore	49.8	27.7	-44	1414	22.22	SIGNIFICANT
5 - Campodoro	63.8	31.5	-51	1339	26.83	SIGNIFICANT

Linear Regression between SULPHATES in air, µg/mc, and time in days.
Calculation and evaluation of the value of "t" of Student

STATION	INITIAL VALUE	FINAL VALUE	DIFF. %	EVENTS	"t"	EVALUATION
3 - Castelfranco	3.44	6.32	84	1248	-19.66	SIGNIFICANT
7 - Erbé	5.05	7.96	58	1089	-13.26	SIGNIFICANT
4 - Caorle	3.89	4.76	22	1143	-6.03	SIGNIFICANT
8 - Badia Polesine	5.07	6.08	20	1205	-6.44	SIGNIFICANT
2 - Monte Cesen	3.92	4.67	19	1295	-5.09	SIGNIFICANT
6 - Carrara S.S.	4.34	4.80	11	1165	-2.85	SIGNIFICANT
0 - Monte Cherz	2.65	2.39	-10	2055	3.17	SIGNIFICANT
1 - Cesiomaggiore	5.00	3.06	-39	1414	16.02	SIGNIFICANT
5 - Campodoro	7.43	4.13	-44	1339	19.63	SIGNIFICANT

Linear regression between SO2 in air, µg/mc, and time in days.
Calculation and evaluation of the value of "t" of Student

STATION	INITIAL VALUE	FINAL VALUE	DIFF. %	EVENTS	"t"	EVALUATION
7 - Erbé	11.37	8.68	-24	1089	7.51	SIGNIFICANT
3 - Castelfranco	14.66	8.99	-39	1248	15.23	SIGNIFICANT
0 - Monte Cherz	1.35	0.78	-42	2055	11.42	SIGNIFICANT
5 - Campodoro	15.38	8.79	-43	1339	15.86	SIGNIFICANT
6 - Carrara S.S.	15.99	8.86	-45	1165	18.46	SIGNIFICANT
4 - Caorle	15.12	6.82	-55	1143	19.55	SIGNIFICANT
8 - Badia Polesine	17.52	7.66	-56	1205	24.80	SIGNIFICANT
1 - Cesiomaggiore	16.11	3.86	-77	1414	38.32	SIGNIFICANT
2 - Monte Cesen	19.08	4.10	-78	1295	36.37	SIGNIFICANT

Critical Values of "t" of Student for Freedom Degree higher than 120:

t=2.58 at probability level of 99%; t=2.58 at probability level of 99%

The difference between the initial and final values extrapolated from the regression line is evaluated as:

SIGNIFICANT, at level of 99%, for "t" > 2.58
INDICATIVE, at level of 95%, for "t" between 1.98 and 2.58
UNSIGNIFICANT, at level of 99%, for "t" < 1.96

Table 8.17. Atmospheric pollution data of SO_4^{2-} and SO_2 : average in µg/m³, molar ratio (SO_4^{2-} / SO_4^{2-} + SO_2) and correlation coefficient between the two series of data : period 1989-1995

STATION	Quarter	n. data	SO_4	SO_2	SO_2 total	M.R.	r
0	I	561	1.6	1.6	2.7	0.42	0.06
	II	572	3.5	0.8	3.1	0.74	0.05
	III	461	3.9	0.8	3.4	0.76	0.34
	IV	491	1.0	1.0	1.7	0.45	0.21
	1989-1995	2085	2.5	1.1	2.8	0.59	0.12
1	I	385	3.5	9.5	11.8	0.24	0.11
	II	360	4.2	9.1	11.9	0.28	0.18
	III	320	5.0	8.0	11.3	0.36	0.18
	IV	349	2.9	7.9	9.8	0.26	0.11
	1989-1995	1414	3.8	8.7	11.2	0.28	0.20
2	I	354	3.4	10.0	12.3	0.24	0.32
	II	376	5.6	9.8	13.5	0.36	0.07
	III	280	5.8	11.9	15.8	0.34	0.03
	IV	285	2.5	8.6	10.3	0.23	0.11
	1989-1995	1295	4.4	10.0	12.9	0.29	0.13
3	I	303	5.2	11.7	15.2	0.28	0.04
	II	401	5.7	12.3	16.1	0.30	0.15
	III	243	5.4	10.3	13.9	0.31	0.08
	IV	301	4.2	10.3	13.1	0.28	0.10
	1989-1995	1248	5.2	11.3	14.8	0.29	0.07
4	I	300	4.7	12.1	15.2	0.28	0.20
	II	262	4.8	9.7	12.9	0.31	0.10
	III	279	4.9	12.5	15.8	0.29	0.12
	IV	302	3.2	7.0	9.1	0.31	0.05
	1989-1995	1143	4.4	10.3	13.2	0.30	0.13
5	I	395	6.5	12.1	16.4	0.32	0.24
	II	287	4.6	11.8	14.9	0.29	0.03
	III	341	5.5	11.3	15.0	0.31	0.05
	IV	316	5.0	10.4	13.7	0.28	0.12
	1989-1995	1339	5.5	11.4	15.1	0.30	0.18
6	I	309	5.3	11.4	14.9	0.29	0.09
	II	332	5.1	12.3	15.7	0.29	0.02
	III	214	4.8	12.0	15.2	0.26	0.29
	IV	310	3.3	10.9	13.1	0.21	0.16
	1989-1995	1165	4.6	11.7	14.8	0.26	0.13
7	I	352	8.0	9.1	14.4	0.41	0.03
	II	239	7.1	10.4	15.1	0.36	0.30
	III	160	6.5	8.5	12.8	0.40	0.31
	IV	338	6.2	9.9	14.0	0.35	0.23
	1989-1995	1089	7.0	9.5	14.2	0.38	0.10
8	I	325	5.9	11.3	15.2	0.30	0.06
	II	322	5.6	10.6	14.3	0.36	0.08
	III	252	6.0	10.2	14.2	0.36	0.01
	IV	306	5.4	12.1	15.7	0.29	0.09
	1989-1995	1205	5.7	11.1	14.9	0.33	0.04

Table 8.17 summarizes, the data of concentration of SO_4^{2-}, SO_2 and total SO_2 (SO_2 + SO_4^{2-} expressed as SO_2) in air on all the stations and the respective values of the M.R. and of the correlation coefficient ("r"). For the "zero" station of Monte Cherz the quarterly concentrations of SO_2 (0.8 μg m^{-3}) in the summer period were slightly lower than those relative to the winter period (1.0-1.6 μg m^{-3}), while the concentrations of SO_4^{2-} are significantly higher in spring-summer (3.5-3.9 μg m^{-3}) in comparison with autumn-winter (1.0-1.6 μg m^{-3}). The Molar Ratio $SO_4^{2-}/(SO_4^{2-} + SO_2)$ changed from 0.74-0.76 in summer to 0.42-0.45 in winter, confirming that the fraction of SO_2 converted to SO_4^{2-} is higher in summer due to the influence of photochemical activity in the atmosphere.

During all the periods of the year, however, the correlation coefficient between the two series of data was particularly low (0.05-0.34) and showed a tenuous relationship between the two pollutants: the presence of SO_4^{2-} in air could therefore be due to factors other than the oxidation of SO_2 to SO_4^{2-}, such as dust removal from the soil. It was noticed finally that the total pollution by SO_2 (SO_2 + SO_4^{2-} expressed as SO_2), was higher in the summer months than in the winter months (3.1-3.4 against 1.7-2.7 μg m^{-3}); this may confirm the hypothesis that the origin of the two pollutants is different from the classic one of emissions from heating plants. Elaboration of data on a monthly basis, carried out to deepen the evaluation, shows that :

— for the "zero" station, the M.R. is clearly higher in the summer months (0.77) than in the winter months (0.4-0.5),

— for stations 1, 2, 6, 7 and 8, the M.R. is slightly higher during the summer months than in the winter ones (0.3-0.4 against 0.2-0.3), and

— for stations 3, 4 and 5 the higher values for M.R. (0.3-0.4) are spread throughout the year.

Correlation between SO_4^{2-} in the rain and SO_4^{2-} and SO_2 in air

The linear correlation existing between the data of concentration of SO_4^{2-} in the rain, in μeq/l, and the data of concentration of SO_4^{2-} and SO_2 in air in μg m^{-3}, measured in the case of rainy events, was calculated. The correlations were carried out for the entire period for each single station and for all the stations together. The results are reported in Table 8.18.

SO_4^{2-}, μeq l^{-1} = 70 + 10.9 SO_4^{2-}, μg m^{-3}; r = 0.30

SO_4^{2-}, μeq l^{-1} = 88 + 2.6 SO_2, μg m^{-3}; r = 0.21

The correlations for all the stations together (2192 events) show that the linear relationship is not very high between the two series of data : this is probably because of the different temporal and meterological factors which can influence this relationship.

SO_4^{2-}, μeq l^{-1} = 50 + 4.4 SO_4^{2-} μg m^{-3}; r = 0.28

SO_4^{2-}, μeq l^{-1} = 55 + 1.1 SO_2, μgm^{-3}; r = 0.21

The correlations obtained using the data concerning events with a volume of rain higher than 10 mm (725 events) give practically the same result considering all the events, thus showing that the volume of rain does not influence the correlation between these data.

The regression lines between SO_4^{2-} in the rain and SO_4^{2-} in air for the single station are similar, the correlation coefficient is always low and varies from 0.18 for Castelfranco to 0.41 for Monte Cherz. The regression line between SO_4^{2-} in the rain and SO_2 in air for the station of Monte Cherz is quite different from those for the other stations, probably due to the different concentration of SO_2 in air existing between the "zero" station and the "background" station.

Table 8.18. Correlation between the concentration of SO_4^{2-} in the rain, µeq/L, and the concentration of SO_4^{2-} + SO_2 in air, µg/m³, in the case of rainy events : period 1989-1995

CORRELATION : SO_4^{2-}, µeq l^{-1} = B + A * SO_4^{2-}, µg/m³

STAZ	CORRELATION				
	n	B	A	r	
0	460	27	9.7	0.41	
1	303	56	8.1	0.36	
2	243	44	5.6	0.36	
3	246	106	7.7	0.18	
4	189	69	10.1	0.31	
5	204	68	12.8	0.39	
6	171	150	13.2	0.24	
7	179	114	6.3	0.24	
8	197	139	12.0	0.21	
0-8	2192	70	10.8	0.30	All the events
0-8	725	50	4.4	0.28	Events with mm > 10

CONCLUSION

The study conducted during the period 1989-1995 on the samples collected at the "zero" station at Monte Cherz and the eight "background" stations of the monitoring network spread all over the regional territory enabled the evaluation of the characteristics of meteorological precipitations and the air quality in the region.

Meteorological precipitations

The most salient feature of the precipitation regime is a temporal and spatial variability. Stations located in the North of the region are more rainy. The rain samples were analyzed for the following ions: SO_2^-, NO_3^-, Cl^-, HCO_3^-, NH_4^+, Ca^{++}, Mg^{++}, Na^+, K^+ and for pH, conductivity and alkalinity. The total volume of each sample was recorded.

The acidity of the rain, expressed by a calculated value of hydrogen ion concentration, varied both in time and space with the lowest values being determined at the stations of Monte Cherz and Monte Cesen (north of the region, a mountainous area). On the whole, however, the acidity of the rain in the entire region can be considered as mean-low. The ionic strength (given by the sum of all the cations and anions) is lowest at the station of Monte Cherz and increases considerably for the stations sited in the south of the region. For all the stations the principal components of the rain samples are the same : sulphates, nitrates and ammonium. As may be expected, sulphates are present in a concentration higher than nitrates in winter and only in south of the region, but not in summer. For the stations at Carrara, Erbe and Badia Polesine concentrations of sulphates, nitrates and ammonium are higher than those measured at the other stations.

Though an inverse correlation between ion concentrations and rain volume may be supposed, it was not clear up to what kind of model it occurs. Hypotheses include the rain-out (influence of the cloud on the content of the rain drop) and the wash-out process (washing of the various pollutants from the air). An influence of marine aerosol of the quality of the rain sampled at all the stations, particularly at Caorle (a seaside location) was noticed. Agriculture and cattle breeding exerted some influence on all the stations except for those located in the mountains (Monte Cherz and Monte Cesen) and at sea side (Caorle). This is particularly evident for the ammonium ion. At Erbe, calcium carbonate is exceptionally important, which can be attributed to the local activities of marble quarrying.

Air pollution

In the "background" stations, the annual mean concentrations varied between 34 and 48 $\mu g\ m^{-3}$ for particulates, between 3 and 7 $\mu g\ m^{-3}$ for sulphates and between 9 and 12 $\mu g\ m^{-3}$ for SO_2. On the "zero" station the data are below the concentrations, measured on the other stations. On the basis of obtained data, the Italian standards for these parameters were found to have been fulfilled throughout the period of the study.

The linear regression between the data of single daily determinations on the air pollutants and time are calculated and correlations between the atmospheric pollutants SO_4^{2-} and SO_2 and the Molar Ratio $SO_4^{2-}/(SO_4^{2-} + SO_2)$ and the correlation coefficient between the two series of data was calculated. The fraction of SO_2 converted to SO_4^{2-} was higher in summer due to the influence of photochemical activity in the atmosphere. The correlation coefficient between SO_4^{2-} in the rain and SO_4^{2-} in air was always low (0.19-0.41).

FUTURE PROGRAMME

The University of Venice and the region of Veneto are trying to agree to extend for the next four years the research on the rainwater and air pollution in the regional territory using the existing monitoring network, taking into account the experience acquired by the personnel working in the laboratory on the analysis of the samples, the data elaboration and the knowledge of the territory. In particular atmospheric trace metal measurements (Al, Cu, Fe, Pb, V, Zn etc.) of air particulates pollution should be carried out, according to the international literature (Infante *et al.*, 1985). Meteorological parameters characterizing the area should be considered and correlated with the influence of transboundary transportation of pollutants on the physical-chemical characteristics of the rainfall (Raynor and Hayes, 1982). Finally an accurate study has to be carried out on the influence of the modification of pollutants emission by urban and industrial sources and in particular by the vehicular traffic.

REFERENCES

APHA, 1985 : *Standard Methods for the Examination of Water and Wastewater*. APHA, AWWA, WPCF, 16th Ed., New York.

Asbury, C.E., F.A. Vertucci, M.D. Mattson and G.E. Likens, 1989 : Acidification of Adironck Lakes. *Environ. Sci. Technol.*, 23 : 362-365.

Barrie, L.A., S.E. Lindberg, W.H. Chan, H.B. Ross, R. Arimoto and T.N. Church, 1987 : On the concentration of metals in precipitations. *Atmos. Environ.*, 21 : 1133-1135.

Basile, G. and G. Gelano, 1990 : Le deposizioni atmosferiche nell'alto bacino del fiume lao. *Inquinamento*, 1 : 47-52.

Bassanino, N., M.A. Brovelli, A. Cavallaro, C. Corradi and G. Tebaldi, 1989 : Analisi multifattoriale del chimismo delle precipitazioni atmosferiche. *Acqua - Aria*, 2 : 147-154.

Bellandi, S., C. Oppo, F. Pantani and R. Cini, 1995 : *Proc. Int. Conf., Mediterraneanchem.*, Toronto.

Bettinelli, M., E. Joannilli and G. Ziliani, 1984 : Influenza delle modalita di campionamento nello studio delle precipitazioni nella Valle Padana. *Sep Pollution Conference Acts.*, 125-144.

Bolze, D. and J. Beyera, 1989 : The Citizens' Acid Rain Monitoring Network. *Environ. Sci. Technol.*, 23 : 645-646.

Brocco, D., A. Cecinato, A. Cerquilini and F. D'Innocenzio, 1983 : Convenzione per lo studio del transporto a lunga distanza degli inquinanti atmosferici, E.M.E.P. Metodi di analisi degli inquinanti, *ISTISAN 1983/40, Istituto Superiore di Sanita*, 1-73.

Brocco, D., M. Rotatori and R. Tappa, 1986 : Valutazione delle deposizioni atmosferiche per lo studio del trasporto degli inquinanti a grande distanza. *Acqua - Aria*, 5 : 445.

Brovelli, M. and M. Bassanino, 1988 : Analisi statistica dell'acidita delle precipitazioni a Milano. *Acqua-Aria*, 5 : 607-615.

Cavallaro, A., G. Tebaldi, C. Corradi, R. Gualdi and M. Bassanino, 1984 : Analisi delle caratteristiche chimiche delle acque meteoriche raccolte a Milano negli anni 1982-83. *Sep. Pollution Conference Acts.*, 155-171.

Cavallaro, A., C. Corradi, R. Gualdi and G. Tebaldi, 1986 : Le caratteristiche chemiche delle acque meteoriche raccolte a Milano e a Cassano d'Adda tra il 1982 e il 1984. *Provincia di Milano*, U.S.S.L., 75/11.

Ciattaglia, L., 1975 : Costituzione della rete OMM per il controllo dell'inquinamento atmosferico di fondo. *Riv. Met. Aer.*, 25 : 236-239.

Ciattaglia, L., 1981 : Programma OMM, Inquinamento atmosferico di fondo : Dati di torbidita dell'aria e chimica delle precipitazioni della rete italiana 1978-1979, *Riv. Met. Aer.*, 41 : 49-57.

Coppi, R. and S. Bolasco (Eds.), 1988 : *Multiway Data Analysis*, North Holland, Amsterdam.

Cosma, E., D. Sepulcri and G. Maffei, 1988A : Stazioni E.M.E.P. "Tipo O" controllo del transporto transfrontaliero degli inquinanti atmosferici : 2 anni di osservazioni sull'atmosfera e nelle precipitazioni. *SEP Pollution, Citta e Ambiente*, 142-154.

Cosma, E., G. Maffei, L. Szpyrkowicz and F. Zilio Grandi, 1988B : Organizzazione di una rete regionale per la valutazione delle caratteristiche chimico-fisich delle precipitazioni atmosferiche. *SEP POLLUTION Conference Acts, Citta' e Ambiente*, 143-155.

Cossu, A., A. Luglie's, R. Mosello and N. Sechi, 1989 : Eventi di deposizione acida nella Sardegna Nord-occidentale. *Acqua-Aria*, 7 : 815-820.

Dell'Atti, A., L. Ruggiero and F. Zuanni, 1988 : Caratteristiche chimiche dell precipitazioni atmosferiche nel territorio di lecca. *Difesa Ambientale*, 11 : 36-38.

Del Turco, A., F. Zilio Grandi and F. Franzin, 1986 : L'inquinamento atmosferico e l'impatto sull'ambiente", 53-68.

Del Turco, A., 1987 : Studio delle precipitazione dell'area di Porto Marghera. C.N.R., *Progetto Finalizzato Energetica*, Sottoprogetto n. 13, Relazione Annuale, 1986/87.

Desideri, A., P. Menghi, G. Bartolini and M. Costa, 1989 : Indagine sulle acque meteoriche nella citta di Prato. *Inquinamento*, 7/8 : 58-62.

EMEP Manual, 1977 : Cooperative programme for monitoring and evaluation of long range transmission of air pollutants in Europe. Manual for sampling and chemical analysis, EMEP/CHEM, 3/77.

Escoufier, Y., 1985 : Objectifs et procedured de l'analyse conjointe de plusiers tbleaux de donnees, *Statistique et Analyse des Donnees*, 10 : 1-15.

Galloway and Likens, 1976 : Calibration of collection procedures for the determination of precipitation chemistry. *Water, Air and Soil Pollution*, 241-251.

Galloway and Likens, 1978 : The collection of precipitation for chemical analysis. *Tellus*, 30 : 71-82.

Giunta Regionale del Veneto, *Delibere* n. 7029 del 18.12.1986; n. 575 del 28.10.1987 e n. 3230 del 5.6.1992.

Greghi, D., M. Gatti and F. Pantani, 1989 : Monitoraggio delle deposizioni meteoriche nell'area fiorentina. *Inquinamento*, 6 : 52-56.

Gruppo di Studio Nord India, 1987 : Deposizioni atmosferiche nel Nord Italia (rapporto finale anni, 1983-1984). *Ingegneria Ambienale, Quaderni*, 6 : 1-63.

Infante, R., A. Carrasquillo, F. Feliciano, M. Hernadez, G.A. Infante and G. Rosado, 1989 : Atmospheric trace metal measurements of air particulates pollution in ponce, Puerto Rico. *Environ. Technol. Letters*, 10 : 687-696.

Lauterwasser, E., 1988 : *Landscape Urban Planning*, 16 : 45.

Lavit, C., 1988 : Analyse conjointe de tableaux quantitatifs. *Masson, Paris*.

Lipfert, F.W., S.C. Morris and R.E. Wyzga, 1989 : Acid aerosols : The next criteria air pollutant? *Environ. Sci. Technol.*, 23/11 : 1316-1322.

Mantovan, P., A. Pastore, L. Szpyrkowicz and F. Zilio-Grandi, 1992 : Rainwater characterization by multivariate analysis of Venice Region Network Data *Simposio Italo-Brasiliano di Ingegneria Sanitaria-Ambientale, Rio de Janeiro (Brasil)*, pp. 42-59.

Mantovan, P., A. Pastore, L. Szpyrkowicz and F. Zilio-Grandi, 1995 : Characterization of rain water quality from the Venice Region Network using Multiway Data Analysis. *Sci. Tot. Environ. Chem.*, 164/1 : 27-43.

MAP3S/RAINE Research Community, 1982 : The MAP 3S/Rain Precipitation Chemistry Network : Statistical Overview, 1976-80, *Atmos. Environ.*, 16 : 1603-1631.

Millan, M.M. and G. Gangoiti, 1988 : Rain scavenging from tall stack plumes : the non-proportionality problem. *Environ. Technol. Letter*, 9 : 877-890.

Morselli, L., A. Carati and F. Mattivi, 1986 : Indagine sulle metodologie analitiche e di campionamento delle deposizioni atmosferiche. *Inquinamento*, 10 : 44-50.

Mosello, R., A. Barbieri, P. Mandriolo, G. Righetti, C. Sabbioni and G. Tartari, 1986 : Relazioni fra alcune variabili chimiche nelle acque di pioggia e loro uso nella verifica dei risultati analitici. *SEP POLLUTION*, 115-123.

Mosello, R., G. Tartari, B. Sulis and A. Boggero, 1988 : Ricerche sulle deposizioni acide e sull'acidificazione delle acque superficiali. *Acqua-Aria*, 1 : 61-67.

Mosello, R. *et al.*, 1990 : Metodologie e controlli di qualita per lo studio della chemica delle deposizioni atmosferiche in Italia, *Documenta n. 23 C.N.R. dell' Istituto Italiano di Idrobiologia - Verbania, Pallanza*.

Mosselo, R., 1993 : Rapporto sull'attivita della rete RIDEP (Rete Italiana per lo studio delle deposizioni atmosferiche) nel quinquennio, 1988-1992. *Documenta* n. 44, C.N.R. - *Istituto Italiano di Idrobiologia*, C.N.R.

Mosello, R. *et al.*, 1995 : Acid rain analysis intercomparison. AQUACON - MedBas Project Subproject No. 6, 1/1994, Joint Research Centre, European Commission, EUR 16332 EN.

Norton, G.A. and Levin, 1989 : Combustion of refuse-derived fuel and oil. *Environ. Sci. Technol.*, 23 : 774-780.

OECD, 1979 : The OECD Programme on long range transport of air pollutants. *OECD*, Paris.

Olem, H. and P.M. Berthouex, 1989 : Acidic deposition and cistern drinking water supplies. *Environ. Sci. Technol.*, 23 : 333-340.

Pierson, W.R., W.W. Brachaczek, R.A. Gorse, S.M. Japar, J.M. Norbeck and G.J. Keeler, 1987 : Acid rain and atmospheric chemistry at Allegheny Mountain. *Environ. Sci. Technol.*, 21 : 679-691.

Possanzini, M., F. De Santis and A. Liberti, 1984 : Aspetti tecnici e critici di valutazione connessi alla misura dell'acidita atmosferica, *Atti delle giornate di studio su "Precipitazioni Acide", Venezia*, 91-111.

Raynor, G.S. and J.V.´Hayes, 1982 : Variation in chemical wet deposition with meteo conditions. *Atmos. Environ.*, 16 : 1647-1656.

Repubblica Italiana, *Suppl. Ord. G.U.* 25.5.1982 : *Legge* 27.4.1982, n. 289, Ratifica ed esecuzione della Convenzione sull'inquinamento atmosferico attraverso frontiere a lunga distanza.

Repubblica Italiana, *DPCM*, 28.3.1983 : Limti mssimi di accettabilita delle concentrazioni e di esposizione relative ad inquinanti dell'aria ambiente esterno.

Repubblica Italiana, *DPR* 24.5.1988 : n. 203 - Alegati I° e II° e Note 1 e 2.

Sah, S.P. and K.J. Melwes, 1989 : Rates of acid deposition and their interaction with forest canopy and soil in two beach forest ecosystems on limestone and triassic sandstone soils in N. Germany. *Environ. Technol. Letters*, 10 : 995-1002.

Schofield, C.L., *Acid Rain/Fischeries* (R. Johnson ed.), American Fisheries Society, Bethesda, MD, p. 57.

Schultz, J.A.M., 1988 : Biological implications of seasonal and urban nitrate (acid) wet deposition. *Environ. Technol. Letters*. 9 : 713-720.

Smith, V.R., 1987 : Chemicalo composition of precipitation at Marion Island (sub-Antartic). *Atmos. Environ.*, 21 : 1159-1565.

Spiegel Murray, R. (Ed.) 1994 : In : *Statistica* Cap. 11, IInd Ed., McGraw-Hill, Milano.

Technicon Traacs, 800, 1986 : *Industrial Methods Manual.*

United Nations, 1979 : Economic commission for Europe, Convention on long-range transboundary air pollution. Geneva, E/ECE/1010.

Voldner, E.C. and M. Alvo, 1989 : On the estimation of sulfur and nitrogen wet deposition to the great lakes. *Environ. Sci. Technol.*, 23 : 1223-1232.

Zilio-Grandi, F. and L. Szpyrkowicz, 1991a : Sampling and analysis of rain : methods and results of the Venice regional network. *Fresenius J. Anal. Chem.*, 341 : 625-630.

Zilio-Grandi, F. and L. Szpyrkowitz, 1991b : Air monitoring network for the Venice Region : preliminary results for the rain quality. *Toxicol. Environ. Chem.*, 29 : 281-296.

Zilio-Grandi, F. and L. Szpyrkowicz, 1994a : Rete di campionamento e caratterizzazione delle precipitazioni atmosferiche della Regione del Veneto. Valutazione dei rilevamenti del biennio, 199-1991. *Ingegneria Ambientale*, 23 : 329-402.

Zilio-Grandi, F., L. Szpyrkowicz, G. Maffei and R. Morandi, 1994b : Progetto EMEP. Rete di campionamento e caratterizzazione delle precipitazioni atmosferiche della Regione del Veneto, 1994 *4° Convegno Naz. Inquinamento dell'aria e Tecnche di Riduzione, Padova*, pp. 381-394.

Zilio-Grandi F. and L. Szpyrkowicz, 1996 : Progetto Sistema Lagunare Veneziano. Caratterizzazione delle precipitazion atmosferiche in localita Giare - Mira (Venezia) - Periodo Aprile, 1992 - Marzo, 1994, *Ingegneria Ambientale*.

Chapter 9

Conservation of the Desert Environment

Ishwar Prakash

The escalation of human and livestock population has greatly changed the ecology of the fragile ecosystem in the Thar desert. Overgrazing and cultivation on marginal lands are severe threats for the survival of typical deserticolous biota. Due to flood irrigation practised for crop raising, the extensive grasslands of an endemic grass, *Lasiurus sindicus* and a number of artiodactyles and carnivores are facing extinction in the 150 mm rainfall zone and these are being replaced by mesic elements. A number of new pests are emerging in the region. A series of suggestions have been made for restoring the desert ecology, conservation of biodiversity and to minimise the impact of human activities in the Thar environment.

INTRODUCTION

The Great Indian or the Thar desert is situated in northwest of India and is spread in four states: Punjab, Haryana, Rajasthan and Gujarat. On the total area (3,17,000 km^2), 61 per cent falls in Western Rajasthan. It is not a man-made desert. Its origin lies with geotectonic and climatic changes during the upheaval of the Himalayas or even earlier. It is contended that the arid conditions must have originated sometime between the end of Pliocene and before the end of the last glaciation (Vishnu-Mittre, 1977). The Palaeolithic man did exist in the desert and the Mesolithic man had settled over the sand dunes or the sandy alluvium thousands of years ago (Misra, 1971). Till recently man has been dwelling in the arid region in harmony with the environment and the over-exploitation of the natural resources started in the post-independent era which has accentuated the desertification processes.

The major purpose of this communication is to focus attention of the policy planners to the utilisation of the desert as a 'future land bank', conversion of vast grasslands into crop fields, myopic developmental activities, which are the causative processes for threatening the survival of xeric biodiversity; besides being apathetic to human health and welfare (Gupta and Prakash, 1975).

THE NATIVE DESERT SCENARIO

The scant and erratic monsoon manifested its utmost influence over the xeric ecosystem, its vegetation, wild denizens and human alike. With the first shower of the rainy season (July to September), the forbidden sea of sand assumes a gay cloak of the most magnificent greenery. The lush monsoon vegetation of the desert was constituted by a number of grasses, especially *Cenchrus ciliaris, C. setigerus* and vast grasslands of *Lasiurus sindicus* in the western Barmer, Jaisalmer and Bikaner districts. The shrubs dominated the landscape with few *Acacia* and *Prosopis* trees. The rangelands were inhabited by a large number of artiodactyles, the black buck, *Antilope cervicapara*; the Indian gazelle, *Gazella bennetti*. The desert fox, *Vulpes vulpes pusilla*, the desert cat, *Felis sylvestris* were the abundant carnivores. In the scrublands the caracal, *Felis caracal* was found. Among the birds, the great Indian bustard, *Ardeotis nigriceps* inhabited the more drier parts in large flocks : Patridge and sandgrouse occurred in innumerable numbers. The migratory birds visited the Thar in large flocks, the Imperial sandgrouse, *Pterocles orientalis*; the houbara, *Chlamydotis undulata* in the central and northern desert and the lesser florican, *Sypheotides indica* in the south-eastern grasslands (Prakash, 1975). Inspite of the fact that man did shoot them in large numbers but it was for sport and table and not for trade of fur and bones. The population of human beings was scattered all over in villages but their pressure on the land or on the natural resources was minimal till the pre-Independence era - their needs were rather limited. The desert dwellers thrived on animal husbandry, fetched drinking water from long distances, lived in hutments and appeared to be contended with the hard life.

THE CHANGING ECOSYSTEM

Impact of human activities

Soon after independence, the human impact on the desert ecosystem became gradually apparent mainly due to the exponential escalation of human population. In twelve desert districts of Rajasthan it increased from 5.5 million to 16.3 million in 1991 (Malhotra, 1977; Prakash, 195). As a consequence, an increase has been registered in the sown area for food crop production, even the marginal lands are being cultivated which are actually not suitable for crop raising. Resultantly, the area under natural rangelands decreased and the productivity per unit area also diminished (Mann *et al.*, 1977), besides increasing the magnitude of soil erosion and loss of soil fertility. The change in landuse pattern has a direct detrimental impact on the larger animals like ghe gazelle, black buck, a large number of carnivores and the great Indian bustard. The presence of human beings everywhere in the desert concerned the wildlife to remote areas where food and shelter are scarce. The desert is fast moving towards urbanisation. This process is being assisted by the tourist pressure on the land. For example, Jaisalmer was a less important town 40 years ago. The author has seen it as a typical desert town, no linkage with tar roads, railway line, very little electric supply, water had to be fetched from dug wells or rain water ponds. Cultivation was limited to rainfed crops like millet. Today it is a reverberating tourist centre connected by excellent roads with Barner, Jodhpur and Bikaner. The broadgauge railway line, the air service draw thousands of people into the heart of the desert. The discovery of fossil water of high quality in the Lathi series rocks and that from Indira Gandhi canal has changed the face of the town and nearabout area. Plantation of exotic trees, mustard and vegetable crops are raised for the benefit of the people. A large number of hotels, guest houses have emerged for satiating the increasing tourist requirements. Electric and water supply has been provided everywhere but the people grow with dissatisfaction and destruction of the environment is rampant. The mining of fossiliferous stone and lime has added to the glory drama. The ecology of region has totally changed. Four decades ago, partridges, sandgrouse (in tens of thousand), flocks of great Indian bustard, herds of gazelles could be seen within 15 km of Jaisalmer town, in the grasslands. At present one watches tourists or the wheat and mustard fields all around! The native endemic plants are being replaced by 'quick-growing' species and the typical desert animals are vanishing.

Irrigated agriculture

The Ground Water Board has been following an ambitious plan of tubewell boring throughout the desert. As a result, irrigated *rabi* crops are being raised everywhere over the highly permeable desert soil in as much as that the water table has sunk to a point of no return (Chatterji and Dubey, 1991). The mighty Indira Gandhi canal, traversing through whole of the western boundary of Rajasthan, upto Ramgarh, has ultimately irrigated 11 per cent of the desert area. In the Gangadhar and Hanumangarh districts, irrigated agriculture had started forty years ago through the Gang Canal. This region has totally transformed into a vast cropland and most parts of Phase I of IG canal are being flood irrigated for growing wheat, mustard, rice, cotton, groundnut, sugarcane and citrus. The venture will fulfil the present and future food needs of escalating human population but from the ecological point of view, the drastic change in landuse pattern and depletion of grasslands have deleterious impact on deserticolous biota which occurs only in this ecosystem and most of them nowhere else (Prakash, 1995). Water logging and rise in water table in the irrigated area have their own manifestations.

Threat to seven grassland

The highly adapted, perennial, tussocky, sewan grass (*Lasiurus sindicus*) is spread over tree-less vast expanse in 100-150 mm rainfall regions of Bikaner, Jaisalmer and Barmer districts in the region of IG canal command area. It has already vanished from Ganganagar and Hanumangarh districts due to irrigation from Gang Canal and from most of the area of IG Canal Phase I. Ultimately this grassland will be transformed into crop fields. *Lasiurus sindicus* is endemic to the Thar desert and unless we take up an extremely nutritive perennial species will be lost due to human intervention in the desert environment.

Diminishing wildlife

The rangelands in the IG canal region harbour specialised forms, adapted to extreme arid conditions. A fair abundance of the solfugid false spider, *Galeodes agilis* has been observed. Their large size is to be seen to be believed. The endemic reptiles of the desert (Sharma, 1996) like *Phrynocephalus laungwalensis, Stenodactylus orientalis, Cyrtodactylus kachhensis* and *Lytorhynchus paradoxus* thriving in dry soil find the perpetually moist soil unsuitable for their survival. The wildlife species like the caracal, *Felis caracal*; desert cat, *Felis sylvestris*; the desert fox, *Vulpes v. pusilla*; the gazelle, *Gazelle bennetti*; the great Indian bustard, *Ardeotis nigriceps* will either perish or migrate from IG canal zone to vulnerable areas due to change in landuse pattern. Moreover, the man and its livestock are everywhere in comparison to 4 humans per km^2 prior to incoming of IG canal water. The habitats of wildlife have been now occupied by man. It is immediately required to formulate an action plan to conserve the native faunal diversity of the arid zone.

Augmentation of weeds

The typical scrub vegetation represented by *Calligonum polygonoides, Leptadenia pyrotechnica, Haloxylon salicornicum, Aerva pseudotomentosa, Calotropis procera, Crotalaria burhia* are sacrificed during levelling of land for irrigation. Their place is taken over by non-desert species of aggressive weeds which are introduced due to transportation of soil, crop and impurities and due to various modes of dispersal. Chatterji and Saxena (1988) have listed 30 species of weeds which are new to this region. In the water-logged areas and depressional marshy lands created due to seepage of water from the canal are colonised by species new to the area. The important ones are *Desmostachya hipinnata, Arudo donax, Saccharum spontaneum, Typha angustata, Phragmites*, etc. (Saxena, 1991). The water hyacinth, *Eichhornia crassipes* has abundantly encroached over the full length of the canal. The ecological concern is that mesic plants are fast replacing deserticolous species and their survival in the IG canal area is severely threatened.

Emergence of new pests

Grassland insects are being replaced by crop pests. Vyas (1996) has carried out intensive study of the problem and has identified them in three categories: (a) minor insect pests assuming status of of major pests (15 species), (b) insects changing host plants (5 spp.) and (c) insect pests new to the region (17 spp.). Some of the insects are extremely injurious and losses inflicted by them reach an alarming level of 65 and 69 per cent of total produce (Vyas, 1996). It is also feared that the vast irrigated land will be ideally suitable for breeding of locust and mosquitoes.

Replacement of desert rodents especially gerbils (*Gerbillus* spp., *Tatera indica* and *Meriones hurrianae*) by mesic forms like the metad, *Millardia meltada* and the bandicoot, *Bandicota bengalensis* in the irrigated crop fields has been studies in detail (Prakash, 1978; Prakash and Mathur, 1979). The latter species are many times more injurious to standing crops and stored foodgrains due to their higher food requirement and hoarding propensity (Sridhara, 1992). The change-over of species has a serious implication as *Tatera indica* and *Meriones hurrianae* are fairly resistant to the plague baccilus, *Yersina pestis* but the invading metads and bandicoots are highly susceptible (Gratz, 1988). The non-specific carrier of the bacillus, the flea, *Xenopsylla chaeopsis* and *X. astia* keep on shifting from one rodent species to other and there is a grave possibility of plague epidemic emerging in human population through the house rat, *Rattus rattus* which comes in contact with the mesic elements.

Over-grazing

Inspite of the low productivity of the arid rangelands, the Thar Desert sustains a high population of livestock animals. Paradoxically with the decline in grazing area during the last three decades, the livestock number has registered a phenomenal increase, from 10.27 million in 1951 to 23.04 million in 1995. More than ten time animals are grazing in the desert which has hardly a carrying capacity of 6-8 heads per hectare. As a consequence of over-grazing, the perennial edible vegetation has been replaced by annuals as the perennial grasses are grazed in their early stages of growth and are unable to reach the seed formation stage. The impact of overgrazing is so intense that Gupta and Saxena (1972) reckon that the normal vegetation succession has been reversed. The abundance of livestock animals provide a severe competition to wild herbivorous birds and mammals resulting into decline of their numbers. On the other hand destructive small mammals - like rodents have taken to degraded scrubland and have increased to a very high density, 400-475 individuals per hectare (Prakash, 1969).

In nutshell the human activities in the desert region are playing a severely deleterious role which can be summarised as follows :

Factors	*Processes*	*Consequences*
1. Pressure of increasing human population	Cultivation of marginal lands	Decline in crop productivity, soil erosion, loss of soil fertility
	Increase in irrigation/cropping	Waterlogging, salinity problems, over-exploitation of groundwater, introduction of new weeds and pests
	Exploitation of woody biomass	Degradation of wood-lot, suppression of natural regeneration, diminution of wildlife
2. Pressure of increasing livestock population	Overgrazing	Degradation of vegetation resource, diminishing livestock production, livestock migration and inducemet of nomadism

After Mann and Prakash (1975)

CONSERVATION OF XERIC BIODIVERSITY

It has become extremely difficult to manage the environment in the present day setup of democracy. Inspite of the well formulated Wildlife Protection Act, even some of the sanctuaries and National Parks are under the control of unfriendly poaching, grazing and tree felling continues unabated. The foresters, scientists and wildlifers are silent on-lookers!, unable to stop these encroachments due to so many well-known reasons. The idea of involving the masses educating them, taking the environment protection to grass root level has fallen flat in practice before the needs and greed of the jungle dwellers and tradesmen. However, the recent growing concern about protecting the environment and conserving biodiversity has raised hopes that some hard decisions will be taken and they will be enforced in totality.

With this wishful thinking a few rather difficult suggestions are presented.

Cultivation of grasses

Due to erraticity of arrival of monsoon, the farmers plough their crop fields during May and June with a view to take advantage of even the first drop of rain water. The bullock driven ploughs have been replaced by tractors. Consequently, the soil is excavated upto 0.3 to 0.5 metre deep. Soon after the first shower millet is sown. The climatological data indicate that in 150-200 mm rainfall zone, the crops yield enough grains only in 1 year out of 5 years. In rest of the years, the yield is so low that the farmer does not get the cost of farming. The stems and leaves are utilised as fodder. In this process, during May and June the soil is blown away by the hot, high speed desert winds, most of the fertile top soil is lost, eroded, and new sand dunes are formed. Space imageries of the Earth show maximum assemblage of suspended dust over the Thar Desert. Often the stratosphere over this arid zone is called the largest 'dust bowl'. It has to be stopped if we are 'really' interested in stopping desertification. Ploughing of land should be totally banned in 150-200 mm rainfall zone by law and grasses like *Cenchrus ciliaris* and *Lasiurus sindicus* should be cultivated by farmers. Their water requirement is low as compared to millet, *moong* or *moth* bean. Another advantage would be that for grass cultivation, ploughing of the land and their seeding every year would not be necessary. The monetary gains would also be higher. This has been experimentally ascertained (Singh, 1991). If the traditional millet growing can be stopped, the potential of grass production is so high that fodder banks can be established throughout the desert to be utilised for livestock production in drought years. Moreover, it will minimise escalation of desert conditions and would stabilise the xeric environment.

Grazing policy

There is no grazing policy effective in the desert. Everyone is absolutely free to keep any number of livestock and to graze them anywhere and everywhere. As a consequence with the increase of animals in proportion of the carrying capacity of land, their productivity levels are going down (Kalla *et al.*, 1977). Moreover, greatest impact of depletion of vegetation and scarcity of forage is seen in seasonal migration of livestock population and inducement of nomadism. Years-long experiments carried out in more than 60 range management and livestock production paddocks by Central Arid Zone Research Institute have yielded sufficient knowledge to formulate a viable grazing policy which should be firmly executed in the desert environment. Likewise a serious thinking is to be given to check exploitation of underground water.

Wildlife conservation

Immediate attention is required for the protection of wildlife in the IG canal irrigated region. It has already been suggested (Prakash, 1994) that regions along the canal, unsuitable for irrigation, should be declared as closed areas so that the wildlife of the region could thrive in them. Translocation of animals from vulnerable areas of sanctuaries should also be taken up. In the desert region not affected by IG canal, serious efforts are to be made by protecting the habit. The Desert National Park is to be upgraded as the Biosphere Reserve (Anonymous, 1988). This proposal of Ministry of Environment and Forests is pending since a decade and its expeditious implementation an immediate requirement. A number of success stories are available from other countries pertaining to captive breeding and their re-introduction in the wild. The Jodhpur and Bikaner zoos should be shaped as captive breeding centres for threatened species and serious research work on tranquilisers, DNA finger printing, artificial insemination etc. should be started. We are losing ground. We have to take bold steps and we have to introduce 'science' in wildlife management (Prakash, 1994a).

REFERENCES

Anonymous, 1988 : *Thar Desert Biosphere Reserve,* Ministry of Environment and Forest, New Delhi, pp. 100.

Chatterji, P.C. and J.C. Dubey, 1991 : Geology and groundwater conditions of IGNP area. In : *Prospects of Indira Gandhi Canal Project.* ICAR, New Delhi, 45-54.

Chatterji, P.C. and S.K. Saxena, 1988 : Canal irrigation in arid zone. In : *Desert Ecology* (Ed. Ishwar Prakash). Scientific Publishers, Jodhpur, 223-258.

Gratz, N., 1988 : Rodents and human disease : A global appreciation. In : *Rodent Pest Management*. pp. 101-170, CRC Press, Boca Raton, USA.

Gupta, R.K. and Ishwar Prakash, 1975 : *Environmental Evaluation of the Thar Desert*, 484 p. English Book Depot, Dehradun.

Gupta, R.K. and S.K. Saexna, 1972 : Potential grassland types and their ecological succession in Rajasthan desert. *Ann. Arid Zone*, 11 : 198-218.

Kalla, J.C., P.K. Ghosh and B.R. Joshi, 1977 : Livestock productivity and desertification in the arid lands of western Rajasthan. *Ann. Arid Zone*, 16 : 360-366.

Malhotra, S.P., 1977 : Socio-demographic factors and nomadism in the arid zone. In : *Desertification and its Control*. ICAR, New Delhi, 310-323.

Mann, H.S. and Ishwar Prakash, 1975 : *Halting the March : Eco-development of the Thar*. Deptt. of Env. and WWF, India, New Delhi, 36 p.

Mann, H.S., S.P. Malhotra and K.A. Shankaranarayan, 1977 : Land and resource utilisation in the arid zone. In : *Desertification and its Control*. ICAR, New Delhi, 89-101.

Misra, V.N., 1971 : Two late mesolithic settlements in Rajasthan. *J. Poona Univ.*, 35 : 39-77.

Prakash, I., 1969 : Eco-toxicology of Indian desert gerbil, *Meriones hurricanae* Jerdon. V. Food preference in the field during monsoon. *J. Bombay Nat. Hist. Soc.*, 65 : 581-589.

Prakash, I., 1975 : Amazing life in the Indian desert. *Illustrated Weekly Annual*, 96-121.

Prakash, I., 1978 : Impact of changing landuse pattern on the rodent communities in the Indus basin. *Proc. Symp. Land and Water Mgmt. in the Indus Basin*, 2 : 481-486.

Prakash, I, 1994a : Biodiversity conservation in the Thar desert. *Indian For.*, 120 : 873-879.

Prakash, I., 1995 : The risk prone impacts of Indira Gandhi canal irrigation in the Thar desert. In : *Desert Resource Management and Development* (Ed. R. Joshi), Institute of Rajasthan Studies, Jaipur, India.

Prakash, I. and R.P. Mathur, 1979 : *Bandicota bengalensis* in Bikaner town, *Rodent Newsletter*, 3 : 12.

Saxena, S.K., 1991 : Impact of canal irrigation on the ecology of arid tract of Rajasthan. In : *Prospects of Indira Gandhi Canal Project*, ICAR, New Delhi, 65-73.

Sharma, R.C., 1996 : Herpetology of the Thar desert. In : *Faunal Diversity in the Thar desert : Gaps in Research*. (Eds. A.K. Ghosh, Q.H. Baqri and Ishwar Prakash), Scientific Publishers, Jodhpur, India, 297-306.

Singh, K.C., 1991 : Production of forage grasses with minimal irrigation in western Rajasthan. In : *Prospects of Indira Gandhi Canal Project*, ICAR, New Delhi, 82-86.

Sridhara, S., 1992 : Production losses due to rodents : Rice. In : *Rodents in Indian Agriculture* (Eds. Ishwar Prakash and P.K. Ghosh), Scientific Publishers, Jodhpur, India, 1 : 211-230.

Vishnu-Mittre, 1977 : Origin and history of the Rajasthan desert - Palaeobotanical evidence. In : *Desertification and its Control*. ICAR, New Delhi, 6-9.

Vyas, H.K., 1996 : Insect pests in Indira Gandhi canal region of the Thar desert. In : *Faunal Diversity in the Thar Desert : Gaps in Research* (Eds. A.K. Ghosh, Q.H. Baqri and Ishwar Prakash), Scientific Publishers, Jodhpur, India, 203-214.

Chapter 10

The Ganges River Dolphin

R.K. Sinha, *K. Prasad and Gopal Sharma

INTRODUCTION

The Ganges river dolphin, *Platanista gangetica* commonly known as 'susu', is a freshwater dolphin distributed throughout the Ganges-Brahmputra-Meghna river system in India, Bangladesh and Nepal (Jones, 1982; Reeves and Brownell, 1989). Cuvier (1836) claimed that it ascended the Ganges in great numbers to as far upstream as river is navigable. Anderson (1878) reported that even in the month of May, when the Ganges is very low, dolphins were present in the Yamuna river as far as Delhi. He also emphasised that the upstream range of this dolphin was apparently only limited by insufficiency of water and by rocky barriers. The last report of a dolphin at Delhi in the Yamuna river was in 1967 when a dead specimen caught in fisherman's net was brought to Delhi Zoo (Personal Communication by Dr. K.S. Sankhla, the then Director, Delhi Zoo). In September, 1994 susu in Ganges was sighted upstream at Nangal (80 km upstream Bijnor) (Pers. communication Raju Kumar). Smith *et al.* (1994) sighted 2-3 susus in the Karnali river in Nepal a few km upstream of the India-Nepal border. Now-a-days susus are mainly confined to the portion of the Ganges in Eastern Uttar Pradesh and Bihar state.

The World Conservation Union (IUCN) has classified *Platanista gangetica* as vulnerable (Klinowska 1991). Susus are found usually alone or in groups of 2-3 (Nath, 1974; Jones, 1982). Infrequent sightings of pairs are presumed to involve mothers and calves (Kasuya and Haque 1972; Haque *et al.*, 1977; Jones 1982). Pilleri (1970) observed schools of 5-10 animals in the Brahmputra river, and Reeves and Brownell (1989) summarized reports of larger aggregations.They have been reported to move from the main channel of a river into its tributaries usually during rainy seasons when the rivers are in spate. The dolphins are normally found downstream of shallow areas or tributary junctions (Kasuya and Haque, 1972). Smith (1993) observed susus in the Karnali river most often in "primary habitats" where stream convergence creates an eddy counter-current system in the mainstream flow.

Evolutionary adaptation by the dolphins to a turbid fluviatile environment has resulted in a regression of the eye (Herald *et al.*, 1969; Purves and Pilleri, 1974) and the development of a sophisticated echolocation system, with two highly directional acoustic fields (Purves and Pilleri, 1974; Pilleri *et al.*, 1976). In captivity susus continuously emit trains of high-frequency (15-150 KHZ) clicks, interrupted only by pauses of 1-60 sec (Herald *et al.*, 1969; Pilleri *et al.*, 1970, 1976); these pauses have been interpreted by Pilleri & Pilleri (1987); to be polyphasic sleep. The dolphins swim almost constantly, often on their sides. Shortly after initiating a dive, the dolphin spins 90° on its lateral axis and 180° on its longitudinal axis to swim on its side, in the direction opposite from the surfacing direction (Smith, 1993). During side swimming, the body is oriented head down at an angle of approximately 10° from the bottom. The head sweeps up and down in a scanning motion and the flipper trails along or slightly above the bottom. The head sweeps up and down in a scanning motion and the flipper trails along or slightly above the bottom (Herald *et al.*, 1969; Pilleri *et al.*, 1970). The flippers are thought to have an important tactile function (Pilleri, 1970, 1974; Pilleri *et al.*, 1976). This probably explains why the flippers almost feel the bottom to identify the habitat by the nature of its bottom.

The body of the Ganges river dolphin is fusiform. Female is larger than male. The maximum length of a female was recorded as 2.5 m whereas that of male only 2.33 m (Sinha, 1994). At the time of birth the length of cafs is about 70 cm. The pectoral flippers are more or less triangular in shape with a rudimentary dorsal fin and horizontally placed tail fluke. Gupta (1986) reported sighting records of 42-46 dolphins from Buxar to Rajmahal in Bihar, 7-9 in West Bengal and 5-6 in Uttar Pradesh. Ali (1992) surveyed the stretch from Buxar to Rajmahal in Bihar again during February 10-18, 1988, May 7-15, 1989 and August 6-14, 1989 in different segments covering a total area of only 50 km^2. He sighted 92 dolphins in 50 km^2 area of Ganga and based on his observation he estimated 2000 dolphins in Ganga in Bihar. The present study was undertaken to evaluate the current status, and conservation of susus in Ganga mainly in Bihar stretch.

It has also been tried to identify the threats to these dolphins and the effectiveness of conservation measures that have been taken to protect the species in the Ganges, especially in the 600 km segment of the river in Bihar was also considered. No systematic study on this animal has been undertaken in India since the classic work of Anderson (1878). The importance of knowing the current status of susus has grown in recent years as the population of the species appears to be dwindling very rapidly (Reeves *et al.*, 1993).

The work was undertaken as a part of the Ganga Action Plan which has a mandate to conserve the rare and endangered biota of the river Ganges. Funds were provided by the Ganga Project Directorate, Govt. of India. The research team consisted of five individuals assisted by local fishermen. Surveys were conducted by two teams of researchers, using country boats powered by sail or paddle. Each team was led by an experienced researcher. The two teams coordinated their activities fully during the survey.

STUDY AREA

River Ganga originates in the Himalayas at an altitude of 4100 m from Gaumukh glacier near Gangotri in India. After traversing the plains of Uttar Pradesh, Bihar and Bengal for about 2525 km, the river falls into the Bay of Bengal. In Bihar the Ganga receives several tributaries originating from the Himalayas in Nepal, such as the Ghaghra (in Nepal called Kamali), Gandak (Narayani in Nepal), Burhi Gandak and Kosi, the Sone from Central India, and several small rivers from floodplains of North Bihar, Nepal, and the plateau of South Bihar. The total area of the Ganga Basin is 861,404 km^2 which is more than 27% of the total area of India (31,66,828 km^2). The variation of the discharge volume between the lowest and highest flood seasons in the Ganga is more than 100 times near Patna. The annual average silt load in the river is 0.25 gm/litre at Patna. The catchment area of Ganga at Patna is 7,44,563 km^2 whereas at Farakkha 9,52,788 km^2. During the monsoon the width of the river can be upto 5-10 km; it is as deep as 40 m at certain places.

The study area extends from Buxar (N 25° 34.029' E 83°57,000') in Bihar to Farakka (N 24°47,979', E 87°54.756') in West Bengal, a stretch of about 600 km. This segment was selected because the maximum population of susus was expected in this stretch. Nine discrete portions of the 600 km were selected. These portions are representative of dolphin habitat and include the single mainstream of the river, shallow areas with sandbars, tributary junctions, reaches downstream of bridge pilings, areas polluted by both domestic and industrial effluents, and reaches both upstream and downstream of barrages. The study segments are Buxar (25 km), Patna (35 km), Mokama (10 km), Munger (10 km), Sultanganj (10 km), Kahalgaon (15 km), Rajmahal (10 km) and Farakkha (10 km) (Fig. 10.1). During 1991-92 the study was undertaken at Koilwar (confluence of Ghaghara, Sone and Ganga) whereas in 1992-93 the Kahalgaon segment was surveyed instead of Koilwar (10 km). These nine segments were surveyed during post-monsoon (October-November), winter (January-March) and summer (April-June) of 1991-92 and 1992-93.

At Buxar two small tributaries, the Karmanasa and the Thora, join the Ganga from the southern side. Two major tributaries, the Sone and the Ghaghra, drain into the Ganga from the south and the north respectively, about 40 km upstream of Patna. At Patna the Ganga receives the Gandak from the north. At Patna untreated domestic effluents flow into the Ganga. Just upstream of Mokama a small rivulet the Baya, joins the Ganga from the North.

Major industrial, both treated and untreated, discharges are made directly into the Ganga from the Barauni-Mokama industrial complex, which includes Bata tannery, McDowell distillery, Barauni Thermal Power station, Hindustan Fertilizer Corporation and Indian Oil Corporation Refinery.

Between Munger and Sultanganj, the river Burhi Gandak, and between Kahalgaon and Rajmahal, the river Kosi, join the Ganga from the north. At Farakka, there is a barrage (commissioned in 1974) and a feeder canal (38 km) to divert Ganga water into the River Bhagirathi which flows into Hooghli. Surveys at Farakkha were conducted in the river Ganga both upstream and downstream the barrage and in the feeder canal (3 km) near the barrage.

SURVEY METHODS

Because of the low number of dolphins in all the segments, and following the recommendations of a panel of experts (Perrin and Brownell, 1989), a direct count survey method was used to estimate dolphin abundance. When dolphins were sighted, was remained in the area for approximately 15 minutes before recording the count. Usually the number was greater near confluences and downstream of bridge pilings. At such sites we devoted at least one hour to observations in order to reduce the chances of counting a single animal more than once or of undercounting when more than one animal was present. At all the nine stations a map was drawn of the area, indicating peculiar geological formations, river courses, and positions of dolphin sightings. In every segment we surveyed both downstream and upstream. Water depths were measured at various points across the river bed where dolphins were sighted using plastic rope and a 20 kg weight. As the current of the river Ganges is usually sluggish in this zone, except during the monsoon, the depth measurement is considered to be reasonably accurate. The water samples of river Ganges were collected, preserved and analysed following Standard Methods (APHA, 1985). Whenever a dead *susu* became available from fishermen, the gut contents were examined (Sinha *et al.*, 1993). Tissue of the dolphins and any fishes collected from the stomachs of dead susus were analysed for heavy metals and organochlorines (Kannan *et al.*, 1993, 1994).

RESULTS AND DISCUSSION

Depending on location where the river flows in single channel and slightly downstream of a confluence of bridge where the velocity of water increases, eddy counter-currents are created. Such areas are preferred habitat of susus as more dolphins were encountered in these areas as compared to the areas where river flows in single channel without eddy-counter currents. The bottom substrate is mainly clay at Buxar, Kahalgaon, Rajmahal and Farakkha (upstream barrage) whereas it is fine silt at Patna Mokama, Munger, Sultanganj and Farakkha (downstream of the barrage). At Koilwar, the bed of the river is sandy. At Buxar, a large deposit of sand and fine silt has been formed just upstream of the confluence of the Thora river and the Ganges. At Patna, Mokama and Munger a relatively shallow, sloping point bar, composed of fine silt, deflects the main flow and creates an eddy-counter current system with a centre pool. At Kahalgaon a small hill jutting out from the right bank of the river creates a big eddy-counter current. At Rajmahal the river has maximum depth (40.5 m). At Farakkha upstream of the barrage a reservoir has formed, changing the river, from a lotic to a lentic environment. In the feeder canal, however, through which water is released, several small eddy-counter current systems are formed near the gates. Downstream barrage, in the main channel of the Ganges, the flow of water is low especially in lean season.

The physical and chemial riverine environment varies little between the segments so far the habitats of dolphins in river is concerned. A definite variation in the physico-chemical characteristics of river water was observed. The water temperature of the river varies between 18.2 and 34°C. Variation in other abiotic factors of dolphin habitat in Ganga has been presented in Table 10.1.

Maximum number of Ganges river dolphins were observed near confluences and downstream of bridges in eddy-counter current systems. The total counts of dolphins sighted in the nine discrete segments in the Ganga has been presented in Table 10.2.

Total and average number of susu during the three seasons in the total stretch surveyed has been shown in Table 10.3. The average density of susu was approximately 1.5 animal per km in 1991-92, whereas the average of two years of survey revealed the average density to be 1.088 per km. In many segments the river fans-out and flows through several channels. During the survey only main channel was covered; even then the total number of susu in 600 km stretch of Ganga can be expected to be $1.088 \times 600 = 652.8$ or 653. Thus even by any conservative method the total population of susu in Ganga including all the channels between Buxar and Farakkha can be estimated to be 700-1000. However, the distribution of susu in river is not uniform. Ali (1992) during his survey in 1988-89 estimated 2000 susu in Ganga between Buxar and Rajmahal.

Table 10.1. Habitat characteristics of Dolphins in the river Ganga between Buxar and Farakkha

S.No.	*Parameters*		*Range*
1.	Air temperature	(°C)	21.5-39.0
2.	Water temperature	(C)	18.2-34.0
3.	Transparency	(cm)	15.2-154.9
4.	Conductivity	(mmhos/cm)	138-429
5.	pH		8.2-9.2
6.	Dissolved Oxygen	(mg/l)	5.6-9.4
7.	Total alkalinity	(mg/l)	90-247
8.	Hardness	(mg/l)	84-189
9.	Chloride	(mg/l)	11.4-37.9
10.	Sulphate	(mg/l)	14.5-46.8
11.	Nitrate-N	(mg/l)	0.13-6.65
12.	Phosphate-P	(mg/l)	0.036-0.380
13.	BOD	(mg/l)	0.40-4.00
14.	COD	(mg/l)	3.4-28

During the survey of both years, maximum dolphins were sighted in post-monsoon season and minimum during the summer season (Table 10.2). Large aggregation of dolphins were observed in winter season especially at Patna and Rajmahal. In summer season, mostly the dolphins were alone or mother was accompanied by calf. Less sighting frequency of dolphins in summer suggests that susus probably get concentrated in suitable habitats during the unfavourable conditions, though the species is considered as non-gregarious. The controlling factors of such concentrations may be the ecological factors namely depth of water, availability of food, water current etc. With the onset of monsoon in July the river water becomes highly turbid. During this period maximum incidental killing of dolphins especially calves and juveniles were recorded at Patna. For about ten days one dolphin was reported to have been caught in a gill-net almost every day.

A temporal and spatial variation in dolphin population was observed during the two years of survey. During 1991-92 the number of dolphins observed were almost double as compared to 1992-93. Between Patna to Rajmahal the density was more than that of Buxar and Farakkha. Minimum number of dolphins were recorded at Farakkha, where the density of the animal varied between 0.1 in summer '93 to 1.5 dolphin per km in post-monsoon '91. Highly density (3.9/km) was recorded at Sultanganj in post-monsoon '91. Sultanganj is the starting point of recently established Vikramshila Gangetic Dolphin Sanctuary in 50 km stretch of the Ganga between Sultanganj and Kahalgaon.

In the stretch of river Ganga between Buxar and Farakkha (about 600 km) there are three bridges; one road bridge each at Buxar and Patna and one rail-cum-road bridge at Mokama. The occurrence of dolphin near the bridges shows large variations. Surprisingly, no dolphin could be recorded near the bridge at Buxar; whereas maximum number of dolphins (10-40) were encountered near the bridge at Patna. Similarly 2-5 dolphins were always sighted near Mokama bridge. It may be emphasised that about 500 m upstream Patna bridge, the Gandak river from north joins Ganga increasing current velocity as well as creating eddy counter-current systems. Due to increased current the availability of fishes is more at this point and hopefully this attracts dolphins near the confluence and bridge. Similarly at Mokama about 750 m upstream bridge, a small river Baya joins Ganga from north resulting in high current, formation of eddy counter-current systems, whereas at Buxar there is no confluence near the bridge. This may be the reason for absence of dolphin near Buxar bridge. Thus confluence is the preferred habitat of susus not the bridge.

Dams and barrages have long been present on rivers of the Indian sub-continent and they have had a major impact on dolphins (Reeves and Leatherwood, 1994). Chila barrage was constructed on the Ganges near the foothill of Himalayas between 1972 and 1976. However, there is no record of dolphins in and around the Chila barrage. In 1974 Farakkha barrage was commissioned. It isolated the dolphins upstream of the barrage. Though, both ways some genetic interchange is possibly during flood season when the gates of barrage are opened and the water level both upstream and downstream

Table 10.2. Census of the Ganges river dolphin, *Platanista gangetica* in nine discrete segments of river Ganga during different seasons in 1991-92 and 1992-93

S.No.	*Survey Stations*	*Year*	*Stretch of Ganga surveyed (km)*	*Dolphin Count*			*Density of Dolphin in Ganga (per km)*			*Maximum Depth of River Ganga (metre)*		
				Pm	*W*	*S*	*Pm*	*W*	*S*	*Pm*	*W*	*S*
1.	Buxar	1991-92	25	17	29	10	0.68	1.16	0.4	13.5	11.3	9.1
		1992-93	25	10	6	3	0.40	0.24	0.12	9.1	8.6	6.7
2.	Koilwar	1991-92	10	10	11	9	1.0	1.1	0.9	7.6	5.8	3.0
		1992-93	10	ND	ND	ND	ND	ND	ND	ND	ND	ND
3.	Patna	1991-92	35	53	85	45	1.5	2.43	1.29	12.1	10.2	9.1
		1992-93	35	30	30	14	0.86	0.86	0.40	16.8	14.2	12.2
4.	Mokama	1991-92	10	36	19	14	3.6	1.9	1.4	12.0	10.4	8.2
	1992-93	10	12	19	6	1.2	1.9	0.6	11.9	9.6	8.2	
5.	Munger	1991-92	10	27	20	16	2.7	2.0	1.6	23.0	21.3	16.1
		1992-93	10	16	10	4	1.6	1.0	0.4	15.2	13.1	9.4
6.	Sultanganj	1991-92	10	39	5	2	3.9	0.5	0.2	10.9	8.0	6.0
		1992-93	10	21	10	6	2.1	1.0	0.6	16.8	10.8	6.8
7.	Kahalgaon	1991-92	15	ND	ND	ND	ND	ND	ND	ND	ND	ND
		1992-93	15	10	3	8	0.67	0.2	0.53	13.7	10.4	8.3
8.	Rajmahal	1991-92	10	20	25	11	2.0	2.5	1.1	17.0	14.0	9.1
		1992-93	10	13	24	6	1.3	2.4	0.6	40.5	38.4	22.5
9.	Farakkha	1991-92	10	15	4	4	1.5	0.4	0.4	20.5	16.0	10.7
		1992-93	10	3	4	1	0.3	0.4	0.1	11.4	9.7	6.0

Pm - Post-monsoon (October-November), W - Winter (January-March), S - Summer (April-June), ND - Not done.

Table 10.3. Total and average number of dolphins sighted in Ganga in 1991-92 and 1992-93

Year	*No. of dolphins sighted*			*Average number of dolphins*	*Total distance surveyed*
	Pm	*W*	*S*		*Km*
1991-92	217	198	116	177	120
1992-93	115	106	48	89.66	125
Average	166	152	82	133.33	122.5

Pm - Post-monsoon (October-November), W - Winter (January-March) and S - Summer (April-June).

barrage is equal. However, it has not been confirmed if susus move upstream or downstream through barrage during the monsoon as the security personnel do not allow any observation close to the barrage. In 1974 Sarda Nagar barrage was commissioned on river Sarda, a tributary of river Ghaghra, Girijapuri. Girijapuri barrage was constructed in 1976 on river Ghaghra to feed the Sarda canal system. Both Gandak and Kosi have barrage at Indo-Nepal border namely Tribeni barrage (commissioned in 1968) and Birpur barrage (commissioned in 1965) respectively. On river Sone, Indrapuri barrage was commissioned in 1965. All these barrages have not only isolated the dolphin population genetically and made the susu vulnerable but also degraded the historical habitats and reduced the food availability as the migratory fishes cannot cross the barrage. At every barrage the lotic environment has changed into lentic environment. Heavy siltation has resulted and infestation of macrophytes has degraded the habitats of susu especially at Farakkha. Except upstream Girijapuri and Farakkha barrage the susu are likely to be extinct by the end of this century in upstream stretch of rivers of all other barrages in India and Nepal. It is a pretty pessimistic conclusion but hopefully right one.

Besides this, 1400 MLD (million litre per day) untreated domestic sewage from 27 class-I cities (population more than 1,00,000) and 73 cities with less population were discharged into the Ganga. 68 Gross Polluting Industries have been identified on the bank of Ganga which discharge 260 MLD untreated industrial effluents. However, under Ganga Action Plan of Govt. of India (started in 1985) more than half of the domestic sewage has been interrupted and diverted and being treated before discharged into the Ganga. All the industries have already set up Effluent Treatment Plants which is mandatory under the law. Thus untreated sewage and effluents are being taken care of.

Fishing activities with monofilament nylon Gill-nets are not only posing threat to dolphin population as the susu are incidentally caught but the availability of food of susu is also getting reduced due to intensive fishing operation using smaller mesh-size nets. The susus feed mainly on fishes of small size (Sinha *et al.*, 1993). Thus the river dolphins are directly competing with human for their food and habitat.

Conervation of the Ganges river dolphins

Although the susu have been supposedly protected under Wildlife (Protection) Act 1972, a lack of awareness has meant that they are still being killed deliberately as well as accidentally, mainly for their oil which is used in fish bait. Mass awareness campaign through mass media of communication, by arranging exhibitions, seminars/symposia etc. was organised to educate the common people including fishermen. Special lectures and exhibitions were arranged for schools and colleges as well as for volunteers and government officials. The efficacy of the Wildlife (Protection) Act 1972 was found to be almost nil. However, with vigorous awareness campaign the message of conservation of susu has percolated to almost all strata of the society including wildlife officials who are now active in implementing the acts to save this animal from extinction.

In 1991 a Vikramshila Dolphin Sanctuary between Sultanganj and Kahalgaon in a stretch of 50 km of the Ganges has been declared by the State Government of BIhar. The management plans have also been formulated for this sanctuary. Ecotourism, in the form of dolphin watching trips has also started, which will provide an important additional income for the fishermen.

A study has been undertaken to popularise the Sardine and Shark liver oil as an alternative of dolphin oil as fish lure. The fishermen are interested in and opting for these alternative sources of fish attractant. During the survey it was realized that local people, who depend on the river, knew very little about the dolphins sharing the river with them.

Intensive and extensive effortsare being made to educate fishermen about the susu's need for protection. It took two years to win over the trust and confidence of some fishermen and develop a close relationship which has proved invaluable in collecting information. Some fishermen have become strong supporters of the project and are now actively helping to spread the word about the need to conserve the dolphins. The fishermen realize that they share many of the problems faced by the dolphins. Pollution, construction of dams/barrages and use of certain types of net have reduced the fish population in the river.

A report for the restoration of fisheries productivity of the Ganga has been prepared in 1991-92 by Fisheries Coordination Committee of the Ganga Action Plan. The pollution abatement measures undertaken under the Ganga Action Plan will also improve the habitats of susu. Thus the conservation efforts are focussing both on the preservation of habitat and basin-wise resource management besides creating awareness among the fishermen.

The cultures of subsistence farmers and fishermen living within the river basin ecosystem are key components to the ecorestoration of Ganga and conservation of dolphin in the river. The socio-economic conditions of fishermen and dwindling riverine indicate the population pressure on the Ganga river system. Alternative strategies are clearly needed on all these fronts. The conservation of river dolphins in the Ganga river needs to be addressed as part of an overall strategy of conservation and development of this largest river ecosystem of India with the local people playing an integral role. Just like Project Tiger there is a need to have Project Dolphin not only to save dolphins but the entire Ganga river ecosystem.

SCOPE FOR FURTHER WORK

The Ganges river dolphin has been referred in the great Indian Epic, the Mahabharat (with origin of the Ganga) and was given legal protection about 2200 years ago under government decree by Ashoka the Great; this has been depicted by miniature painting in Babur Nama in 16th century but discovered into the scientific world of Roxburgh (1801). After the classic work of Anderson (1878), which mainly deals with taxonomy and biology of the animal, for the first time systematic study on status and conservation of the susu was undertaken as a part of the Ganga Action Plan. The study was mainly undertaken in Bihar, eastern part of U.P. and Bhagirathi and Hooghli in West Bengal. However, many areas remained uncovered and need thorough survey especially upstream barrages in all the tributaries of the Ganges. This will help in formulating conservation plan for such isolated and threatened smaller population of the animal.

REFERENCES

Ali, S.M., 1992 : The Gangetic dolphin. *MYFOREST*, 28 : 245-250.

American Public Health Association, 1985 : *Standard Methods for the Examination of Water and Waste Water*. APHA, AWWA, WPCF, 16th Edn., New York.

Anderson, J., 1878 : Anatomical and Zoological researches : Comprising an account of zoological results of the two expeditions to western Yunnan in 1868 and 1875; and a monograph of the two cetacean genera *Platanista* and *Orcella*. *B. Quaritch*, London, two volumes.

Cuvier, F., 1836 : Del'histoire naturelle des Cetaces on recoeil et examen desfaits dont se compose. Roret, Paris, 252 p.

Gupta, P.D., 1986 : The Gangetic dolphin, *Platanista gangetica* (Lebeck, 1801). In : "*Wildlife Wealth of India* (Resources and Managements) (Ed. T.C. Majpuria), 553-562." Teepress Service, L.P. Bangkok.

Haque, A.K.M. Aminul, M. Nishiwaki, T. Kasuya and T. Tobayama, 1977 : Observations on the behaviour and other biological aspects of the Ganges susu, *Plantanista gangetica*. *Sci. Rep. Whales Res. Inst.* 29 : 87-94.

Herald, E.S., Jr. R.L. Brownell, F.L. Frye, E.J. Morries, W.E. Evans and A.B. Scott, 1969 : Blind river dolphins : first side-swimming cetacean. *Science*, 166 : 1408-1410.

Jones, S., 1982 : The present status of the Gangetic susu, *Platanista gangetica* (Roxburgh), with comments on the Indus susu, *P. minor* Owen. FAO Advisory Committee on marine Resources Research Working Party on Marine Mammals. *FAO Fish Ser.*, 5 : 97-115.

Kannan, K., R.K. Sinha, S. Tanabe, H. Ichihasi and R. Tatsukawa, 1993 : Heavy metals and organochlorine residues in Ganges river dolphins from India. *Mar. Poll. Bull.* 26 : 159-162.

Kannan, K., S.,Tanabe, R. Tatsukawa and R.K. Sinha, 1994 : Biodegradation capacity and residue pattern of organochlorines in Ganges river dolphins from India. *Toxicol. and Environ. Chem.* 42 : 249-261.

Kasuya, T. and A.K.M. Aminul Haque, 1972 : Some informations on distribution and seasonal movement of the Ganges dolphin. *Sci. Rep. Whales Inst.*, 24 : 109-115.

Klinowska, M., 1991 : "Dolphins, Porpoises and Whales of the World". *The IUCN Red Data Book* IUCN, Gland, Switzerland and Cambridge, UK.

Nath, B., 1974 : On some aspects of habit and habitat of the Gangetic dolphin *Platanista gangetica* (Lebeck) in the river Ganges at Patna. *Naturalist* (*Bulletin of the Bihar Natural History Society*, Patna, India), 1 : 6-7.

Perrin, W.F. and Jr. R.L. Brownell, 1989 : Report of the workshop. In : *Biology and Conservation of the River dolphins*, (Eds. W.F. Perrin, Jr. R.L. Brownell, Zhou Kaiya and Liu Jiankang) Occ. Pap. IUCN Species Survival Commn., 3 : 1-21.

Pilleri, G., 1970 : Observations on the behaviour of *Platanista gangetica* in the Indus and Brahmputra rivers. *Invest. Cetacea* 2 : 27-60.

Pilleri, G., M. Gihr and C. Kraus, 1970 : Feeding behaviour of the Gangetic dolphin, *Platanista gangetica* in captivity. *Invest. Cetacea*, 2 : 69-73.

Pilleri, G., 1974 : Side swimming, vision and sense of touch in *Platanista indi* (Cetacea, Platanistidae). *Experientia*, 30 : 100-104.

Pilleri, G.M. Gihr, P.E. Purves, K. Zbinden and C. Kraus, 1976 : On the behaviour, bioacoustics and functional morphology of the Indus River dolphin *Platanista indi* (Blyth, 1859). *Invest. Cetacea*, 6 : 11-141.

Pilleri, G. and O. Pilleri, 1987 : Indus and Ganges river dolphins *Platanista indi* (Blyth, 1859) and *Platanista gangetica* (Roxburgh, 1801). *Invest. Cetacea*, 20 : 2-33.

Purves, P.E. and G. Pilleri, 1974 : Observations on the ear, nose, throat and eye of *Platanista indi*. *Invest. Cetacea*, 5 : 13-57.

Reeves, R.R. and Jr. R.L. Brownell, 1989 : Susu *Platanista gangetica* (Roxburgh, 1801) and *Platanista minor* (Owen, 1853). In : "*Handbook of Marine Mammals*" IV. *River Dolphins and the Larger Toothed Whales*". (Eds. S.H. Ridgway and S.R. Harrison), Academic Press London.

Reeves, R.R., S. Leatherwood and R.S. Lal Mohan, 1993 : Report from a *Seminar on the Conservation of River Dolphins of the Indian sub-continent*, 18-19 August, 1992. New Delhi, India. Whale and Dolphin Conservation Society, Bath, Avon, UK.

Reeves, R.R. and S. Leatherwood, 1994 : Dams and River Dolphins : Can They co-exist? *Ambio*, 23 : 172-175.

Roxburgh, W., 1801 : An account of a new species of Delphinus, an inhabitant of the Ganges. *Asiatick. Res.*, (Calcutta), 7 : 170-174.

Sinha, R.K., N.K. Das, N.K. Singh, G. Sharma and S.N. Ahsan, 1993 : Gut-content of the Gangetic dolphin *Platanista gangetica. Invest. Cetacea*, 24 : 317-321.

Sinha, R.K., 1994 : Bioconservation of the Gangetic dolphin *Platanista gangetica. no. J-3901/5/91-GPD. Technical report* (April 1991 - March 1994). Zoology Department, Patna University, Patna, India.

Smith, B.D., 1993 : 1990 Status and Conservation of the Ganges river dolphin *Platanista gangetica* in the Karnali River, Nepal. *Biol. Conserv.*, 66 : 159-169.

Smith, B.D., R.K. Sinha, U. Regmi and K. Sapkota, 1994 : Status of Ganges river dolphins *Platanista gangetica* in the Mahakali, Karnali, Narayani and Saptakosi rivers in Nepal and India. *Marine Mammal Science*, 10 : 368-375.

Chapter 11 Environmental Implications of the New Pest Management Concepts : Trends and Policies in the New Century

R.N. Sharma

Radical changes in concepts and trends in Pest Management must necessarily influence future policies. Mitigation of environmental pollution by synthetic organic insecticides by resort to IPM in continuing improvements of latter, forms the bulwark of modern trends in the area. Different components of IPN, including most recent and futuristic ones are described. The search for ecofriendly alternatives is emphasized, and current developments, as well as future projects are outlined. Universal stress on preservation of environmental quality by increasing incorporating of biorational and ecocompatible agrochemicals in holistic systems of management is highlighted. Shape of future policies and programmes is discussed.

INTRODUCTION

Insect pests and vectors have commanded an important segment of human effort and ingenuity ever since the dawn of agriculture (McEvan, 1978) more than 10,000 years ago. Advances in science and technology in the present century laid the formulations of efficient pest control ushering in the green revolution. However, economically serious inroads were made on the planetary environment by the synthetic organic insecticides, once considered as a panacea for all pest problems. Concerns for environmental quality have led to quantum changes in the overall scenario of pest control concepts and practices in especially latter half of the 20th century. It is inevitable that one may except logical termination and fruition of these radical changes in the new millennium. The present essay recapitulates the extent pest control scenario and introduces newer concepts, policies and projections which may form the 21st century vistas of universal pest management (Japan Pesticide Information, 1987).

CONTROL VERSUS MANAGEMENT

The first major change in perception has been the abandonment of the policy of total extermination of pest species. Such complete elimination of a biological entity is not only scientifically incorrect but it can be brought about mainly by non-specific bio-poisons such as the broad spectrum conventional synthetic insecticides. Consequences of continued, widespread and often indiscriminate use of the latter are quite well understood. Damage to the ecosystem by indiscriminate destruction of not only the pest but other beneficial predator/parasite species, induction of resistance in target pest species, resurgence of minor pest complexes etc., defeat the primary purpose of 'control'. In addition, residues, often fairly long lasting of chemicals etc., defeat the primary purpose of 'control'. In addition, residues, often fairly long lasting, of chemicals of high hazard to life and environment in general, have all militated against the conventional synthetics (WHO, 1972). Concomitantly, the concept of pest management (Metcalf and Luckman, 1975), incorporating reduction in pest populations below economic injury thresholds also emerged (Beirne, 1969 and Sharma, 1983).

OVERT HAZARDS AND COVERT EFFECTS

Many conventional synthetic insecticides still in global use today are known to exhibit high acute toxicity, as well as long persistence in many cases. Some are known to be carcinogenic as well. Even more disturbing is not so well known aspects viz. covert/chronic effects of sublethal residues of some insecticides on not only insects but a wide gamut of non-target species. Although not actually pinpointed, it may be valid to assume that many such toxicants exercise not so readily apparent, short or long term deleterious effects on humans also. Non-specific ill health or disease may well include such properties of conventional pesticides in aetiology.

The natural consequence of the foregoing has been an intensive search for alternative products and systems which would reduce or eliminate the use of such hazardous chemicals.

INTEGRATED PEST MANAGEMENT (IPM)

The philosophy of combining conventional chemical control with other non-insecticidal strategies, devices and products gave rise to now well accepted doctrine of integrated pest management. IPM seeks to reduce, as far as possible use of highly toxic, broad spectrum and non-specific synthetic organic insecticides by resorting to methods such as cultural control, enlightened field practices, encouragement of alternative strategies such as biological control and use of botanical and microbial pesticides (Duke, 1986; Duke and Lydon, 1987; Cutler, 1988; and Duke, 1990). Scientific progress in this field has also yielded its crop of non-hazardous synthetics such as the insect hormones and the pheromones. These also have become an integral part of new IPM protocols wherever these are sought to be applied.

NEWER FRONTIERS

Scientific

The most exciting scientific developments in the field of pest management have come from the area of insect-plant interaction. Plant secondary metabolites (Beck, 1965; and Kennedy, 1972), the allelochemics, have been recognised as the primary framwork of plant resistance to insect herbivory (Sharma, 1993), obviously evolved over aeons of insect plant co-evolution. To data, a multitude of such allelochemics have been identified (Whittekar, 1970). Efforts are now underway to use them in numerous ingenious strategies for managing pest insect populations. Thus at the primary conventional level, formulations of such principles alone or in synergistic combinations are being attempted.

Technological

Modern biotechnology is also seeking to use allelochemics or other pest anti-pest principles by incorporating them in cloned crops. The most prominent example of this has been the *Bacillus thuringiensis* (Bt) and its endotoxin which has been used to produce transgenic plants (Vaeck *et al.*, 1988) showing high resistance to depredatory pests such as *Spodoptera* and *Helicoverpa* sps. Production of newer transgenics with the more potent allelochemic principles is obviously a very innovative and hopefully bioefficient step.

The bio-pesticides

This term is used here as an inclusive one for both botanical and microbial pest control agents (Perlak *et al.*, 1988). Whereas the Bt and Nuclear Polyhedrosis Virus (NPV) have gained some measure of acceptance and utilisation, by and large focus has been on botanicals, or plant based/derived principles/ formulations. The most prominent among the botanicals is, of course, the Neem. From the mid-century, discovery of antifeedant property of Neem extracts against locusts (Pradhan *et al.*, 1962), led to identification of its principal constituent allelochemic, Azadirachtin, Neem has been steadfast as the first among the plant products. Sharma (1983) had cited abundance, cost, technical feasibility and broad spectrum, acceptable bioefficacy as some of the basic requirements for plant products to succeed as commercially viable Pest Control Agents. Neem meets many of these, and therefore continues to rule the roost. However, vigorous attempts are being made to add to this desirable arsenal.

CURRENT TRENDS

In summing up, it is evident that, despite present widespread use, the conventional synthetic organic insecticides are being phased out. The chlorinated hydrocarbons - DDT, Aldrin, Dieldrin, Chlordane, Heptachlor etc. are already completely banned in many countries (Edwards and Mill, 1986), and only restricted use of some is allowed in highly specialised situations e.g. DDT for Malaria Control; Heptachlor for building termite control etc. In forseeable future, this trend will continue as more effective and much less persistent molecules such as the OPs ,and the pyrethroids take their place. The accent now is on IPM, the use of biological control, as well as the newer "biorational" synthetics such as the IGRs, and the botanicals or microbials where possible, along with, or even in place of the synthetics (Menn and Henrick, 1981).

FUTURE SCENARIO

Among the conventional synthetics, the most innocuous pyrethroids, followed by the relatively milder OPs may still continue to be used, albeit with extreme circumspection. Controlled/slow release formulations or dispensors for still safer and more effective application are also likely to become more popular. It is conceivable that still more safer molecules than the pyrethroids are synthesised, in which case they will definitely take over. Although Juvenile Hormone Analogues have failed to find universal use due to several limitations including high costs, some other insect growth regulators notably anti-ecdysials e.g. Dimilin are being used in wider areas of application including agriculture, forestry, public health and horticulture (Mulla *et al.*, 1975). Many groups are also fast coming up with Diflurobenzuron like molecules without the limitations of the latter. These newer IGR's are active on body contact and do not need to be ingested. It is obvious that these newer molecules will replace the older ones. Attempts are also being made to find more and novel synthetic analogues with anti-juvenile and other developmental inhibition activities (Ho Chan Mei *et al.*, 1990; Pawar *et al.*, 1995 and Sawaikar *et al.*, 1995). The advantage of all these lies in their high specificity of action, and absence of significant deleterious effect on other non-target species. The other class of biorational pest control agents are the botanicals. Despite a host of neem formulations still lie in the future. Synthesis of the azadrachtin molecule, or even pinpointing still more active sites on it will again be an endeavour for the future. There are several other principles in neem itself which may lend themselves to exploitation. Likewise, several allelochemics in many plants have been shown to possess significant bioactivity singly or in combination, these are bound to emerge as viable pest control agents of the future. Transgenic plants incorporating such principles, are of course going to increase and many possibly become a major component of the holistic IPM of the future.

A relatively new trend which is slowly picking up and may well become a major ingenious strategy of futuristic pest management is the deployment of tritrophic principles for enhancement of conventional biological control efficiency. Thus the utilization of parasite/predator attractive chemicals can increase parasitization or predation. For this intensive studies of tritrophic interrelationships and the chemical influences governing this are already being pursued vigorously in many parts of the world. The use of lures, baits, attractants, sex or aggregation pheromones in combination with suitable trapping systems is also emerging as a management strategy in addition to its established monitoring potential (Vartak *et al.*, 1944a and b; and Vartak *et al.*, 1995).

PROBLEMS, PITFALLS AND SOLUTIONS

The newer ventures, ideas and concepts which are mushrooming in the field of modern pest management are not without their individual handicaps. Thus, with an average 10,000 alien molecules being synthesized and introduced into the planetary environment, a ceiling, especially on the overtly toxic ones may become mandatory. Like the CFGs, the synthetic organic insecticides, whatever their technical worth, may have to be phased out due to environmental concepts becoming more and more rigorous.

Notwithstanding optimism to the contrary, insects have been able to maintain their one-up stance on mankind by coming up with ingenious defences against the arsenal of man. An example is resistance against their own hormones - the JHAs. The microbials and botanicals including Neem, have at best a record of mild and moderate bioefficacy, and seem to be good supplements or complements in well designed IPM protocols, rather than pest control agents in their own, individual capacities. This limitation, though perhaps unfair in comparison to the broad spectrum, non-specific conventional synthetics, plagues most biopesticides. User education and acceptance, generous government support and subsidy will be essential to overcome these intrinsic handicaps of the biopesticides. The saving grace seems to be the generallyincreasing awareness and resistance of both farmers and the public against perils and evils of continued conventional pesticide use. Traditional cultural management by development and introduction of resistant cultivars, or the modern biotechnological equivalent of developing transgenic plants are both high in time and cost. Unfortunately, despite attractive claims of dramatic efficacy, their advantages may well be short lived due to time honoured insect adaptive genius. The evidence is also coming in of unexpected problems such as human allergy to the introduced genomes' phenotypic expressions.

CONCLUSION

Radical changes tantamount to revolution are apparently going to dominate the Pest Management Scenario in the coming years. The overall trend of a conscious struggle against further/continuing deterioration of environmental quality is inescapable.

Modern and futuristic pest management policies will therefore veer towards mitigation of environmental and human health hazards. IPM will obviously take on newer dimensions of non-hazardous, non-insecticidal, alternative products and protocols. It is difficult to envisage a complete cessation of conventional synthetic organic insecticides, but the middle of the next century may well usher in an era of largely non-hazardous and eminently eco-compatible pest management strategies, designed to reduce and eventually negate the planetary overload of toxic residues accumulated in the 20th century. Futuristic Pest Management is likely to be a more user and environment friendly system of pragmatic strategies and products, in which the scientist, technologist, manufacturer, farmer and the citizen are all likely to be intimately involved with obvious advantages to all.

REFERENCES

Beck, S.D., 1965 : Resistance of plants to insects. *Ann. Rev. Ent.*, 10 : 207-232.

Beirne, B.P., 1969 : *Pest management.* Leonard Hill Books.

Cutler, H.G., 1988 : Natural products and their potential in agriculture. *Am. Chem. Soc. Sym. Ser.*, 380 : 1-22.

Duke, S.O., 1986 : Naturally occurring chemical compounds as herbicides. *Rev. Weed Sci.*, 2 : 15-44.

Duke, S.O. and J. Lydon, 1987 : Herbicides from natural compounds. *Weed Technol.*, 1 : 122-128.

Duke, S.O., 1990 : Natural pesticides from plants. In : *New crops* (Eds. J. Janick and E. Simon). Timber Press, Portland, 511-517.

Edwards, R. and A.F. Mill, 1986 : *Termite in Building.* Rentokil Ltd., East. Grinstead, U.K.

Expertise on pesticides in the twenty first century 1987 : *Japan Pesticide Information*, 50 : 1-19.

Ho Chan Mei, Wu Shy Hueg and Wu Chinchen, 1990 : Evaluation of the control of mosquitoes with IGRs. *K. GoL' Siung IH Sureh K'o H. Such Tsa Chin.*, 366-377.

Kennedy, J.S., 1972 : Host plant relationships. In : *Reading in Entomology* (Eds. P. Barbosa and T.M. Peters), W.B. Saunders Co., Philadelphia, London, Toronto, 140-144.

Mc Even, F., 1978 : Food production, the challenge for pesticides. *Bioscience*, 28 : 773-777.

Menn, J.J. and C.A. Henrick, 1981 : Rational and biorational design of pesticides. *Phil. Transaction of the Royal Soc.* London. B. 259 : 57-71.

Metcalf, R.L. and W.H. Luckman, 1975 : *Introduction to Insect Pest Management.* Wiley-Inter Science, Publ., New York, YSA.

Mulla, M.S., G. Majori and H.A. Derwazch, 1975 : Effects of the insect growth regulators Dimilin or TH 6040 on mosquito and some non-target organisms. *Mosq. News*, 35 : 211-216.

Pawar, P.V., S.P. Pisale and R.N. Sharma, 1995 : Effect of some new insect growth regulators on metamorphosis and reproduction of *Aedes aegypti, Indian J. Med. Res.*, 101 : 13-18.

Perlak, I.J., M.G. Obrkowicz, L.S. Wartrud and R.J. Kangman, 1988 : Development of Genetically engineered microbial biocontrol agents. *Ibid* : 284-296.

Pradhan, S., M.G. Jotwani and B.K. Rai, 1962 : The neem seed deterrent to locust. *Indian Fmg.*, 12 : 7-11.

Sawaikar, D.D., B. Sinha, G.D. Hebbalkar, R.N. Sharma and S.A. Patwardhan, 1995 : Products active on mosquitoes. Part VII Synthesis and biological activity of longifolene derivatives. *Ind. J. Chem.*, 34B : 832-835.

Sharma, R.N., 1983 : Development of pest control agents from plants. A comprehensive working strategy. In : *Natural pesticides from the neem and the other tropical plants.* (Eds. H. Schumutterer and K.R.S. Ascher), G.T.2, FRG : 551-563.

Sharma, R.N., 1993 : Pest resistant crop varieties : A case for reconsideration. *Curr. Sci.*, 64 : 550.

Vaeck, M., A. Reynaerts, H. Hayte and H.V. Mellaert, 1988 : Transgenic crop varieties resistant to insects. In : *Biotechnology for crop protection* ACS symposium series. 379 : 280-283.

Vartak, P.H., V.B. Tungikar and R.N. Sharma, 1994a : Advise for monitoring of mosquitoes by behaviour. *Ind. J. Exp. Biol.*, 32 : 662-664.

Vartak, P.H., V.B. Tungikar and R.N. Sharma, 1994b : Comparative repellent properties of certain chemical against mosquito, houseflies and cockroaches using modified techniques. *J. Com. Dis.*, 26 : 156-160.

Vartak, P.H., V.B. Tungikar and R.N. Sharma, 1995 : Laboratory evaluation of an ovipositional trap for *Aedes aegypti J. Com. Dis.*, 27 : 32-35.

Whittekar, R.H., 1970 : The Biochemical ecology of higher plants. In : *Chemical ecology* (Eds. Sonelhimer and Simeane) Academic Press, New York, 43-70.

WHO/FAO, 1972 : *Evaluation of some pesticide residues in food.* FAO/AGP/1971/M/9/1, WHO Pesticide residues series no. 1.

Chapter 12

Sustainable Development : Concept and Achievement in India

G.S. Roonwal

Our future is linked to controlling environmental damage, increasing energy and food resources and stabilizing population. This is necessary to give the people of India the quality of life that the founders of the modern India envisaged.

INTRODUCTION

The Brundtland report of the World Commission on Environment and Development, 1987, titled "Our Common Future" is significant because it is accepted as the new order for sustainable development. This report now forms the basis to all our policy makers at the government and non-government level. The report is based on the idea that poverty leads to environmental degradation. Mrs. Indira Gandhi once said "Poverty is the worst form of pollution". She said this in particular context to Indian cities and in general to the developing world. But the prosperity of the industrialized countries produce so much waste in air, water and soil that "environmental protection" has mostly to do with cleaning "environmental dirt" (Seibold, 1993). News such as from Bangalore where enterprising engineers are trying out methods to convert garbage into manure and energy produce are encouraging innovations. Concept of sustainable development has been accepted in a number of countries such as New Zealand, the Netherlands, the UK and Canada (Glasby, 1995).

THREAT TO MANKIND - WORLD SCENE

There are three major threats to mankind (Brown, 1981) as a result of our present day unsustainable development : (a) the erosion of soil, (b) the deterioration of ecological system, and (c) rapid depletion of oil resources. Each of them adversely effect food production (Glasby, 1995). Poverty, population growth and environmental destruction are various connected aspects of the dangerous circle in which we live. It is worth considering that since 1990, the world population has increased by three fold. In the same period the world economy has increased by 20 times. The petroleum consumption has increased by 30 times, and the industrial production has increased by 50 times (MacNeill, 1989).

According to the World Resource Institute (1990), and MacNeil (1989), the economic activity is required to be increased in the order 5 to 10 fold to meet the need of the increasing population. The economic growth has been the most important consideration in the Bruntland Report in which environment has been considered as an important factor.

INDIA : QUALITY OF LIFE

The focal theme of the 80th session of the Indian Science Congress held in Goa, under the general presidentship of Professor S.Z. Qasim was "Science and Quality of Life". This session addressed several questions on sustainable development for the country. India with her size and diversity is unique. This has been deliberated by Glasby and Roonwal (1994). Substantial improvement in human well-being has been noted since 1947 in several ways. Life expectancy has increased from 32 to 58 per cent, literacy rate from 17 to 57 per cent, per capita income has doubled, number of hospitals have increased four folds. But the population increase from 362 to 860 million has almost marginalized our achievements. Today nearly 28 per cent of the people still live in poverty, which comprise 80 per cent in the rural areas and 18 per cent in the urban areas. The population figures are of interest because while the birth-rate has decreased from 40 to 28.5 per thousand of population since 1951, the death rate has decreased from 27 to 9.2 per thousand of the population. It is this lowering of the death rate which has also significantly contributed to the 233 per cent increase in population (Glasby and Roonwal, 1994).

The various contributions in the book "Science and Quality of Life" (Qasim, 1994) bring out that there are about 51 million people who still live in slums, 2 million are homeless, 35 million mentally ill - alcoholics and, or drug addicts. In terms of international comparison, India has a GDP of US $ 350 which is 8 per cent of the world's average of US $ 4,200. Thus we already come in the category of low income economies.

URBAN MIGRATION AND FEAR OF NATURAL DISASTER

An equally great challenge of the late 20th century is the urban migration of the population. The cities world over are growing. But in the developing world this phenomena is very strong. In India, the growth of the cities has made such rapid an expansion that slums are now so common a sight in all open spaces in most cities. Delhi has grown from an administrative capital 50 years back to a city of 11 million. It was 8.8 million in 1990. Mumbai (Bombay), with a population of 11.2 million and Calcutta with a population of 11.8 million figure in the list of large world cities. Projected population of Delhi, Mumbai, Calcutta and Bangalore in the year 2000 are 13.2, 15.4, 15.7 and 8.0 million respectively (UNFPA, 1993). Such a large concentration of population always has dangers of unsurmountable damage and human misery in the event of natural disasters. Think of the earthquake which took place in Latur in Osmanabad District in Maharashtra in 1993. It was indeed not a relatively strong earthquake, nowhere compared to the ones happening in Japan. But the human loss of more than 10,000 lives was very large. This happened because of poor awareness of dangers of natural calamity, as also general ignorance and casual attitude in allowing dwelling to be built with heavy stones on roofs. Each year the cyclones along Orissa, Andhra Pradesh and Tamil Nadu coast inflict human misery. Likewise the human, crop and property damage due to floods following the monsoon rains are familiar to all of us.

In a small way, we are carrying out a seismological-geological-tectonic study of Delhi and the surrounding area.The seismic activity of magnitude 2 is recorded almost periodically in the Delhi region. Regular monitoring of seismicity has helped in preparing of zonation maps. Such information shall be of much use in planning further growth of the Delhi region, and the designs of civil engineering projects to meet the needs. Similar studies would be necessary to other fast growing cities as well. Full adherence to the engineering norms have to be adopted to prevent or minimize the unfortunate and unpredicted seismic events of high magnitude. Here the role of an earth scientist becomes significant because we are expected to foresee dangers of damages through natural disasters such as earthquakes, landslides, floods, but also space, time and extent. We are expected to act or produce remedial measures or even prevention, at least to keep the damage to a minimum. While these new challenges must be faced, earth scientists are required to continue to hold responsibility for supplying new materials to the growing need of all industries - steel, aluminium base metal, nuclear energy, building stone for habitation and construction materials. A continuous search for new hydrocarbon deposits is a great challenge.

WATER SUPPLY

The oceans cover nearly 70 per cent of the globe, but sweet water suitable for drinking, irrigation for crops, washing and industrial use is limited. India may be considered one of the wettest countries in the world and thus has enormous water resources because of the monsoons. But the uneven spread both seasonally and geographically in the country results in large surface run-off and problems of both drought and floods. India today remains an inefficient user of water. To illustrate this point take the case of national capital Delhi where 2.1×10^9 litres of water per day are produced for use (Glasby and Roonwal, 1994). This is the largest per capita water availability in any major Indian city. Yet this is insufficient for the needs of the city. Most residential areas in Delhi receive 2 to 3 hours of water supply per day. Life for the slum people is unimaginable. But where we need to improve is the fact that of this filtered and treated water, about 15-20 per cent is lost due to leakage, taps left open etc. Of our 150,000 villages the per capita water availability is less then 40 litres per day. Water management needs immediate attention. Contamination of water including groundwater is an equally alarming fact of the Indian life because this leads to several water borne diseases, throughout the year. The data presented at the UN Water Resource Conference in 1979, led the UN to declare the 1980's as the International Drinking Water Supply and Sanitation Decade. After more than a decade, we, the developing nations of the world, are closer than ever to a water crisis. For millions of our people in Rajasthan, Gujarat and elsewhere, the long distance track for a pot of water continues. The groundwater table in India is falling low each year at an alarming rate of almost 1 metre a year in some areas. Attempts really need to be directed to integrate environmental concern and basic needs of local communities, because each eco-system will have different needs.

The problem of the domestic sewage is also great. Mumbai discharges 365 million tonnes into the Arabian sea, and Calcutta 396 million tonnes sewage annually into the Hooghly river respectively. The Ganga Action Plan may do some good to the Kanpur area. Upgraded sewage treatment facilities can then reduce water pollution. These aspects have been discussed elsewhere (Glasby and Roonwal, 1995).

FUEL AND ENERGY

Commercial energy consumption (coal, gas, hydropower) amounted to 158 million tonnes of oil equivalent in 1987. Non-commercial energy (fuel wood, dung cake, agricultural waste) amounted to 115 million tonnes of oil equivalent. But this is not sufficient to meet the demand. The presence on fuel wood is so much that we have lost forest area from nearly 45% of land in 1947 to 12% today. Most importantly since the use of biomass for fuel is well above sustainable yield, damage to ecosystems and soil is great. We cut 13,000 km^2 forest area each year. In broad terms we cut 4 trees a year and grow only 1 tree for regeneration of our forests.

Sectors such as transport are vital which require our attention. In Delhi alone, 4.2 million trips were made per day in 1988-89 by 6000 buses and 1.4 million personal vehicles. By 2001 this is estimated to increase to about 11.4 million trips by 16,000 buses and 4 million personal vehicles. The average speed is expected to decrease from 20-30 km per hour in 1988-89 to 10-15 km per hour in 2001. This means we should reckon 100 per cent more time for the distance to be covered in the year 2001. Delhi already has more cars than Mumbai, Madras and Calcutta put together. Fuel efficient buses and mass transport system is needed to keep Delhi livable, to control unending vehicle emitting 1400 tonnes of pollutant every day where atmospheric SO_2, CO and suspended particulate matter contents exceed acceptable levels. Already people complain of "Delhi throat".

POST SCRIPT

In conclusion, one would say to achieve sustainable development, population stabilization, and environmental protection need to be balanced, to achieve a better life for the present and the future. Kandall (1994) has pleaded for possibilities for a sustainable development. He suggests factors such as (a) controlling environmental damage, (b) resources for food and energy, (c) stabilizing population, and (d) reducing conflicts, as some important ways. An emphasis on study of environment is necessary. One finds here that while this is partially adopted in our education system it has at present emphasis on biology. Complete overlooking role of geology in environment is unfair. In fact, environmental geology is the key to the study of environment. One has only to see through the geological ages. Even the orographic changes which have taken place in the recent past are quite clear. Look at the wandering of the rivers in Punjab, the growth of Thar desert, the sea level changes, the burried river channels. A study of present environment is a page in the long history of the Earth. Past environment was as significant as the environment of today - the difference being that today the environment is more relevant because we dwell in here.

REFERENCES

Brown, L.R., 1981 : *Building a sustainable society*. W.W. Norton and Co. New York, 433 p., USA.

Glasby, G.P., 1995 : Concept of sustainable development : a meaningful goal? *The Sci. of the Tot. Env.*, 159 : 67-80.

Glasby G.P. and G.S. Roonwal, 1994 : India and quality of life. *Current Sci.*, 66 : 172-173.

Glasby, G.P. and G.S. Roonwal, 1995 : Marine pollution in India : an emerging problem. *Current Sci.*, 68 : 495-497.

Kandall, H.W., 1994 : Global prospects - the next 50 years : population, environment, and resources. *UN International Conf. Population and Dev.*, Cairo, MS.

Mac Neill, J., 1989 : Strategies for the sustainable economic development. *Sci. Amer.* 261 : 104-113.

Qasim, S.Z. (Ed.) 1994 : *Science and Quality of life*. Offsetters, New Delhi, 621 p.

Seibold, E., 1993 : Offensive and defensive geology in our environment. *Nat. Res. and Dev.*, Tubingen, 37 : 98-109.

UNFPA, 1993 : *The state of World Population*. United Nations Population Fund, New York, 54 p.

World Commission on Environment and Development, 1987 : *Our Common Future*. Oxford Univ. Press, Oxford, 400 p.

Chapter 13

Impact of Urbanization on Climate and Air Quality : A Case Study of Delhi

Prof. B. Padmanabhamurty

INTRODUCTION

Urbanization and industrialization increases demands on energy, air, water and land. Energy production can be Thermal, Hydro, Nuclear or by Non-conventional sources like wind, solar and tidal. In India, Thermal Power accounts for more than 72%, Hydro 24%, Nuclear 2 to 3% and non-conventional 0.5%. Thermal Power is produced by burning fossil fuels and on account of abundance of coal, in India, it forms the main source. Coal burning leads to extensive environmental pollution and even climate change.

Pressure on land leads to deforestation, conversion of fertile agricultural land into industrial areas, extinction of biological species etc. Increasing population as a consequence of industrialisation renders efflux of population from rural areas into urban complexes in search of employment which in turn increases the demands for more potable water and at the same time deteriorates the environment. As the surface water sources get exhausted underground sources are to be tapped and in doing so the overburden crumbles. Further, these depleted sources are not replenished faster and repeated overdrawing of the groundwater deteriorates the fertility of the soil.

In tropics air is bountiful and is so far deemed to carry out the effluents released into the atmosphere. However the carrying capacity of the air also has a limit and of late some of the tropical cities experience high levels of atmospheric pollution necessitating restriction or control of emissions.

URBANIZATION AND CLIMATE

Urbanization and industrialization modify the microclimate in the urban fabric in comparison with rural environment. Urbanization increases temperature, cloudiness, precipitation, contaminants, fog, haze, precipitation, cloudiness and even causes acid rain and reduces relative humidity, solar radiation, wind speed and visibility. Extensive studies have been made on local climatic changes in Delhi and some are reported below (Tables 13.1, 13.2 and 13.3).

Table 13.1. Heat intensity in various cities during winter season

City	*Heat island intensity °C in winter*
Delhi	6.0
Mumbai	9.5
Calcutta	4.0
Chennai	4.0
Bhopal	6.5
Pune	10.0
Visakhapatnam	0.6
Vijayawada	2.0

Table 13.2. Annual mean concentration of pollutants in Delhi (μg/m³)

Pollutant	Urban Delhi Daryaganj	Rural Delhi J.N.U.	CPCB standard
Total SPM	456	207	200
Nox	74.2	17.2	80
SO_2	38.7	7.8	80

Table 13.3. Acid precipitation in Delhi (1991)

February	6.80
April	6.50
May	6.41
June	6.85
July	5.74
August	5.02*
September	4.68*
October	6.45

* Neutral precipitation will have pH = 5.6. If pH > 5.6 the rain is said to be basic and if pH < 5.6 the rain is said to be acidic.

The differences in the heat balance of urban and rural areas both by day and night in almost all the cities in India result in warmth at the centre compared to rural environment. The differences in precipitation, evaporation, radiation etc. between urban and rural settings cause different conditions leading to differences in comfort conditions both diurnally and seasonally.

Even on a local scale there are large differences in temperature, humidity and wind under exposed and shaded environments. Climatic elements differ largely at urban complexes comared to rural environment. Green areas or forest microclimate reduces appreciably the temperature as well as wind.

Table 13.4. Daytime noise levels in Delhi

Location	*Noise level (dB)**
Chandni Chowk (Urban) "C"	82
JNU (Rural) "R"	57
Near hospitals "S"	
Sri Ganga Ram	66
Moolchand Khairati Ram	65
Ram Manohar Lohia	70
Safdurjang/AIIMS	70
ESI	73
Bara Hindu Rao	70
Sucheta Kriplani	74
LNJP/GB Pant	62
Holy Family	62
Deen Dayal Upadhyaya	60
Army	62

"C" = Commercial, "R" = Residential, "S" = Silent Zone
* Noise standards (daytime) for commercial, residential and silent zones according to Ministry of Environment and Forest are 65, 55 and 50 dB respectively.

URBANIZATION AND AIR POLLUTION

Delhi which had a population of 3.6 million in 1970, with a growth rate of 50.64% is likely to cross 15 million by the turn of the century. The industrial units which are mainly located in the west, south and southeast and were 26,000 in 1971 swelled upto 73,000 in 1991 and are expected to cross the million mark by the turn of this century. Delhi has the highest density of population of 6319/km^2.

The pollution trend is upwards in Delhi during the last few years. Delhi stands 4th in the world and 1st in India in respect of Suspended Particulate Matter. It is 27th in the world and 2nd in India in respect of Sulphur dioxide. Delhi's pollution load is more than 900 tonnes of SO_2. The sources of pollution are industries. Owing to poor dispersion, maximum concentrations of SO_2 occurred during night. Vehicular pollution is not far behind in Delhi. It contributes 55-60% of the pollution load. With a meagre 1.7 million vehicles in 1971 the number is expected to cross 40 million by the year 2001 with 10,000 vehicles being registered every month. The total number of vehicles registered in Delhi are more than the combined number of vehicles registered in the other three metropolitan cities viz. Mumbai, Calcutta and Chennai put together. With increasing population, by 2001, 12 million person trips every day are expected. This increases the pressure on the roads which will become inadequate. Of the vehicles that ply in Delhi, large number have Uttar Pradesh and Haryana registration numbers. The per capita vehicular trip which was 0.49 in 1969 is likely to cross 1.25 by 2001 with the number of trucks entering and leaving Delhi by that time would be 65000-70000. The most polluted intersections due to vehicular traffic (CO) are Shahdara Chowk, Daryaganj, Safdarjung Hospital, Inter State Bus Terminus, ITO, Tis Hazari and Lothian Bridge and of these Shahdara Chowk and Lothian Bridge stand significantly. In early hours between 5 and 7 A.M. the levels of CO exceeded sometimes 1.8 ppm.

URBANIZATION AND HUMAN COMFORT

Human comfort and health are weather dependent. The climatic parameters that control human comfort are air temperature, humidity, wind, cloudiness, sunshine duration and barometric pressure. Of all these, temperature is dominant, however, humidity changes the sense of comfort leading to the concept of "Effective Temperature" (ET). ET is that temperature which saturated air would provide the same comfort as does the actual temperature and humidity. Wind and sunshine extend the range of conditions which are comfortable. High effective temperature causes physical discomfort/distress and affects all mental activities. This depends on the degree of acclimatization, age etc.

Indoor climates are dependent on outdoor climate. By suitable building design, town planning and microclimatic amelioration, indoor climate can be made comfortable for living. The thermal environment of a human being is influenced by the balance between the incoming radiation from the sun and the atmosphere on the human body and that emitted by the human body into the surroundings. Additional effects relate to humidity of the air and the winds.

That the temperature of an urban centre exceeds the rural environment is a recognised fact and is often termed as "heat island effect". Heat island effect is always observed at all urban locations, though of different magnitudes. The heat island intensity (excess of urban temperature over the rural) depends upon time of the day (maximum at minimum temperature epoch and minimum at maximum temperature epoch), proximity to water bodies, population density, building material, wind speed, temperature profile etc. Studies in Delhi showed that the heat island intensity has seasonal variation with 6°C in winter/post-monsoon and reduces to 3-4°C as pre-monsoon approaches and during monsoon. During the day too, the heat island exists at Delhi with excess of urban temperature of 2-3°C. The humidity field showed an inverse relationship with temperature field except at locations close to the water bodies. The heat and humidity islands attain an early evening peak followed by an early morning peak and of these two, the latter is intense. Wind, evaporation, total radiation, net radiation at urban Delhi are found to be higher than at rural locations. The climatic classification of urban Delhi tended to be more moist compared to rural, where no change is observed. The rainfall pattern is observed to follow the urban heat island location coupled with wind flow and occasions of higher rainfall down wind of heat islands also are noticed. Spatial distribution of net radiation and pollution indicated pairs of sinks and sources of pollution suggesting the necessity of greeting Delhi to mitigate pollution problems. Rural Delhi has higher tendency to record lower temperature. Lower calm periods at urban areas compared to rural locations are found. At rural locations winds blow from all directions but in urban areas winds have preferred directions leading to channeling of winds. Lower temperatures causing condensation coupled with re-entrainment of soil dust due to high winds in rural Delhi reduced visibility compared to urban Delhi. The microclimates also responded to urbanization. In urban complexes, temperature relief in the shade at midday in urban, suburban and rural complexes, urban forest cover and cropped farmlands were 2.5, 2.0, 1.5, 2.5 and 1.0°C respectively.

The studies suggest that to mitigate adverse heat island effect, urban areas should be planned to have green fields, parks, ponds, vegetative covers appropriately spaced. Also microclimate amelioration can be achieved by developing vegetative covers, urban forests etc. to reduce the heat stress on the residents and make urban living more comfortable.

URBANIZATION AND NOISE POLLUTION

Noise pollution has assumed significance in Delhi primarily due to increased number of vehicles followed by industries and community. Noise is as much as environmental pollutant as any other gaseous or solid waste or water pollution. Community noise is mainly due to transportation. Narrow streets and high rise buildings increase noise by multiple reflections and channeling. The vehicle operators and occupants are also subjected to community noise. Other forms of community noise are due to domestic appliances, electric and electronic audio-visual equipment, loud conversations, street hawkers, barking dogs etc. Mechanical services and equipment like air conditioners, elevators and refuse handling chutes make noise that disturb tenants. People living in multi-storeyed flats in groups often complain noise transmission between units which affects living of conversation and other activities. Noise is more annoying in the evening and nights when people relax than at other times.

Noise produces temporary hearing loss and sometimes permanent damage to hearing. Noise changes heartbeat rates resulting in fluctuations in arterial blood pressure and vasoconstrictions of peripheral blood vessels. Noise produces temper tantrums, headache, fatigue and even nausea. Constant noise causes blood vessels to contract, skin to become pale, muscles to constrict and adrenaline (responsible for excitatory and inhibitory responses in living beings) to be shot into blood streams. Intermittent startling noise affects subliminally. Such cumulative subliminal interruptions cause irritability, anxiety, nervousness and sudden emotional breakdown. At unexpected and unwanted noise, pupils dialate, skin pales, mucous membranes drain, intestinal spasms develop and biological organism is disturbed. Shreik of siren or roar of a jet engine produces internal wreckage like gastric ulcers and thymus gland atrophy.

Noise was measured at 40 identified places in Delhi in 1987 and 1993. The survey indicates that the noise levels increased at many places by more than 1-10 decibels (dB). [Decibel is a unit of noise measurement which is a ratio of intensity of noise at any particular time and place to that of a reference intensity at 1 kHz (represents the threshold of hearing) expressed in logarithmic scale]. This is due to increased vehicular/human activity during the above period. In 1987 the noise levels are in the range of 80-86 dB while in the 1993 the range rose to 85-90. The trends of noise in Delhi is continually increasing as the number of vehicles and other man made activities are increasing every day. The most distinct feature of noise survey is the existence of high noise levels (83-86 dB) at Mool Chand, Sucheta Kriplani and Jai Prakash Narayan hospitals. Even during the night, these hospitals have high noise levels much to the discomfort of patients in the hospital complexes. The most noisy commercial areas are Chandni Chowk, Karol Bagh, Sadar Bazar and Shahdara. The residential areas of Vishwas Nagar, Tri Nagar etc. also have high noise levels of 80 to 89 dB respectively due to anthropogenic sources. Improved vehicle design and maintenance, improved road construction to avoid rolling noise at the point of contact between tyre and road surface, more flyovers and subways at busy road intersections, bushes and shrubs on the road dividers can mitigate the noise problem in Delhi.

Hospitals, Educational institutions, Courts etc. are declared "Silent Zones" by the Government of India. Use of vehicular horns, loud speakers and bursting of crackers is prohibited in these zones. Noise levels in these zones should not exceed 50 and 40 dB during day (6 AM to 9 PM) and night (9 PM to 6 AM) respectively. In order to assess how silent are our silent zones, noise levels were monitored at 12 hospitals in Delhi viz All India Institute of Medical Sciences, Safdarjung, Moolchand, Holy Family, Deen Dayal Upadhyaya, E.S.I., Army, Sri Ganga Ram, Bara Hindu Rao, Lodnayak Jayaprakash Narayan, G.B. Pant, Guru Tegh Bahadur, Ram Manohar Lohia and Sucheta Kripalani hospitals. Surprisingly the noise levels (Table 13.4) within the premises of all these were much higher than the stipulated standards. Norms laid down for silent zones were blatantly violated. These high levels of noise are due to their location and the density of traffic outside as well as inside the precincts. The noise environment can be improved in these regions by diverting the traffic either by fly-overs or sub-ways rendering free flow, relocating bus stops within 500 m from the hospitals and prohibiting the vehicular traffic within the hospital premises and planting bushes, shrubs and heavy canopy trees between the hospitals and the traffic arteries.

CONCLUSION

To sum up, rural environment have lower minimum temperature, less calm periods, poor visibility, high turbidity, less relief of temperature and higher humidity in shade. Urban areas should be planned in such a way that they are not drastically different from their rural counterparts. This could be achieved by developing green areas, peaks, ponds, vegetative covers appropriately spaced so that the microclimate slides smoothly from urban atmosphere into the rural setting without much discomfort to others. Such a rural environment prevents urban migration which in future exerts less pressure on urban resources like housing, transport, water, infrastructural facilities and minimize crime in urban areas. Urban Meteorologists will therefore have to be associated with the Town Planning Boards so as to ensure that the microclimates are designed and developed for the comfort of Urban Delhites in future.

REFERENCES

Achuta, N. 1992 : "*Noise levels in some silent zones in Delhi*", M.Sc. Project Report, School of Environmental Sciences, Jawaharlal Nehru University, New Delhi, 67.

Padmanabhamurty, B., 1984 : "Some aspects of urban climates of India "*Proc. Tech. Conf.*", "*Urban Climatology and its applications with special regard to tropical areas*" WMO - No. 652 : 136-165.

Padmanabhamurty, B. and D. Bandopadhyaya, 1994 : "Radiation Balance in a Tropical City - Delhi (India). *Boundary Layer Met.* 70 : 197-210.

Padmanabhamurty, B. and J.S. Bisht, 1995 : "Noise Pollution in U.T. of Delhi" *Bull. of Pure & Appl. Sci.* 47D : 45-51.

Ravichandran, C. and B. Padmanabhamurty, 1994 : "Acid Precipitation in Delhi", *Atm. Env.* 28 : 2291-2297.

Chapter 14

Problem of Oil Pollution in the Indian Ocean - An Overview

Parvin Farshchi, Z.A. Ansari* and S.A.H. Abidi

INTRODUCTION

The Indian Ocean as a whole, unlike the Pacific and Atlantic, has its northern boundaries closed with land mass. The western region of the Indian Ocean could best be covered, by division into three zones: the Arabian Sea and the two off-shores; the Gulf of Oman/Persian Gulf extension and the Gulf of Aden/Red Sea extension. After the discovery of oil in the Middle East, Suez Canal became a clock point because it gave access to the oil rich Gulf region. At present the Gulf supplies the lion's share of oil import of the industrialised countries of the world. The Persian Gulf connects the world's single largest site of oil reserves and production with the world market.

PETROLEUM—A CRITICAL SOURCE AND THE INDIAN OCEAN

Petroleum is the most critical source of energy and a major determinant of development at the national level and its availability and demand are indicators of international economic status. By the year 2085 AD, the world population is likely to be around 12000 million. The rate of growth is higher in the developing countries, Socio-economic development has become increasingly sensitive to food requirement and also to energy needs. The consumption of commercial energy in developing countries has tripled between 1970 and 1990. It now accounts for 27% of the world's total.

Fossil fuel constitutes the major source of commercial energy. Statistical review of the Asia's energy is demonstrated in Table 14.1. The general trend of expanding production and recoverable reserves would no doubt have some effect on the prospects of oil/gas resources in the Indian Ocean region. It seems that with current rate of production, oil resources would be exhausted within the next 25 years in South and South-East Asia. Indonesia, Malaysia, Australia and Pakistan are believed to be in future dependent on West Asia and new areas of oil exploration. West Asian oil resources are based on the Persian Gulf. World demands for petroleum by the year 2000 will likely to be more than 77 million barrels per day and by the year 2010, the demand may be 95 million barrels per day (Teitelbaum, 1995). A major part of this will be coming from Persian Gulf and Indian Ocean will play an important role in the transportation.

Table 14.1. Asia's energy production and consumption, 1991

	Production (barrels)	*Oil consumption (per day)*	*R/P ratio (year)*
China	2,810	2,405	22.6
Indonesia	1,515	675	12.2
Malaysia	655	240	12.5
India	645	1,200	25.6
Australia	539	680	8.0
Brunei	165	NA	33.2
Vietnam	72	NA	19.0
Pakistan	57	210	7.8
New Zealand	57	100	11.1
Japan	13	5,295	11.6
South Korea	—	1,185	—
Taiwan	—	570	—
Thailand	—	440	—
Singapore	—	380	—
Philippines	—	225	—
Bangladesh	—	—	—

Source : BP, Energy Economics, EIU.

By 1970 the oil accounted for 2/3 of world's energy requirement. The exploration and extraction of oil increased the seaborne shipping of oil from the Persian Gulf (Singh, 1992). Historical clock point at Suez Canal, Strait of Hormuz and Malacca assumed critical importance. Since these chock points lies in the Indian Ocean, it suffers maximum level of oil pollution. Even in the Indian Ocean, the Arabian Sea has the highest concentration of floating oil and tar in the world. This is because the only route of the large oil tanker carrying the oil cargoes from the Gulf countries is through the Arabian Sea. Annual statistical data of the British Petroleum indicates that more than half of the global marine transport of oil was shipped from the Gulf countries during 1973-1980. The total oil transport across the Arabian Sea in 1982 was 579×10^6 tonnes (Sen Gupta and Qasim, 1985). During 1981 and 1987 the marine transport decreased considerably and a consequent reduction in oil pollution was noticed (Qasim and Sen Gupta, 1988). Oil transport from Gulf through the Indian Ocean is shown in Fig. 14.1. Oil tankers passing through this route go to different parts of the world. They turn into

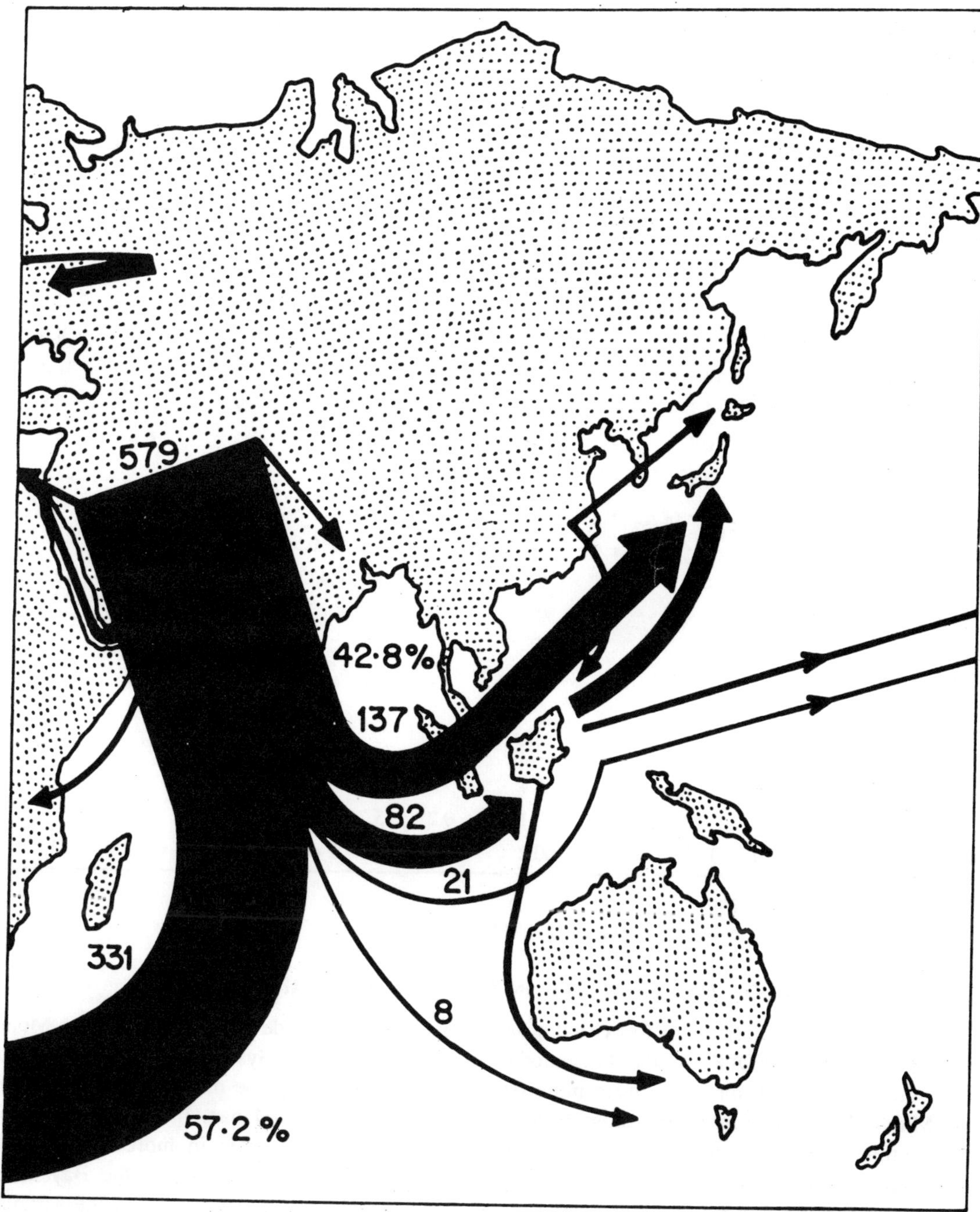

Fig . 14.1. Movement of oil from Middle-East countries by sea, 1982 (million tonnes).

Bay of Bengal for South Asia, Japan, Latin America and to other countries. Table 14.2 indicates the inter-area total oil movement in 1992. It can be seen that of the total oil movement of 1625.1 mmt, the contribution of oil movement from the Middle East countries was 686.2 mmt.

Table 14.2. Inter-area total oil movement in 1991 (Values are in million tonnes)

From To	*USA*	*Canada*	*Latin America*	*OECD Europe*	*Africa*	*Asia**	*Japan*	*Austra-lasia*	*Rest of the world*	*Unidenti-fied*	*Total*
USA	-	3.4	4.5	12.2	-	-	0.2	0.9	0.9	19.8	47.9
Canada	50.2	-	0.2	1.1	0.1	0.8	0.3	-	-	2.1	54.8
Latin America	113.9	4.7	-	33.8	-	-	7.9	-	0.1	8.5	168.9
OECD Europe	20.1	17.4	0.3	-	2.9	0.2	0.1	0.3	6.5	31.4	79.2
Middle East	97.4	4.7	15.8	192.7	18.3	165.8	163.7	9.3	17.6	0.8	686.1
North Africa	13.5	0.4	-	103.3	4.1	-	1.0	-	6.0	9.2	137.5
West Africa	41.0	3.1	-	52.7	5.1	-	0.6	-	-	21.9	124.4
East-South Africa	-	-	-	-	0.7	-	-	-	0.1	-	0.8
South Asia	-	-	-	0.1	-	-	0.5	-	2.8	-	3.4
Other Asia	6.9	0.1	-	0.1	-	4.9	47.1	4.3	3.6	13.6	80.0
Japan	-	-	-	0.1	-	2.5	-	0.4	0.2	-	3.2
Australasia	1.5	0.1	-	-	-	5.1	2.6	-	-	1.5	10.8
Ex-USSR	1.9	0.5	0.2	71.9	-	6.2	0.6	-	30.9	-	112.2
Other non-OECD Europe	0.1	-	-	3.7	-	-	-	-	0.1	2.5	6.4
China	5.0	-	0.2	0.4	-	9.1	13.3	0.1	-	-	28.1
Unidentified	34.2	0.8	-	28.1	0.2	15.7	-	-	2.4	-	81.4
Total	385.7	35.2	21.2	500.2	31.4	209.7	243.9	15.3	71.2	113.3	1625.1

* Excluding China and Japan

Source : B.P. Statistical review of world energy, 1992.

VISUAL OBSERVATION OF OIL SLICKS IN THE INDIAN OCEAN

By the end of 1978, more than 8500 visual observations have been made by Marine Pollution Monitoring Pilot Project (MAPMOPP) of the Integrated Global Ocean Station System (IGOSS), over much of the word's oceans. Visible slicks were more frequently reported along the major tanker routes between the Middle East and Europe and those between the Middle East and Japan. Indeed, the plot of these observations clearly delineated the major tanker routes. Oil was reported in more than 10% of the observation made along the East coast of Africa and West coast of India, across the Bay of Bengal and throughout the South China Sea. These areas coincided with tanker lane between the Middle East, Europe and Japan.

The MAPMOPP data on the distribution of tar in the Indian Ocean suggested that the concentration along the West coast of India was substantially higher than along the East. This was undoubtedly a consequence of the tanker route which passes from the Persian Gulf across the Arabian Sea and then across the Bay of Bengal to the Strait of Malacca enroute to Japan. The MAPMOPP data for dissolved/dispersed petroleum residues in the Indian Ocean contained no values less than 10 μg/L, while some exceeded 300 μg/L. The geometric mean of 8.8 μg/L suggested a much higher level of pollution in this region than was indicated for any other area of the world ocean. The data demonstrated a tremendous difference between the values obtained along the west coast (GM = 86.4 μg/L) and those obtained along the east coast (GM = 0.7 μg/L) of India. This is presumably a reflection of the influence of the tanker lane that passes along the west coast of India and across the Bay of Bengal.

It is interesting to note that in the final report of MAPMOPP (IOC/WMO, 1981), 85 μg/L dissolved/ dispersed oil was measured along the west coast of India and this range was considered to be highest value found so far in the world oceans. Much higher values for dissolved/dispersed oil in the surface water along the east coast of the Red Sea have been reported (Awad, 1988).

SURFACE DISTRIBUTION OF OIL POLLUTION IN THE PERSIAN GULF AND GULF OF OMAN

The sea area of the Persian Gulf and Gulf of Oman is divided into three zones according to the sources and transport mechanism in oil pollution.

First zone of surface water

This zone is the shipping route in the two Gulfs occupying the open sea area between the land masses. The surface water at the Strait of Hormuz and the coastal waters of Emirate of Sharjah are included in this zone. The measured concentration of dissolved/dispersed hydrocarbon in this surface water changed gradually from a mean of 27 ppb in the Gulf of Oman to a mean of 22 ppb in the seawater inside the Persian Gulf (El-Samra and El-Deeb, 1988). The higher level of oil pollution in the Gulf of Oman as compared to that of Persian Gulf was attributed to the deballasting operation which are carried out in the Gulf as a regular practice. Fondekar and Alagarswamy (1984) estimated oil concentration upto 41.6 ppb in the area of oil tanker route in the Arabian Sea while El-Samra *et al.* (1986) reported oil concentration around 7 ppb in the area of the oil tanker routes in front of Saudi Arabia.

Second zone of surface water

This is the coastal waters of the Arab countries side. Oil concentration of this zone were generally lower than 20 ppb. The coastal water in front of UAE did not exceed 17 ppb, while that in front of Southern part of the State of Qatar did not exceed 11 ppb. Values less than 5 ppb have been given for the coastal water of Kuwait, [5.7 ppb for Bahrain, 3.8 ppb for UAE and 7.5 ppb for Oman, respectively (National Report of Kuwait, 1985)]. According to published data these coast are considered as the primary treat area for coastal impact of oil.

Third zone of surface water

This zone is the area in the Persian Gulf where offshore oil fields are found. These are justifiably considered as fixed point sources of oil pollution (hot spot) in the Gulf. Pollution can arise as a large oil spill from a pipeline break or well blow out, or chronically from continuous or intermittent operational discharges. In 1986 a value of oil pollution upto 40.9 ppb was measured near the biggest centre of oil fields of UAE at Das Island and a value of 68.6 ppb was measured near Halul Island oil field belonging to Qatar. The highest value of oil pollution (546.4 ppb) was measured in the Persian Gulf at the area of Ras Tanura where the largest centre of oil production and refinery in the world is found in front of Saudi Arabian coast (El-Samra *et al.*, 1986). It has been reported by El-Samra and El-Deeb (1988) that there exist a subsurface movement of pollutant and the water of high density is directed from the Persian Gulf into the Gulf of man through the sill-less Strait of Hormuz.

DISTRIBUTION OF FLOATING AND DISSOLVED/DISPERSED HYDROCARBONS IN THE NORTHERN INDIAN OCEAN

Consolidated picture of oil pollution in the Northern Indian Ocean has been reported by Sen Gupta and Kureishy (1981). Data on observations of oil slick, floating tar and concentration of dissolved/dispersed hydrocarbons have been reported. The data over the entire Indian Ocean down to 40° S latitude, have been sorted out for every 5° square because of the large number of observations available. The positive number of oil sightings ranged from 51% to 96%. The frequency of oil slicks along the trade route from the Eastern to the Western hemisphere through the Red Sea was also of same order. The number of oil sightings increased away from the source of oil. Even in area away from the tanker and trade route, the percentage of oil sightings was almost of a similar order. Tanker disasters have not been very significant so far. Therefore, it could be concluded that oil is released in this region mainly from the ballast and bilge washings of the ships and that Northern Indian Ocean is a well known area for the occurrence of slicks.

Tar balls

The tar balls are a common scene during the monsoon period on all sandy beaches. The concentration of tar balls ranged from 0 to 6.0 mg/m^2 with a mean of 0.59 mg/m^2 in the Arabian Sea. The total amount of floating tar balls on the surface of the Arabian Sea at any given time is about 3700 tonnes (Sen Gupta and Kureishy, 1981). For the Bay of Bengal the concentration of tar balls ranged from 0 to 69.75 mg/m^2 with a mean of 1.52 mg/m^2. Total quantity of floating tar balls in the Bay of Bengal amounts to 1100 tonnes (Sen Gupta and Kureishy, 1981). Accumulation of tar balls entirely depends on the surface currents. Annual deposition of tar balls on the beaches along the west coast of India has been estimated to be 750-1000 tonnes (Sen Gupta and Qasim, 1985). The floating tar ball has a residence time of 60-90 days.

Dissolved and dispersed hydrocarbon

There is a large variation in the distribution of dissolved/dispersed hydrocarbon in the Indian Ocean. Statistical analysis upto May 1979 indicated a mean value of 60.10 μg/kg with a standard deviation of 92.66. A summary of the data is presented in Table 14.3.

The average concentration of dissolved/dispersed hydrocarbon in the upper 20 m of the tanker route in the Arabian Sea and the Bay of Bengal were 42.8 and 28.22 mg/kg, respectively. The quantity of petroleum hydrocarbon present in the upper most 20 m of the water column is estimated to be 5×10^6 in the entire Arabian Sea and 0.4×10^6 tonnes in the Bay of Bengal tanker route (Sen Gupta and Qasim, 1985). An average of 7.1 mg/L of dissolved/dispersed hydrocarbon ranging from 0-41.6 mg/l in the upper 20 m in the Northern Indian Ocean is reported (Sen Gupta, 1990). The concentration of petroleum hydrocarbon in zooplankton and in the sediment of the Arabian Sea was found to range from 19.5-83.3 mg/g and 4.8-8.5 mg/g, respectively (Fondekar *et al.*, 1980).

The Indian West coast is vulnerably exposed to any oil spill on the Arabian Sea. Analysis of the data indicates that the trade and tanker routes across the Arabian Sea are more contaminated as compared to similar routes over the Southern Bay of Bengal. In the near-shore areas of the Bombay high oil field, the concentration of dissolved/dispersed hydrocarbon range from 2-46 mg.dm^3 in the water column and from 4-32 mg/g in the sediment. In the Bombay Harbour region, the concentration of hydrocarbon after a tanker fire accident, ranged from 27-105 mg/dm^3 at the surface and from 36-59 mg/dm^3 at 5 m depth (Fondekar, 1988).

Table 14.3. Dissolved/dispersed hydrocarbons in the upper 20 m of the Arabian Sea and the Bay of Bengal

	Arabian Sea			*Bay of Bengal*		
	Transport	*Concentration*		*Transport*	*Concentration*	
Year	*Mt*	*Range*	*Mean*	*Mt*	*Range*	*Mean*
1978	975	0.9-42.5	24.31	323	0-28.2	17.14
1979	1010	10.4-41.6	24.48	351	—	—
1980	869	2.4-9.0	5.28	308	1.2-27.4	12.47
1981	725	—	—	247	0-2.8	1.40
1983	513	0-17.7	5.02	222	—	—
1984	489	—	—	252	0-3.4	1.70
1985	447	0.65-31.0	7.64	232	—	—
1986	—	1.0-23.5	7.50	—	—	—
Net decrease (%)	54	—	—	28	—	—

Source : Qasim and Sen Gupta (1988).

OIL SPILL IN THE INDIAN OCEAN

Sources of pollution emerging from the hydrocarbon sector are largely associated with the following activities : (1) Drilling, (2) Crude oil production, (3) Transportation, and (4) Flaring of natural gas. These operations if carried out properly will lead to insignificant environmental implications. There can also be spill during the transportation of crude to cargo vessels or due to the overflow from the group gathering stations. Leaks or burst in the pipes can lead to sub-surface pollution. Continuous oil production from an installation can give rise to oil seepage into the ground. In the offshore platforms, leaks in the pipeline or sudden break down of submarine pipeline can result into sea pollution. However, due to stricter regulations and improved environmental protection, there has been a considerable reduction in the incidents of oil spillage (Qasim, 1995).

The greatest concern of the oil industry is the oil spills caused by the oil tankers during the transportation of different types of oil by the sea route to different countries. The Gulf war oil spill is the world's largest oil spill. The estimates of this oil spill varies widely from 160 to 340 million gallons. The range of total input of oil from all sources is estimated to be between 1.7 and 8.8 million tonnes per year (Table 14.4). Out of many oil spills in the Indian Ocean few instances which are on record (1970-1995) are reported in the coming paragraphs.

Table 14.4. Input of oil into marine environment

Source (mln tonnes)	Probable range (mln tonnes)	Best estimate
Natural sources		
Marine seepage	0.02-2.0	0.2
Sediment erosion	0.04-0.06	0.05
Offshore production	0.04-0.06	0.05
Transportation		
Tanker operation	0.4-1.5	0.7
Drydocking	0.2-0.05	0.03
Marine terminals	0.01-0.03	0.02
Bilge and fuel oil	0.2-0.6	0.3
Tanker accident	0.3-0.4	0.4
Non-tanker accident	0.02-0.01	0.02
Atmosphere deposition	0.05-0.5	0.3
Other sources		
Municipal water	0.4-1.5	0.7
Refineries	0.06-0.6	0.1
Industrial waste	0.1-0.3	0.2
Urban runoff	0.01-0.2	0.12
River runoff	0.01-0.5	0.04
Ocean dumping	0.005-0.02	0.02
Total	1.7-8.8	3.2

Source : Doerffer, 1992.

In 1970, a Greek oil tanker 'Ampuria' ran aground off Gulf of Kutch, in the Arabian Sea with a full load of 15,622 tonnes furnace oil. About 3500 tonnes of oil leaked out in the sea. In 1973 oil tanker, 'MT Cosmos Pioneer' went aground in the north west coast of India releasing 18,000 tonnes of light diesel in the Arabian Sea. In 1974, oil tanker "Transhuron' carrying 18500 tonnes of furnace oil ran aground on Kiltan, one of the atolls of Lakshadweep, spilling 5000 tonnes of navy special fuel oil in the Arabian Sea (Qasim *et al.*, 1974). In 1978, the tanker 'Sealift Mediterranean' ran aground on an island off the northern tip of Sumatra, spilling 1000 tonnes of crude oil. In 1992, in the Strait of Malacca an EXX on tanker got damaged and started drifting towards the Great Nicobar Island, spilling 40,000 tonnes of crude oil.

In 1993, three cases of oil spills in the Indian Ocean were reported (Anon, 1994). (I) On 21st January 1993, the Danish tanker 'VLCC Maersk Navigator' collided with another ship near the Strait of Maiacca, spilling 18000 tonnes of oil in the Andaman Sea. (II) The rupture of feeder pipeline in Bombay High on 17th May 1993 resulted in an oil spill of 3000-6000 tonnes. (III) The mishap involving 'MV Challenger' caused a leakage of 270 tonnes of fuel oil off Mangalore on 1st August, 1993. In 1995, on 8th January, Pasarlapudi gas field in the East Godavari in Andhra Pradesh blew out, 49 stands of drill pipe were thrown out and the blowing gas caught fire.

Besides these, considerable quantities of oil spills occur from trans-shipping, washing, ballasting, deballasting of oil tankers. Other kind of ship also cause numerous oil spills, due to the large amount of bunker oil they may be carrying. Usually these spills are not as large as those caused by tankers. The level of risk of a blow-out or pipeline spill occurring and causing oil pollution is far elss than is experienced from the transportation of oil at sea (Welch *et al.*, 1991).

The causes of oil spillage are—design fault in terminals, design fault in tankers, breakage and mechanical failure, incorrect operating procedure and human error. The International Maritime Organization (IMO) is taking care to overcome these problems. They have prohibited release of oil and its products within 50 nautical miles from the shore. Using proper tagging techniques, it would be possible to match an oil spill to one of the reference samples and the source of pollution could then be traced. In the latest amendment to the safety of life at Sea Convention, the IMO has deemed that vessel reporting scheme, meeting IMO guidelines shall be mandatory. The greatest benefit will come when ship reporting is not only mandatory but automatic. This would involve new levels of regulation, cooperation and technology and as "stewards of the world environment, we have to bear this responsibility."

REFERENCES

Anonymous, 1994 : *Annual Report of National Institute of Oceanography*, 15-18.

Award, I.I., 1988 : Oil in Sudanian Red Sea Territorial waters. *Mar. Poll. Bull.*, 19 : 287-290.

El-Samra M.I., H.I. Emara and E. Shunbo, 1986 : Dissolved petroleum hydrocarbon in the north western Arabian Gulf. *Mar. Poll. Bull.*, 17 : 65-68.

El-Samra M.I. and K.Z. El-Deeb, 1988 : Horizontal and vertical distribution of oil pollution in the Arabian Gulf and the Gulf of Oman. *Mar. Poll. Bull.*, 19 : 14-18.

Fondekar, S.P., S.R. Topgi and R.J. Noronha, 1980 : Distribution of petroleum hydrocarbon in Goa coastal water. *Indian J. Mar. Sci.*, 9 : 286-288.

Fondekar, S.P. and R. Alagarswamy, 1984 : Petroleum hydrocarbon contamination along oil tanker routes in the Arabian Sea. *Indian J. Mar. Sci.*, 13 : 181-183.

Fondekar, S.P., 1988 : *Studies on petroleum hydrocarbons in the marine environment around India*. Ph.D. thesis, University of Bombay, India.

IOC/WMO, 1981 : *Global oil pollution, results of MAPMOPP*. The IGOSS pilot project on marine pollution (petroleum) monitoring, IOC, UNESCO.

National Report of Kuwait, 1985 : *Marine monitoring and research programme in the State of Kuwait*. ROPME sysmposium, UAE University, Al-Ain, 8-11 December, UAE.

Qasim, S.Z., P.N. Nair and P. Sivadas, 1974 : Oil spill in the Laccadives sea from the oil tanker Transhuron. Mahasagar, *Bull. Natl. Inst. Oceanogr.*, 7 : 83-89.

Qasim, S.Z. and R. Sen Gupta, 1988 : Some problem of oil pollution in India. *Mar. Poll. Bull.*, 19 : 100-106.

Qasim, S.Z., 1995 : *Some environmental implications of the Indian hydrocarbon sector.* Keynote address, International Conference - Petrotech, 12 January, 1995, New Delhi, India.

Sen Gupta, R. and T.W. Kureishy, 1981 : Present state of oil pollution in the northern Indian Ocean. *Mar. Poll. Bull.*, 12 : 295-301.

Sen Gupta, R. and S.Z. Qasim, 1985 : The Indian Ocean : an environmental overview. In : *The Ocean Realities and Prospect*, (Ed. R.C. Sharma), 7-40.

Sen Gupta, R., 1990 : State of the marine environment in the south Asian seas region. UNEP Regional Seas Report and studies No. 123.

Singh, J., 1992 : Critical resources and the ocean. In : *Indian Ocean challenges and opportunities*, New Delhi, 29-30 Sep., : 183-193.

Teitelbaum, R.S. 1995 : Your last big play in oil. *Fortune*, Oct. 30 : 106-108.

Welch, J., A.M. Stolls and D.S. Etkin, 1991 : *Worldwide oil spill trend.* Proceeding of International 1991 oil spill conference : Prevention, behaviour, control, cleanup, March 4-7, San Diego, California.

Chapter 15

Molecular Mechanisms in Mercury Detoxification in Bacteria

Arif Ali*, Girish Sharma, Imtiyaz Murtaza and Subhash Mishra

* *Gene Expression Laboratory, Centre for Biosciences, Jamia Millia Islamia, New Delhi, India*

INTRODUCTION

The presence of heavy metals in the environment has received a great deal of attention due to their high toxicity, translocation and subsequent magnification through the food chain. The discovery of high level of methyl-mercury in fish and shellfish in Menamata bay (Japan) brought the problem of mercury pollution into limelight (D'Itri, 1972), and the source of mercury was found to be the fertilizer plants, polymer industries where mercury is used as a catalyst in the production of vinyl chloride. High levels of methyl mercury have also been detected in fish from the great lake region of North America.

The sources of mercury in the environment are both natural and anthropogenic in origin. In nature, cinnebar (HgS) and metacinnebar (black HgS) are the most important mercury containing ores. Anthropogenic sources of mercury include those associated with its use in the chlor-alkali, paints agriculture, pharmaceuticals and paper and pulp industries where they are used as a disinfectant, catalyst and fungicidal agents. Industrial mining activities and burning of fossil fuels are believed to be the major source of mercury released into the environment. Over 3000 tonnes of mercury per year is released through the burning of coals and an additional 10,000 to 60,000 tonnes are released from crude oil (Joensuu, 1971). Therefore human activities are estimated to account for 2.104 to 7.104 tonnes of mercury per year, being released into the atmosphere and water supply. Sewage treatment facilitates a widespread source of both inorganic and organic mercury compounds (HgO, Hg^{2+}, MMA, Dimethyl

It is therefore apparent that human activities have significantly contributed to the increased levels of mercury in the biosphere. Mercury containing compounds are abundant in many aquatic and terrestrial ecosystems as a result of either anthropogenic or natural geological processes (Jefferies, 1982). Bacteria have, however, evolved mechanisms for mercury detoxification as mercury resistance is a common phenotype among bacteria isolated from soil and water environment (Rochelle *et al.*, 1991).

MERCURY POLLUTION AND PROKARYOTIC INTERACTION

Problems related to water pollution by mercurial compounds have been primarily handled at managerial levels in India. Very scanty literature is available dealing with mercury removal from the aquatic bodies. The studies carried out are mainly related to surveys of incidence of mercury in and around Mumbai region, but there is not a single report found in the literature which deals with the elucidation of mercury detoxification mechanisms at the molecular level. Though Rajni and Mahadevan (1994) have reported the cloning and expression of mercury resistant genes from marine *Pseudomonas* sp., they found that there existed no DNA homology between the *mer* operon of *Pseudomonas* sp. strain MRI and the characterized determinants of TN501 *mer* DNA, suggesting the possibilities of existence of genetic variability at the molecular level.

There are two types of mercury pollution. One is caused by the presence of inorganic mercury compounds and other is by organic mercurial compounds and consequently, nature has evolved two types of organisms to counter the toxic effects of mercury. Organisms which have evolved mechanisms to detoxify inorganic mercury compounds are classified under narrow spectrum and exemplified by *Pseudomonas aerogenosa*, and microbes which have adopted strategies to detoxify organo-mercurial compounds are categorized under broad spectrum *i.e. Pseudomonas* sp. K62. Most of the work carried out to date has been on the narrow spectrum form of mercury resistance which confers resistance to inorganic mercury [$HgCl_2$, $Hg(NO_3)_2$]. Studies have shown that the proportion of bacteria isolated from various mercury-contaminated soil and aquatic environments are resistant to inorganic mercury in the range of 0.007 to 49% (Rochelle *et al.*, 1991) and their resistance to the more toxic organic mercurial compounds such as methyl mercury and phenyl mercuric acetate (broad spectrum resistance) is less common.

BIOCHEMISTRY OF MERCURY METAL

Mercury is the only metal which exists in liquid state. It forms a variety of electrovalent compounds (metallic nature) as for example, $HgCl_2$, $Hg(NO_3)_2$, $HgSO_4$, Hg_2SO_4 etc. and covalent compounds as for example, phenyl mercury acetate (PMA), alkyl mercury acetate (AMA), fluorescein mercury acetate etc., due to the presence of vacant outer d-orbitals. Much like any other non-radioactive element, the amount of mercury in the biosphere also remains constant. It is only the form which changes and completes a cyclic process in the environment which is mediated by both geological and biological processes. Unlike cobalt, zinc, cadmium, magnesium and iron, mercury has no significant biological role and is in fact toxic even at ppm levels. Hg^{2+} has high redox potential and so it can reduce disulphide bond in proteins which has both the structural and functional importance in almost all proteins. It has also been noted that Hg^{2+} binds with negatively charged membrane phospholipids, like thiol (SH), thioester (–S–) and imadazol group which renders the proteins and the biomembrane (phospholipids) inactive disrupting the vital cellular integrity and activity. Regarding organo-mercurial compound, it is volatile and their toxicity is very high when compared to inorganic mercury. Exact mechanism is not known but it is argued that they act upon electron transport system, cause neural toxicity besides the general deleterious effects on membrane, membrane bound proteins and other proteins.

GENETIC DIVERSITY IN *MER* OPERON

Bacterial cells can activate mechanisms to counteract the toxic effect of mercury ions and organo mercurial compounds. These resistance mechanisms are often encoded on plasmids or transposons and are highly specific for Hg^{2+} ions or mercury covalently bonded to an organic counterpart as for example, phenyl mercury acetate (PMA), methyl mercury acetate (MMA) (Summers, 1986). Presently one of the best understood metal resistance loci is that conferring resistance to mercury compounds (Walsh *et al.*, 1988).

The emergence of recombinant DNA and DNA hybridization technology has helped us towards accessing and understanding the occurrence of specific genetic determinants in the environment rather than just monitoring phenotypes. These technologies have enabled us to understand the underlying genetic mechanisms involved in the detoxification of the mercurial compounds. In essence, the biochemical pathways utilized in narrow and broad spectrum resistance mechanisms are similar. The cardinal difference in broad spectrum resistance is the involvement of another gene encoding lyase. It cleaves the covalent bond between the alkyl or aryl and mercury. The pathway is given in Fig. 15.1.

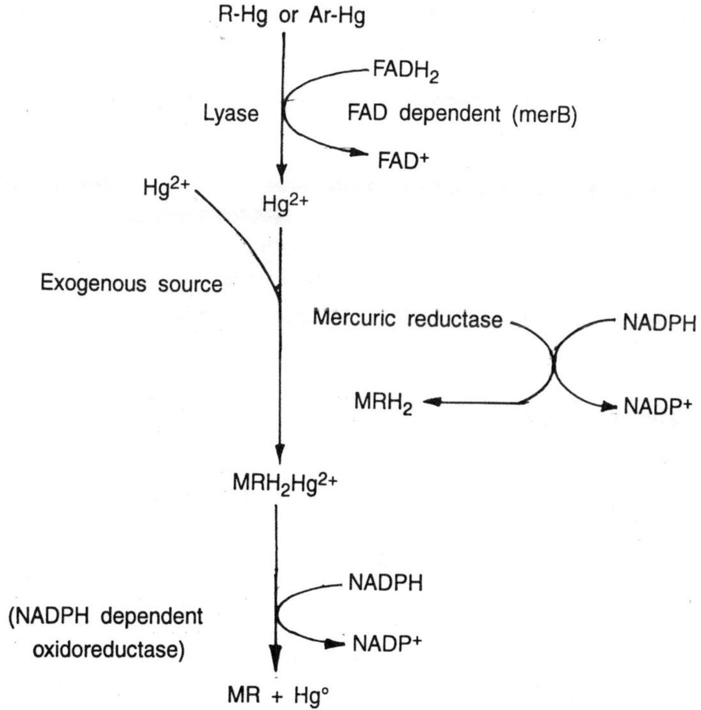

Fig. 15.1. Biochemical pathway utilized in the detoxification of mercurial compounds.

DNA probes have been used to detect the occurrence of mercury resistant, genotypes in environmental isolates (Bridges, 1979; Barrineau and Summers, 1983; Barrineau *et al.*, 1984; Foster and Ginnity, 1985 and Beglay *et al.*, 1986). Genes responsible for transport and detoxification of mercurial compounds are present in a cluster called *mer* operon. The *mer* operons are present either on chromosomal DNA e.g. in *Bacillus* sp., or on plasmids like R-100 (*E. coli*), pVs1 R-831, and pDU1358 originally from *Shigella* (Tn21), *Pseudomonas aerogenosa* (Tn501), *E. coli* and *Serretia* sp. respectively. Out of these, R-100 and pPVS1 carrying Tn21 and Tn501 which code for narrow spectrum resistance whereas R-831 and pDU1358 encode broad spectrum mercury resistance (Silver and Misra, 1988). Besides these operons, there are reports for other DNA sequences and plasmids coding for resistance genotype but they are not well characterized.

The general scheme of the gene cluster of *mer* operon can be designed in the following manner. The gene cluster of the Gram negative bacteria show seven open reading frames (ORFs) coding for *mer*R, *mer*T, *mer*P, *mer*C, *mer*A, *mer*B and *mer*D. DNA sequences between *mer*R and *mer*T is a promoter/operator region which transcribes divergently (Fig. 15.2). *mer*R, *mer*T, *mer*P, *mer*A, *mer*D are common to all plasmids borne *mer* operons coding narrow spectrum resistance. *mer*B is present only in plasmids coding for broad spectrum resistance. *mer*C is present only in R-100 and *Thiobacillus ferroxidans*. *mer*C has been sequenced by Misra *et al.* (1985) from R-100. The sequence homology studies and polypeptide sequence analysis revealed that it is a hydrophobic protein, therefore, he surmised that it is a membrane bound protein but Summers (1992) is of the opinion that it is a regulatory protein and is not directly involved in detoxification since it is not found in Tn501 and pDU358. The

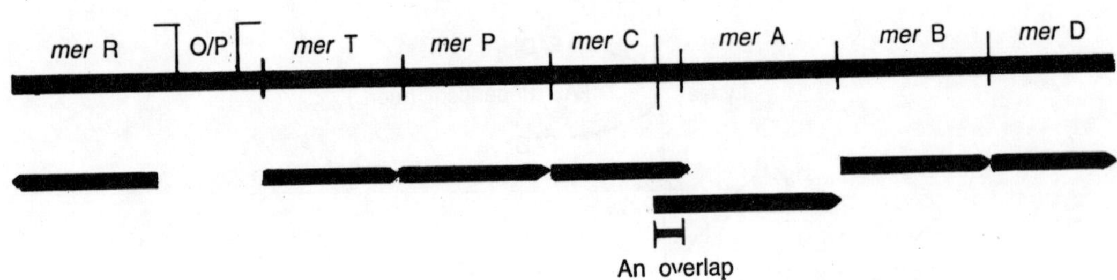

Fig. 15.2. General design of the gene cluster of *mer* operon in Gram negative bacteria.

Functions of different genes:

mer R :	Repressor / regulatory protein	144 amino acids
mer T :	Transport protein	116 amino acids
mer P :	Periplasmic protein	91 amino acids
mer C :	Function unknown. *mer* C mutants are still able to detoxify mercury	140 aa in R-100 and 143 aa in *Thiobacillus ferroxidans*
mer A :	Mercury reductase (detoxifies mercury)	564 aa in Tn*21*, 561 aa in Tn*501*
mer B : (dimer)	Lyase. Breaks the covalent bond between mercury and organic compounds	100 aa in monomer
mer D :	Regulatory protein involved probably in feed back inhibition	120 aa in Tn*21* and Tn*501* (pDU 1358)

situation of gene cluster of *mer* operon in gram positive bacteria is still unclear. Fig. 15.3 illustrates the general design of the *mer* operon found in Gram positive bacteria. There are eight ORFs but the function could not be assigned to four ORFs, *i.e.* ORF 2, 3, 4, 5.

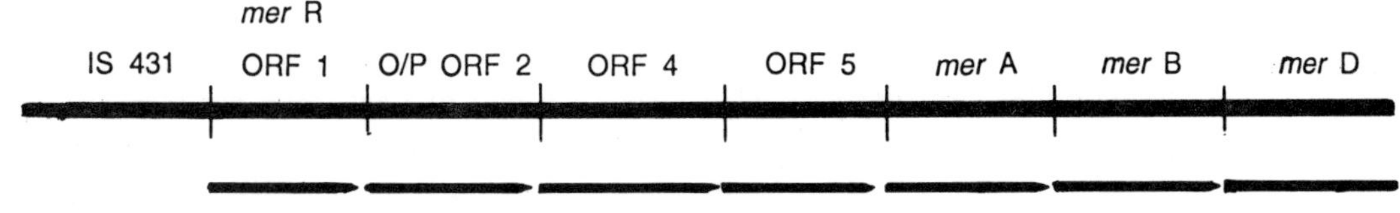

(Size of the cluster from ORF 1 to *mer* B = 6.4 Kb)

Functions of different genes :

IS 431 :	Transposase	
ORF 1 :	*mer* R different from Gram negative	
ORF 2 :	Function unknown	135 amino acids
ORF 3 :	May be involved in transport, may be membrane bound	
ORF 4 :	Like *mer* T and twice the size of *mer* T	232 amino acids
ORF 5 :	Mercury transfer	128 amino acids
mer A :	Mercury reductase but different from Gram negative. (shows 34% similarity)	631, 547 amino acids
mer B :	Lyase. Breaks the covalent bond between mercury and organic compounds	100 amino acids (22.4 KD)

Fig. 15.3. General design of the gene cluster of *mer* operon in Gram positive bacteria.

MICROBIAL PROCESSORS OF MERCURY DETOXIFICATION

Biotransformation of mercury involves the conversion of more toxic into less toxic forms. It mainly occurs in two processes, as follows :

(i) Enterobacteriaceae are known to convert Hg^{2+} into dimethyl mercury by excreting methyl cobalamine which methylates Hg^{2+} non-enzymatically. The methyl mercury is more toxic at the same time it is more volatile and escapes into the environment.

(ii) Some bacteria including Enterobacteriaceae and H_2S producing bacteria which, when come in contact with Hg^{2+}, forms insoluble mercuric sulphide (HgS) and hence the effective mercury concentration decreases. But these two processes of biotransformation are not involved in specific response to mercury and its compounds have arisen gratuitously from a metabolic process with an independent function other than mercury resistance. So far as mercury resistance is concerned, there are genetic systems found exclusively to detoxify mercury. These genetic determinants may be present on chromosome (*Bacillus* sp.) or on plasmids. A plasmid borne determinant may be flanked by with or without transposon. Mercury resistant phenotype can broadly be classified based on two parameters :

(a) Based on the nature of mercury compounds :

(i) Broad range : They detoxify both organic and inorganic mercurial compounds.

(ii) Narrow range : They detoxify only inorganic mercury.

(b) Based on Gram staining :

(i) Gram negative : Most of them are plasmid borne and show a general pattern of mercury resistance. They may be broad or narrow spectrum resistance. Gram negative have been studied extensively by Summers (1992) and Misra *et al.* (1985).

(ii) Gram positive : They may either be plasmid borne or resistance genes may be located at chromosomes (*Bacillus* sp.). They may be a part of transposon but in many cases it is without it. Gram positive have not been studied in detail although the first mercury resistance system was identified in gram positive strain (*Staphylococcus* sp.).

GENE REGULATION OF *MER* OPERON

Gram negative system : Tn*21* (R–100), Tn*501* are narrow spectrum resistance while R831b is broad spectrum resistance system. Their loci and arrangement of genes are illustrated in Fig. 15.2. *Mer* operon is controlled by overlapping divergent operator promoter which is cis acting and transcribes *mer*R gene. It is 432 bp and translated into regulatory protein of 144 amino acids and is trans acting. MerR is a dimeric protein and binds with one Hg^{2+} per molecule. In the absence of Hg^{2+} the synthesis of structural genes is repressed, while in its presence it favours the formation of open complex of RNA polymerase and promoter template (Fig. 15.3). While initiating the transcription of structural gene MerR protein shows autogenic regulation and always repress the synthesis of it's own mRNA. MerR protein is a high affinity intracellular receptor for an environmental signal and an allosterically modulated transcriptional activation. In the absence of RNA polymerase, MerR and Hg-MeR binds to the same region proved by nuclease footprinting analysis (O'Halloran, 1989) from -37 to -10P, a 17 bp and a 19 base pairs spacers required in TN*501* (Parkhill and Brown, 1990), generally it is 15 bp. This places the binding domain (Summers, 1986) out of phase of 70°, as it might provide better alignment for the formation of an open complex. This site occupancy is unique as most of the repressor protein binds upstream of -35 except for l c11 and p22 which bind at -35 consensus region. In the absence of MerR protein, RNA polymerase competes for P_I and P_R transcriptional promoters and transcription is dominated mainly by P_R. As the concentration of MerR increases, both the promoters gets occupied, thereby inhibiting the transcription of either of them (Summers, 1986). But when Hg^{2+} is present at or above the threshold level (5.0-10.0 mM) MerR protein binds to it (Jeffery *et al.*, 1991) and there is allosteric modulation which diminishes MerR affinity for the operator of P_R and thus allosterically modified MerR and RNA polymerase synergistically forms an open complex at P_T to transcribe structural genes located in the rightward direction.

These stated conclusions are based on transcriptional runoff, gel shift assay and detailed protection data when treated with $KMNO_4$ (Rubin and Schmid, 1980), DMS and fenton reagent (Fe^{2+}/H_2O_2) by O'Halloran (1989). MerR has four important domains Helix turn helix (HTH), motif present at N-terminal, interface domain for the interaction of two subunits, Hg^{2+} binding domain and C-terminal for organo mercurial regulation (Nucifora *et al.*, 1989a).

There is unique similarity between MerR and MerD, both have HTH motif at N-terminal, and have atleast three cysteine to form trigonal structure with Hg^{2+} (Helmann *et al.*, 1990). MerR binds from -8 to +22 while MerR from -10 to +25 (relative to *mer*R initiation codon demonstrated by nuclease SI protection assay). Apparently, when MerD concentration reaches a critical level in the cell, it binds with the operator which deactivates P_R and P_{TPCAD} partially suggesting MerD is involved in fine tuning expression of the *mer* operon and acts as a co-regulator (Mukhopadhyay *et al.*, 1991 and Nucifora, 1989b).

Biosynthesis of structural proteins : The first gene transcribed after receiving mercury stimulus is *mer*I which is a transport protein of 116 amino acids. It has extensive hydrophobic regions and hence it is membrane bound and crosses the membrane thrice, it has atleast two pairs of cysteine, one facing the periplasmic and the other cytosolic side. Second gene which is transcribed is responsible for the synthesis of *mer*P (91 amino acids), which is periplasmic and has a pair of cysteine (Lund and Brown, 1987). The third gene is *mer*R [140 amino acids in Tn21 and 143 amino acids in *Thiobacillus ferroxidans* (R100)]. In both Tn21 and R100 the position of the cysteine coding region is conserved. *mer*C locus is exclusively found in *E. coli* and is related to facultative anaerobic bacteria that has limited host range (Gilbert and Summers, 1988). The fourth gene is *mer*A responsible for mercury reductase enzyme which catalyzes the conversion of Hg^{2+} to Hg^0 (Hg^{2+} -Hg^0). It's size varies from 474 amino acids in *Streptomyces lividans* to 631 amino acids in *Bacillus* sp. It is an oxidoreductase flavo protein and the reduction is NADPH dependent. Monomeric polypeptide unit of MerA protein possesses following features :

(a) Transient mercury binding site at N-terminal,
(b) NADPH binding site,
(c) FAD binding site,
(d) Subunit interface domain (hydrophobic),
(e) Redox disulphide side (enzyme active site).

NADPH binds to the NADPH binding site, the electrons after the reduction of NADPH, bind to the FAD binding site and then is transferred to the enzyme active site, at the same time Hg^{2+} is transferred from N-terminus to the cysteine present at the enzyme active site where Hg^{2+} accepts electrons and is reduced to Hg^0 (Fox and Walsh, 1982 and 1983). Fifth gene which is *mer*B is present only in broad spectrum resistance, is of 22.4 Kd and codes for monomer lyase enzyme which cleaves covalent bond between Hg and organic moiety *via* SE^2 mechanism (Begley *et al.*, 1986). It is this gene which is able to furnish Hg^{2+} from organo-mercurial compound and Hg^{2+} follows normal pathway as in narrow spectrum resistance. Last gene of the operon is *mer*I, which is 13.5 Kd (120 amino acids in Tn21 and 121 amino acids in Tn*501* and pDU1358).

Mercury ion transport : As mercury or organo-mercurial compounds are toxic, bacteria trap this mercury with the help of cysteine pairs present in the MerP protein and does not leave the free mercurial compounds over its surface and cytosol. As mercury in the environment increases, it enters the cell (mechanism is not known but mostly they are passively transported). Once it reaches a minimal concentration in the cytosol, it binds with MerR and triggers the transcription of structural genes (*mer*I, *mer*P, *mer*C, *mer*A, *mer*B and *mer*D and translates into proteins. The mercury trapped over the surface by the MerP is then transferred to cys pair of MerT facing periplasmic surface. This pair of cystein of MerT Hg is transported across the membrane and is transferred to another pair of cys facing cytosolic surface. From here the Hg and its compounds are delivered to MerC protein (if present) or is directly given to cys pair to MerB lyase protein (if the compound is organic and the organism is of broad spectrum resistance). From here Hg^{2+} is transferred to cysteine pair of MerA protein present at N-terminal. In narrow spectrum resistance, the Hg^{2+} is transferred from MerT to N-terminal of MerA directly via MerC protein.

Gram positive systems : In Gram positive bacteria the mercury resistance may be plasmid or chromosomal borne, for example p 1258 studied by Ladagga *et al.* (1987) from *Streptomyces aureus* where *mer* operon is present in 6.4 Kb Bg/II fragment. The same fragment is present in different *S. aureus* plasmids (Weiss *et al.*, 1977) and with chromosomal mercury resistance system. The mercury operon is flanked by either IS 431 or IS 257 on either side. IS 431 at the left has 800 bp and 78 bp at the right, bracketed by 22 or 14 bp inverted repeats between 1st and 46th base. This sequence include 234 amino acids which is homologous to chromosomal transposons of *Proteus vulgaris* (IS26) (Bearbaris *et al.*, 1987) which might be helping in transposition. In this, operon ORF2 is *mer*R (Helmann *et al.*, 1989) and structural genes are transcribed in the same direction. Deletion of ORF1 made the expression of the operon constitutive proving it's negatively controlled in contrast to the positive control in gram negative bacteria (Foster, 1983).

BIOSYNTHESIS OF STRUCTURAL PROTEINS

After canonical -35 and -10 and dyad symmetrical O/P there are four open reading frame (ORF2, ORF3, ORF4, ORF5) which are quite different from gram negative system. ORF2 codes 135 amino acids soluble protein. ORF3 has a leader sequence suggesting that it is membrane bound or extended protein, ORF4 is hydrophobic and twice the size of *mer*T. ORF5 is 128 amino acids protein with central hydrophobic region and is like MerT. The amino acids between position 47 to 61 is similar to amino acids from position 24 to 38 of *mer*T of R100 (Tn21). ORF5 is similar to *mer*T. After this ORF, there is *mer*A gene of 547. It is 66% dissimilar to R100 *mer*A product (Ladagga *et al.*, 1987), size of MerA in gram positive (*Bacillus* sp.) is 3.8 kb. However, the essential region like FAD, NADPH and enzyme active site N-terminus Hg^{2+} binding cystein site shows 90% similarity with gram negative bacteria suggesting the conservation during evolutionary drift. After *mer*A, there is *mer*B gene the N- and C-terminals 50 amino acids is barely related to gram negative MerB protein (size of gene is 2.9 kb in *Bacillus* sp.). However, the central 50 amino acids show 62% similarity including a stretch of 18/19 amino acids and a cystein pair. In *Bacillus* sp. RC-607, *mer*A shows 63% homology with p1258, *mer*B lies 2Kb distal to *mer*A (Wang *et al.*, 1987) similar to *E. coli* R831b (Barrineau *et al.*, 1984) MerA is 631 amino acids (Wang *et al.*, 1987) with a repeat of 79 aa at N-terminus means it has two binding sites of Hg (only one site is present in MerA of p1258).

Streptomyces lividans seems to be another interesting gram positive microbe where there are four divergent reading frames, ORF4 of p1258 is similar to the fourth one but others are quite dissimilar, MerA is 474 as long as has no mercury binding domain at N-terminus, like lipoamide dehydrogenase. Lyase shows 55% similarity with p1258.

CONTEMPORARY STATUS OF MERCURY REDUCTASE (MERA)

It is the most important protein of the *mer* operon. Size of the mercury reductase varies from 474 amino acids in *S. lividans* to 631 amino acids in *Bacillus* sp. But most common is of 561 amino

It is the most important protein of the *mer* operon. Size of the mercury reductase varies from 474 amino acids in *S. lividans* to 631 amino acids in *Bacillus* sp. But most common is of 561 amino acids. In *S. lividans*, the Hg^{2+} binding domain of about 90 aas is absent while this domain has duplicated in *Bacillus* sp. In Tn*501*, R100 (Tn*21*), pDU1358 they are 561 or 564 aas long. Basic features of the proteins are conserved in all the cases (both silent and replacement mutation minimum) for example the NADPH binding site, FAD site, redoxactive site and the interface domain. There is no sequence homology of gram positive and gram negative bacteria however, among gram negative strains the sequence homology is quite significant. In terms of similarity of this reductase with other systems of the same class have been well studied, it shows significant similarity with lipoamide dehydrogenase, glutathione reductase and tryptathione reductase shown by dot matrix analysis (Misra *et al.*, 1985). Limited homology of 457-478 at the C-terminus to several membrane bound protein suggests its transient association with membrane or it might help in subunit interaction.

the NADPH binding site, FAD site, redoxactive site and the interface domain. There is no sequence homology of gram positive and gram negative bacteria however, among gram negative strains the sequence homology is quite significant. In terms of similarity of this reductase with other systems of the same class have been well studied, it shows significant similarity with lipoamide dehydrogenase, glutathione reductase and tryptathione reductase shown by dot matrix analysis (Misra *et al.*, 1985). Limited homology of 457-478 at the C-terminus to several membrane bound protein suggests its transient association with membrane or it might help in subunit interaction.

Mercury reductase is cytoplasmic but in *E. coli* it is associated with inner membrane. Cytoplasmic location is advantageous for an enzyme requiring NADPH whereas association with membrane places it in close proximity of the substrate. It may be that the two forms of the enzyme have a biological role in increasing the efficiency of the system. Mercury reductase require a thiol or cysteine residue for ensuring reduced enzyme active site and EDTA (role not known). NADPH dependent violatalization of Hg^{2+} shows sigmoidal or biphasic rather than Michaelis-Menton Kinetics. The highest homologies with lipoamide dehydrogenase glutathione reductase is found in NADPH binding region. Lys 66 present in lipoamide dehydrogenase glutathione reductase is replaced by Ser at 143, Lys at 67 and Glu at 201 forms a salt bridge with conserved residue corresponding to 144 and Glu 283 which helps to expel NADP+ when produced. Remaining region of NADPH site between residue 217 to 224 is corresponding to 299 to 304 of R100 (*E. coli*) and residue 205 to 212 of lipoamide dehydrogenase. In FAD binding region, the sequence of b sheet is strongly conserved and the FAD interacts in the same way. Pyrophosphate binding region from position 104 to 109 of R100 enzyme deviates slightly. Substrate (Hg^{2+}) facing (si) face of the isoalloxazine are highly conserved in all three enzymes. The redox active site having Cys and the region of glutathion reductase Gly 55 to Pro 65 is corresponding to Gly 13 to Pro 142 of *mer* reductase.

In glutathione reductase tyrosine 197 protects reduced flavin from solvent and O_2. This residue is replaced by Ile in Lipoamide dehydrogenase while in mercury reductase it is replaced by Val which cannot protect flavin, therefore, presence of some other mechanism of protection can not be ruled out. Remaining homology around Tyr 197 is quite strong. C-terminal amino acids of different MerR is highly conserved (65 aas) but differ from glutathione dehydrogenase and lipoamide dehydrogenase. This region helps in subunit interaction and catalytic binding site Cys 561 and Cys 562 in R100.

EVOLUTIONARY CONSIDERATIONS

Homology of MerR with glutathion reductase and lipoamide dehydrogenase clearly indicates their evolutionary relationship (Misra *et al.*, 1985; Begley *et al.*, 1986; Brown *et al.*, 1986; Ladagga *et al.*, 1987). We can hypothesize that the glutathione reductase fused with mercury binding domain N-terminal extension of about 80 amino acids to produce *mer* reductase. This N-terminal extension shows 35% identity with 71 aas of MerP, Misra *et al.* (1984 and 1985) hypothesized that the ancestral gene of mercury binding polypeptide duplicated and one of them fused with shorter NADPH dependent oxidoreductase while other sequence fused with determinant of the membrane transporting leader sequence to produce the current MerP. Alternatively N-terminal of MerA may have duplicated and have fused with leader sequence to give rise to MerP and MerC. However, we can not rule out their independent origin as *mer* has region of similarity with *His*M, *His*Q (histidine transport system) and *Mal*F (maltose transport system) where though, the similarity might be signifying their anchorage in membrane which needs hydrophobic region. *Mer*C which is present in R100, *Thiobacillus ferroxidans* shows strong homologies with N-terminal of MerR. Excluding the first 22 aas of MerC which is a signal peptide they show homology with first 68 aas of *mer* reductase. Thus, the evolution of MerC may like MerP, is not universal and that is why its precise role is not yet established.

CONCLUSION

Mercury and organomercurial compounds are highly toxic. Their solubility in lipids as well as their binding to sulphydryl groups of protein in membranes and enzymes accounts for their cytotoxicity. Methyl mercury is 100 times more toxic than inorganic mercury and has been found to be mutagenic under experimental conditions. Investigations revealing the similarities and differences between the *mer* loci of Tn*21* and Tn*501* have facilitated studies of their genetic elements. With the exception of MerA, the *mer* gene products are relatively small molecules of the functional classes of proteins with which they are identified. Although there is much which is yet to be discovered about *mer* operon, the present knowledge accumulated so far can lay the firm foundation towards a prospective future which asks broader questions of its regulation, enzyme mechanism, ion transport, membrane structure, function and epidemiology of mercury resistant phenotypes.

REFERENCES

Barrineau, P. and A.O. Summers, 1983 : A second positive regulatory function in the *mer* (mercury resistance) operon. *Gene* 25 : 209-221.

Barrineau, P. *et al.*, 1984 : The DNA sequence of the mercury resistance operon of the Inc. FII plasmic NRI. *J. Mol. Appl. Genet.*, 2 : 601-619.

Bearbaris, M. *et al.*, 1987 : IS431 a Staphylococcal insertion sequence like element related to IS26 from *Proteus vulgaris*. *Gene* 59 : 107-113.

Begley, T.P. *et al.*, 1986 : Bacterial organomercurial lyase : overproduction, isolation and characterization. *Biochem.* 25 : 7186-7192.

Bridges, K., 1979. Gentamycin and silver resistant *Pseudomonas* and a burns unit. *Br. Med. J.* 1 : 446-449.

Brown. N.L. *et al.*, 1986 : The nucleotide sequence of the resistance operon of plasmids R100 and transposon Tn501 : further evidence for *mer* genes which enhance the activity of the mercuric ion detoxification system *M.G.G.* 202 : 143-151.

D'Itri, F.M., 1972 : *The environmental mercury problem.* CRC Press Boca Raton, Florida, USA.

Foster, T.J., 1983 : Plasmid determined resistance to antimicrobial drugs and toxic metal ions. *Microbiol. Rev.* 47 : 361-409.

Foster, T.J. and F. Ginnity, 1985 : Some mercurial resistance plasmids from different incompatibility groups specify MerR regulatory functions that both repress and induce the *mer* operon of plasmid R100 *J. Bact.*, 162 : 773-776.

Fox B. and C.T. Walsh, 1982 : Mercuric reductase purification and characterization of transposon encoded flavoprotein containing and oxidation reduction act like disulfide. *J.B.C.*, 257 : 2498-2503.

Fox, B. and C.T. Walsh, 1983 : Mercuric reductase homology to glutathion reductase and lipoamide dehydrogenase : iodo acetate alkylation and sequence of the active site peptide. *Biochem.*, 22 : 4082-4088.

Gilbert, M.P. and A.O. Summers, 1988 : The distribution and divergence of DNA sequence related to the Tn*21* and Tn*501* *mer* operon plasmid. *Plasmid*, 20 : 127-136.

Helmann, J.D. *et al.*, 1989 : Homologous metalloregulatory proteins from both gram positive and gram negative bacteria control transcription of mercury resistance operons. *J. Bact.*, 171 : 222-229.

Helman, J.D. *et al.*, 1990 : The MerR metalloregulatory protein binds mercury ion as tri coordinate metal bridged dimer. *Science*, 247 : 946-948.

Jefferies, T.W., 1982 : The microbiology of mercury. *Prog. Microbiol.*, 16 : 21-75.

Jeffery, S.S. *et al.*, 1991 : Transcriptional analysis of the *Staphylococcus aureus* plasmid p. 1258. Mercury resistance determinant. *J. Bact.*, 7 : 5234-5238.

Joensuu, Q.C., 1971 : Fossil Fuels as source of mercury pollution. *Science*, 172 : 1027-1028.

Ladagga, R.A. *et al.*, 1987 : Nucleotide sequence and expression of the mercurial resistance operon from *Staphylococcus aureus* plasmid. p. 1258. *PNAS (USA)*, 84 : 5106-5110.

Lund, P.A. and N.L. Brown, 1987 : Role of the *mer*T and *mer*P gene product of transposon Tn*501* in the induction and expression of resistance to mercuric ions. *Gene*, 52 : 207-214.

Misra, T.K. *et al.*, 1984 : Mercuric ion resistance operons of plasmids R100 and transposons Tn*501*, the beginning of the operon including the regulatory region and the first two structural genes. *PNAS (USA)*, 81 : 5975-5979.

Misra, T.K. *et al.*, 1985 : Mercuric reductase structural genes from plasmids R100 and transposon Tn*501* : functional domains of the enzymes. *Gene*, 34 : 253-262.

Mukhopadhyay, D. *et al.*, 1991 : Purification and functional characterization of MerD a coregulator of the mercury resistance operon in gram negative bacteria. *J.B.C.*, 266 : 18538-18542.

Nucifora, G.L. *et al.*, 1989a : Mercury operon regulation by the *mer*R gene of the organomercurial resistance systems of plasmid pDU1358. *J. Bact.*, 171 : 4241-4247.

Nucifora, G.L., 1989b : Down regulation of *mer* operon function by the production of *mer*D gene of the plasmid pDU1358. *M.G.G.*, 220 : 69-72.

O'Halloran, T.V., 1989 : The *Mer*R heavy metal receptor mediates positive activation in a topologically novel transcription complex. *Cell*, 56 : 119-129.

Parkhill, J. and N.L. Brown, 1990 : Site specific insertion and deletion mutants in the *mer* promoter-operator region of Tn*501*; the 19 bp spacer is essential for normal induction of the promoter by MerR. *Nucleic Acid Res.*, 18 : 5157-5162.

Rajni, R. and A. Mahadevan, 1994 : Cloning and expression of the mercury resistance genes of marine. *Pseudomonas* sp. strain MRI plasmid pMRI in *E. coli., Res. Microbiol.*, 145 : 121-127.

Rochelle, P.A. *et al.*, 1991 : Distribution of DNA sequences encoding narrow and broad spectrum mercury resistance. *App. Env. Microbiol.*, 57 : 581-589.

Rubin, C.M. and C.W. Schmid, 1980 : Pyrimidine specific chemical reaction useful for DNA sequencing. *Nucleic. Acid Res.*, 20 : 5613-5619.

Silver, S. and T.K. Misra, 1988 : Plasmid mediated heavy metal resistance. *Ann. Rev. of Microbiol.*, 42 : 717-743.

Soldano, B.A. *et al.*, 1975 : Air borne organomercury and elemental mercury emissions with emphasis on central sewage facilities. *Atmos. Environ.*, 9 : 941-944.

Summers, A.O., 1986 : Organization, expression evolution of genes for mercury resistance. *Ann. Rev. Microbiol.*, 40 : 607-634.

Summers, A.O., 1992 : Untwist and shout : a heavy metal responsive transcriptional regulator. *J. Bact.*, 174 : 3097-3101.

Walsh, C.T. *et al.*, 1988 : Molecular basis of bacterial resistance to organomercurial and inorganic mercuric salts. *FASEB J.*, 2 : 124-130.

Wang, Y. *et al.*, 1987 : Cloning and expression in *E. coli* of chromosomal mercury resistance genes for *Bacillus* sp. *J. Bact.*, 169 : 4848-4851.

Weiss, A.A. *et al.*, 1977 : Mercury and organomercurial resistance determined by plasmids in *Staphylococcus aureus. J. Bact.*, 132 : 197-208.

Chapter 16

Environmental Degradation in Indian Rivers : A Biological Perspective

V.V. Sugunan and G.K. Vinci

INTRODUCTION

A large number of rivers, small and big, with their ramifying tributaries cascade down the various hill ranges of the country, draining divergent geo-climatic regions and dissecting the Indian landscape into a panoramic trelliswork of land and water. The river systems of the country comprise 14 major rivers each draining a catchment of 20,000 km^2 or more, 44 medium rivers having an average catchment between 2,000 and 20,000 km^2 and the innumerable small rivers and desert streams that have an average drainage of less than 2,000 km^2. While the major rivers carry an annual runoff of 140.6 million hectare meters (mhm) of water, the medium and small rivers discharge 11.2 mhm and 12.7 mhm respectively. Thus, the major rivers share among themselves, 85% of the total surface drainage of the country leaving 7% and 8% to the medium and small ones respectively. The combined linear length of all the rivers in the country along with their tributaries and distributaries is estimated at 45,000 km which carry 6% of the total drainage of the surface of the globe.

Harmony between man and nature is the essence of India's religious-cultural ethos and thus, the rivers in the country have been venerated and worshipped since Vedic times. Among them, Ganga, the presiding river, is not only the lifeline of the country's economy, but it is also considered as the cultural mainstream, epitomising a supreme symbol of purity. But ironically, the very same Ganga and many other rivers in India are subjected to desecation of the worst order, causing concern among all the right thinking people across the country. Apart from receiving untreated sewage and a variety of industrial and agricultural wastes, the riverine environment is subjected to a number of man-made changes in the hydrodynamics and the physiography, that have far-reaching ecological implications. Environmental degradation in the riverine ecosystem is a universal phenomenon which, if left uncontrolled, can upset the natural balance and threaten the very survival of human race on this planet.

BIOLOGICAL IMPACT OF ENVIRONMENTAL PERTURBATIONS IN THE RIVERINE ECOSYSTEM

The environment and biotic communities together constitute an ecosystem. While community refers to the living part of the ecosystem, the environment embodies the total physical, chemical, geological and meteorological factors which exert an effect on these living assemblages. Any community of animals and plants, under normal conditions, will be in a state of dynamic equilibrium. Some will die, some will be born and so on, but this general picture of the community as a whole will remain much the same. Harvestable living resources in the rivers comprise the populations belonging to reckton which form only a section of the community at large. The abundance and distribution of these target populations depend on a very complex community metabolism by which the solar energy trapped at primary producer level passes through different communities of organisms before a fraction of it finds its way to the terminal phase. Community succession is a series of rhythmic by a number of habitat variables. Environmental stress inflicted upon any component populations affects the community metabolism and succession and is reflected in the entire community and ultimately impairs the production at fish level. Therefore, the habitat constraints which have no direct bearing on fish can also affect the fish productivity indirectly. Conservation of the whole community, rather than the specific economic species, is therefore imperative in preserving the biological wealth from the rivers.

ENVIRONMENTAL CONSTRAINTS

Basin modifications due to anthropogenic reasons and their ill-effects on the riverine environment are well-chronicled, especially from public health, aesthetic and economic angles. Common sources of environmental constraints and their effects are portrayed in Table 16.1. The man-induced environmental stresses in the riverine ecosystem, having a direct bearing on the biological production functions, can be broadly grouped into five major categories *viz.*, water abstraction, sedimentation, river training, dams and effluxion.

Table 16.1. Environmental constraints and their effects on aquatic ecosystem

Constraints	*Source*	*Effects*
Nutrient loading	Fertilizers, Sewage	Algal blooms, marine life destruction
Chlorinated hydrocarbons : Pesticides (DDT, PCBs etc.)	Agricultural runoff, Industrial wastes	Contaminated and diseased finfish and shellfish
Petroleum hydrocarbons	Oil-spills, industrial discharge, urban runoff	Ecosystem destruction
Heavy metals : arsenic, mercury cadmium, copper, lead, zinc	Industrial wastes, mining	Diseased and contaminated fishes
Silt load/Particulate matter	Soil erosion, poor basin management, fly-ash	Smoothers benthic communities, destroys juvenile stages of finfish and shellfish, affects recruitment blocks light needed by aquatic flora
Plastics litter	Household wastes	Strangles and destroys natural habitats
Temperature	Thermal plants	Fish kill, destruction of other biota, eutrophication
Reduced flow rate	Water abstraction	Acceleration of sedimentation rate, destruction of breeding grounds, impeding migration, irrational fishing

WATER ABSTRACTION

Although man has been harnessing the river water for irrigation and domestic purposes for centuries, these water abstractions were too insignificant in the past and affected the discharge rate in the rivers. However, modern technology had made it possible to tame even largest of the rivers through gigantic dams and huge storage reservoirs for retaining or diverting water for hydropower generation, irrigation, industries, drinking water supply and a number of other water oriented activities. Large scale diversion of water for off-stream purposes and the resultant diminutive flow in the main channel pose serious threat to the stability of local communities in many river basins of the world.

Environmental changes associated with water abstraction manifest themselves through velocity, depth, width of the channel and the solid transport, each of them having its own impact on the riverine biotic communities. The most important among the abiotic factors influenced by the flow regime is the velocity. The fluviatile biocoenos are highly oriented to currents of water. A fall in velocity may physiologically upset the communities especially adapted for the rapids, which may give way to populations adapted for lesser velocities (like the ones living in the marginal pools). This may upset the normal primary community succession and lead to a decrease in species diversity. Elimination of flooding due to river storage facilities has had major adverse affects on many marshes, estuaries and stream-associated vegetation. A scanty flow encourages colonization of aquatic macrophytes which upsets the normal phytoplankton-based chain of energy transformation and thereby retards fish productivity. There are instances of stabilized low-flow leading to luxuriant growth of water hyacinth in Tuolumne river in California.

Discharge rate plays a vital role in the dilution of contaminants thereby ameliorating the hazards of pollution as a reduced flow can aggravate the situation in the affected stretches. In many advanced countries, flow augmentation has been used as a tool to mitigate problems related to water pollution.

The amount of discharge also determines depth in a channel of a given configuration. At least in some of their life habits most lotic species are linked in their preference for depth of water. Low depth increase the photosynthetic activities through better penetration, especially in the pool zones, to upset the normal balance of species. Conversely, drastic reduction in depth can lead to direct fish kill, as fish being the largest member of nekton, require maximum amount of water.

Width of a channel of given configuration is similarly determined by the discharge and can be important for the life-cycle of fish in terms of their spawning habits, production of fish food organisms, etc. Flow regime also affects the temperature, transport of organic material, sediment transport and other water quality factors. For example, drastic changes in the salinity gradients in the deltas and estuaries due to abnormal river discharge sometimes become cause of serious concern.

Effects on the biota

One of the direct impacts on fish populations stems from the fact that lotic environment is a specialised habitat, the inhabitants of which are geared to centuries of evolutionary adjustments to the seasonal variations of their flowing water environment. Many riverine fishes have a preference for particular velocities and any man-made changes in the stream flow regime can upset the physiological rhythm of fishes. Many fish populations are dependent upon annual flooding for food and spawning.

Stream-flow rate has a direct impact on the migratory habit of fishes. Discharge can cause migration to commence, create barriers at high or low flows, cause delays, disrupt normal routing and change the speed of travel. In many cases, it is the nuances in discharge rates that trigger the migratory instincts. Distinct correlations have been established between different aspects of migration in case of pink salmon and the discharge rates. Reymond (1969) has proved that juvenile chinook salmon used to migrate from Ice Harbour Dam to the Dalles Dam in 14 days @ 18 km per day before the John Day reservoir was formed. After the dam and the consequent fall in discharge rate, the fish took 22 days to cover the distance @ 11 km per day. The stimulus provided by natural freshets can be a significant factor in the migration of juveniles. According to Chapman (1965) freshets cause the downstream movements of juvenile coho salmon.

Considerable knowledge is available on the spawning behaviour of salmon which has bearing on the change in stream velocity and depth. Velcoity of stream flow is an important factor in salmon and trout redd construction, fertilization of eggs and oxygen supply to the eggs. Salmon appears to select gravels with an adequate oxygen supply by sensing a current or upwelling of flow through the gravels. Investigations have revealed that salmon and trout have a rather narrow tolerance to velocity and depth when choosing spawning areas (Fraser, 1972).

Rearing capacity of the river, determined by the survival at hatching and nursery phase of the offspring, depends, to a large extent, on the stream flow. Some fish eggs require a flow of well-oxygenated water through the gravels in which they are incubating. Discharge also influences the fish food species composition and total production as well as the availability of shelter. Most stream fishes exhibit a strong territorial orientation (Onodera, 1962, Allen, 1969) and this territoriality is often recognised in terms of velocity and current. Kelleberg (1958) noted that with reduced flow regimen, fishes were forced to select less desirable feeding stations because of the expanded territories of the more aggressive individuals. Competition for food and space forces the less aggressive groups to remain in smaller size units over longer periods thus exposing them to predators. As a result, some fish populations get eliminated because of the territorial influence. Flow regime also affects the harvestability of fish. Reduce flows or excessive discharge can negate the use of boats, nets and other means which have commonly been used. One of the negative influences often noticed, relates to the indiscriminate and irrational modes of harvest practised in rivers taking advantage of the diminished water flow.

SEDIMENTATION

The catchment of Indian rivers is characterised by a prolonged dry season followed by a turbulant monsoon and flood discharge, causing high rate of natural sedimentation. This sedimentation rate is further accelerated by the removal of vegetative cover to the catchment area and the subsequent erosion of top soil in almost all the rivers and their tributaries. Erosion of top soil in the catchment area

Table 16.2. Sediment rate in some rivers of India

River	*Site*	*Sediment rate (m^3/sq.km/year)*
Chambal	Gandhi Sagar	365
Ganga	Farakkha	560
Kosi	Barahakshetra	2,000
Teesta	Anderson Bridge	5,148
Damodar	Panchet	1,075

is the main man-made factor that leads to increased sediment load in rivers. Vegetative cover on the slopes acts as an adherent of top soil during the surface runoff. Removal of forest cover in the slopes for logging, grazing of cattle, road making or for human settlements make the soil susceptible to erosion, thus adding to the sediment load in the river. Rivers in the world are known to carry as much as 3 billion ton of material in solution and 10 billion ton of sediment every year. Sediment load of different rivers are shown in Table 16.2.

RIVER TRAINING

The main purposes of river training is to allow the movement of bed and suspended load without damaging the banks. Protection of banks becomes necessary to control floods, to safeguard towns, roads, railways etc., and to facilitate navigation. The most common river training devices are the guide banks, spurs (groynes) and river revetments. The spurs are essentially the devices to nudge the stream flow to a desired channel, whereas the river revetments smoothens the banks by masonary or other structures to prevent the river from excoriating its banks. Almost all forms of river training have been practised in India for various purposes. Guide banks are generally constructed upstream of barrage and canal headworks. A typical bank protection work through stone spurs in the Ganga can be seen at Mansi. A more extensively used flood control measure in the Ganga is bank reveting or channelisation as has been done in Varanasi and Patna.

River training, in various forms, upsets the energy expenditure balance in rivers. The embankments decrease the roughness on friction along the river banks, but this leads to increased velocity and accelerated transport of bed material through the protected stretch. Such disequilibrium created due to bank embankments tends to readjust at points where stream channelisation ends and destruction of bank and bed resumes with added vigour.

One of the direct impacts of river embankment is the increase in velocity which upsets the lifehabits of organisms. River training may also lead to deepening of channels and they often severe the natural links with oxbow lakes and deep pools upsetting the biological cycle of many organisms. It destroys the benthic and littoral habitats, the two key links in the trophic events leading to fish productivity. Increased velocity, depending on beds and the attendant turbulance, can adversely affect the migratory behaviour of fishes.

HYDRAULIC STRUCTURES

Effects of dams, barrages, weirs and other hydraulic structures on riverine ecosystems are manifested in three ways *viz.* :

(i) reduced discharge,

(ii) habitat destruction due to impoundment, and

(iii) obstruction of migratory pathways of fishes.

One of the immediate effects of impoundment is the sudden transformation of the lotic habitat into a lentic one. This cataclysmic habitat change is akin to the classical secondary community succession that takes place during earthquakes, wild fires, devastating floods, etc. A good number of planktonic, benthic and periphytic forms along with small nekton perish. Fishes and bigger nekton either escape to more congenial environs or try to adjust themselves to the changed conditions. Thus, the transformation totally upsets the lotic community structure, succession and metabolism. The lentic community that eventually develops invariably shows lesser diversity factors and are often accompanied by single species bloom and the resultant decrease in evenness. Low species diversity and high dominance factor accompanied by low evenness values in man-made lakes, compared to the lotic system has been reported by Sugunan (1991).

Impoundments of river water behind dams destroys the spawning grounds by inundation. Even if the fishes manage to breed in stimulated breeding grounds, the large-scale mortality of the eggs and spawn is common due to the lentic conditions in the reservoir. The impact of dams can be summed up as under :

(i) general reduction of preferred habitat,

(ii) changes in the plankton, benthos, periphyton and other fish food communities,

(iii) alterations in breeding grounds and shelter areas, and

(iv) alterations in water quality characteristics and sediment load.

Construction of barrages across the river causes drastic changes in the discharge rate, water quality and sediment load pattern in the downstream stretch with attendant biological implications. Changes in flow pattern due to water abstraction fail to inundate the spawning grounds of major carps. The spawn output was seriously impaired in river Sone, consequent to the construction of Rihand Dam a barrage at Indrapuri on this drainage (Shetty *et al.*, 1971). Fall in carp spawn output has also been reported due to construction of Konar and Tilaiya reservoirs.

Upstream migrants like hilsa are known to congregate below the dams which obstruct their migratory path and are prone to over-exploitation. The reduction in flow-rate tends to reduce the growing period of hatchlings in nursery areas. Erratic discharge rate reduces the velocity of water downstream, which leads to rise in average temperature affecting the species sensitive to temperature.

Water released from the reservoir is by and large stripped of its sediment load which distorts the energy dynamics downstream resulting in deepening of river bed. Sediment load plays an important role in transport of nutrients downstream. Since sediments pick up toxic substances by absorption, they have an ameliorative effect on the waste phase. Deep reservoirs are known to have klinograde oxygen distribution and the water released from deeper layer is, more often than not, deficient in oxygen. Thus, the reservoirs can cause oxygen depletion in the river downstream. But, quick oxygenation through diffusion from the atmosphere often corrects the situation.

Effects on migration of fishes

One of the direct effects of dams on fishes is obstruction in migratory pathway. Hilsa is a classical example of anadromous fishes getting affected due to obstruction in their up-river migratory path. Hilsa used to constitute a substantial riverine fishery in the Ganga, the Godavari, the Krishna, and the Cauvery. Upstream fisheries of hilsa in all the above rivers have collapsed due to construction of barrages at Farakkha, Dowleswaram, Vijayawada and Mettur respectively. Freshwater eels are known to undertake long seaward migration for breeding and their offsprings ascend the river to return home. Barrages prevent the adult eels to migrate to the sea for breeding and the elvers fail to negotiate the homeward route. Apart from these fishes, freshwater prawns, like *Macrobrachium rosenbergii* and *M. malcolmsonii* which normally live in freshwater, migrate down to the estuaries for breeding. Dams and barrages threaten to obliterate these freshwater prawn populations due to constant recruitment failures. Similarly, *Pangasius pangasius*, the riverine catfish is known to migrate downstream for breeding and their population is also adversely affected due to construction of dams across the rivers. But this remarkable fish has the capacity to adapt itself to the changed environment. In some reservoirs, it has not only started breeding within the reservoir but has also become the most dominant fishery.

EFFLUXION

Rivers, in the past, were considered to be the ideal places for waste disposal (effluxion), perhaps prompted by the irrational belief and misplaced sense of safety that the dirt is carried away by the running water. Thus, the rivers were increasingly subjected to the onslaught of industrial, municipal and agricultural wastes, commensurate with the rapid pace of industrialisation and urbanization since the days of industrial revolution, about 75 years ago. As the human activities in the river basins escalated, it became apparent that the running waters of the river did not solve the problems of pollution, but only shifted them to other areas. In more recent years, on account of the reduced discharge rates and balconisation of rivers, offending wastes even stopped moving downstream, but staying in the vicinity to punish the offenders.

Biological impact of industrial pollution

The industrial effluents include a wide variety of chemical toxicants and heavy metals apart from contributing substantially to the BOD load. Such effluents also include large quantities of pesticides which are used in processing the raw materials in many industries. In addition to the sub-lethal chronic effects on the environment, certain direct impacts are also discernible. Tannery, textile and other mixed organic wastes cause depletion of dissolved oxygen and high BOD load (61 t/day) at Kanpur. Apart from direct fish kill, the fertilizer wastes at Allahabad adversely affect the populations of carps, catfishes and murrels. Plankton and benthic fauna are known to disappear upto a stretch of 300 km downstream due to high pH and ammonia toxicity. Ammonia is known to lead to hyperplasia of gill epithelium and at a concentration of 1.2 ppm or above, it is toxic to marine diatoms (Natarajan, 1970).

The oil-bearing wastes at Barauni affect the major and minor carp populations and cause periodic mortalities. The enormous 1134×10^3 of industrial wastes discharged into the Hooghly from the 96 units in and around the Calcutta metropolis contain lignin, various types of chemicals, heavy metals, pesticides, detergents etc. Calcutta effluents cause a high BOD load (106 t/day) and reduce the DO level. Metals and pesticides are accumulated in the sediment phase and there is a definite decline in the rate of primary productivity. The impacts on aquatic life include reduction in plankton population, absence of benthos around the outfall and bioaccumulation of Zn, Cu, Cr, DDT and BHC in fish and other aquatic organisms.

Effluents received by the Yamuna at Delhi contain chemicals, DDT factory wastes, oil and greese, heavy metals and detergents causing a high BOD load of 30 t per day, organic pollution and DDT residues in water (0.6-3.5 ppb) and soil. Bioaccumulation of DDT is detected in fish and other biota. While at Mathura effluents are characterised by oil bearing and acidic wastes, Agra receives trade and tannery wastes. Free chlorine washed down to the Sone at Amla causes total depletion of dissolved oxygen. Industrial effluent load in the Gomti, detected at Lucknow to the extent of 95,670 m^3/day contains toxic chemicals causing depletion of dissolved oxygen and high BOD load. Extensive fish mortality is common in the Gomti due to industrial wastes. Sindri, Bokaro and Asansol-Durgapur belt on the Damodar is subjected to heavy load of effluents from fertilizer plants, steel plants and collieries. The pollutants include alkalies, chromates, ammonia, cyanide, phenol, naphthalene and fine coal particles from coal fisheries. Chemical pollution becomes acute during summer leading to mass mortality of fishes. The BOD load reaches 43 t/day and the total load of coal particles is estimated at 43 t/day. Regular fish mortality is reported from a 3 km stretch of Suvaon at Balrampur due to high BOD, low DO and generation of H_2S caused by sulphates and sugar factory wastes. Industrial wastes at Bulandshaher in the Kali; originating from sugar, distillery, paints and rayon etc., lead to blackening of river bed and anoxic conditions in water leading to fish mortality. Deoxygenation and formation of ferric hydroxide in water result from the presence of iron in the river bed at Siwan. This leads to heavy mortality of small-size fishes due to choking of their gills by ferric hydroxide.

Toxicity tests with industrial effluents revealed that cotton textile effluent is most toxic to *Macrobrachium* sp. under continuous exposure as indicated by 96 hr LC_{50} value (6.5% by volume) followed by paints and varnishes (19-20%), rubber (32%) and rayon effluents (31-52%). The distillery wastes are toxic to *P. sophore* at 6.7-8.1% and to *M. vittatus* at 11%, while 96 hr LC_{50} values for pulp and paper mill effluents for *P. sophore* and *M. vittatus* lie within 20.6 to 34.7%. Cycle rim factory effluents are highly toxic to *C. catla* and *L. rohita* (24 h LC_{50}, 0.6% and 0.85%). Coal tar effluents have been found to be most toxic to fish food organisms like *Heliodiaptomus viddus* (96 hr LC_{50} : 0.243%). Fluorine at 4.84 ppm is toxic to *Catla catla* causing 50% mortality (when used as NaF) in 96 hrs. Experiments with radionuclide ^{65}Zn have indicated higher concentration factor in marine fish (5000) compared to the freshwater ones (1000). Daphnia absorbs these materials at a higher rate.

Crude oil and oil fractions may form coating over the gills leading to direct fish kills. Even highly reined diesel engine lubricating oil is known to induce tumours in the digestive tract of animals. Phenolic substances discharged from petrochemical complexes and washings from coal mines may cause paralyses of nervous system and cardiovascular congestion in fish. With the addition of detergents to phenol, the toxicity of the latter is increased due to accumulation.

Biological impact of municipal wastes

The major adverse effects of sewage pollution are deoxygenation, high BOD load, rapid eutrophication and accumulation of heavy metals in the environment. Sharp fall in dissolved oxygen of water puts the biotic communities under severe stress. While some species can tolerate a wide range of dissolved oxygen, many communities are highly sensitive to this parameter. Chronic effects of elimination of some component populations from the riverine community due to oxygen depletion may cause far reaching changes in the trophic cycle. For instance, complete absence of zooplankton during January to August and its reappearance in September represented by *Keratella* sp., associated with abundance of phytoplankton like *Microcystis* sp., *Oscillatoria* sp., *Hormidium* sp., and *Nitzschia* sp., have been observed in the downstream of the Ganga and Yamuna. The outfall area is dominated by *Chironomus* sp., followed by oligochaets (*Tubifex* sp., and *Nais*) in both the rivers while areas below the outfall are characterised by the dominance of *Chironous*, followed by gastropods and bivalves.

Apart from affecting the organisms at lower trophic levels, intensive rate of pollution from municipal sources often causes direct fish kills especially in smaller streams, where the problem gets aggravated due to reduced water flow rate. In the United States, more than 50% of the fish kills from 1960 to 1968 due to pollution were attributed to oxygen demanding non-toxic waste. The organic pollution in rivers is monitored through :

1. indicators of organic pollution like BOD, COD, nitrogen and dissolved oxygen;
2. bacterial indicators;

3. major mineral constituents like specific conductivity, pH, chloride, sulphates, sodium, calcium, magnesium, hardness and alkalinity; and
4. physical parameters like temperature, velocity and turbidity.

Some degree of self-purification has been ascribed to the Ganga water in terms of bacterial load. It has been reported that 65% of the *E. coli* organisms, 75% of the fecal *streptococci* organisms and 99.8% of the *Vibrio cholerae* are killed within 24 hours in the water samples (at 30°C ± 2°C) collected from the Ganga at Patna. On the contrary, these microorganisms were found to have survived much longer in the waters drawn from Sone, Gandak, dug wells and tube wells in Patna. They survived two to four times longer in the Sone water and nearly twice in Gandak water. Owing to the increased use of synthetic detergents for domestic purposes, their incidence in the sewage effluents is going up. Synthetic detergents being absorbed into the body system of fish impair their growth and reproductive capacity. Detergents mixed with oil may be 60 times more toxic than the oil alone. Synergistic action of detergents with insecticides has also been recorded. Their sub-lethal concentration causes thinning and elongation of respiratory epithelial cells. Sodium lauryl sulphate is more toxic to freshwater teleosts, compared to alkyl sulphonate (13-60 ppm).

The maximum amount of water is consumed by the agricultural sector which draws freshwater to the tune of 134,484 million cubic meter. A substantial portion of this water is retained as moisture, and a portion is lost through evapo-transpiration. Still, there would possibly be a substantial balance of nearly 20% of total supply which goes off from the irrigated tracts by surface seepage and overland flow as wastewater gravitating towards the natural drainage system (Das Gupta, 1984). The agricultural sector, thus generates a wastewater discharge of 26,896 mcm per year. The agricultural runoff affects the riverine environment mainly in three ways viz.

1. increase in salt and alkali levels in water,
2. increase in nutrient load due to the residual effects of chemical fertilizers, and
3. the accumulation of pesticides in the environment.

Continued irrigation has helped possible building up of the salt and alkali levels in the cultivated soils. In the entire irrigation command area, especially wherever the drainage is poor, the salinity level and alkali status in the soils rise to an appreciable degree. From these areas, there is constant sub-surface seepage. The flow of waste water here is charged with salts and alkalis, which eventually find their way to the river waters in the basin. Excess salt coming from irrigation waters is largely responsible for raising the salinity level in the Yamuna especially near Agra. In the main Ganga too, a similar trend of raising salinity is observed from Haridwar to Kachla bridge. There is also a consistent upward trend in the alkalinity status of the Ganga. Various chemical inputs in the agriculture comprising fertilizers, insecticides, pesticides and weedicides are continuously being added to the soil of various river basins with their impact on the water quality of the respective rivers. It is estimated that 10-15% of the nutrients added to the soils eventually find their way to the surface flow (Alderfer and Lovelance, 1977).

Among the pollutional hazards from agricultural sector, the damage caused by the pesticides is the most lethal and interminable to the environment. Of the three types of pesticides viz., organochlorides, organophosphates and carbamates, used in agriculture, organochlorines are lipophilic, extremely toxic and non-biodegradable. Like heavy metals, they assume alarming proportions as they are prone to biomagnification and accumulation in fish tissues posing serious threat to the fish eating public.

Most of the commonly used pesticides in India like DDT, BHC, endosulfan, ethyl parathion, dimethorate, phosphamidon, carbaryl and 2,4-D have been screened to evaluate their toxicity to fish and fish food organisms. It has been found that plankton and benthos are more sensitive to these chemicals than fishes. Thus, they seriously hamper the natural succession of fish food communities leading to low fish output. Studies have also indicated that DDT was acutely toxic to *Daphnia* sp., and *Chironomus* larvae (24 hr. LC_{50} 31 and 80 ppb respectively). The LC_{50} values obtained for the fish, *Colisa fasciata* and *Oreochromis mossambicus* were 230 ppb and 520 ppb respectively. Among the organophosphorus insecticides ethyl and methyl parathion are highly toxic to *Daphnia* sp. (24 hr. LC_{50} : 0.11-1.32 ppb) compared to fish *Colisa fasciata* (860 ppb). Similarly, malathion was found to be toxic to plankton and chironomid larvae. The toxicity (24 hr. LC_{50}) to *Daphnia carinata* has been in the order of methyl parathion (1.4) > DDT (4.8) > Carbaryl (17.8) > BHC (33.9 ppb). Endosulfan and quinalphos, the two commonly used pesticides in the paddy fields of West Bengal are known to show high toxicity to estuarine tiger prawn, *P. monodon*. Exposure to endosulfan and quinalphos, the two commonly used pesticides in the paddy fields of West Bengal are known to show high toxicity to estuarine tiger prawn, *P. monodon*. Exposure to endosulfan has led to enlargement of the epithelial cells with pyknotic nuclei, hypertrophy of the pillar cells and swollen erythrocyte cells in the gills of *Puntius stigma* (Wagh and Khillare, 1987).

Sub-lethal concentration of DDT and BHC adversely affect the fish at tissue levels. Damage of liver cells besides decline in growth, RBC count, Hb, and PVC level have been noticed in *Oreochromis mossambicus* at an exposure rate of 0.15 ppm of BHC for 90 days. Similar effects in the liver cells of *L. rohita* and *C. mrigala* were observed when exposed to DDT (0.005 ppm) for 30-60 days. Chronic exposure to low levels of DDT (5.2 ppb) and BHC (150 ppb) has resulted in lower growth and fecundity rates in *O. mossambicus*. Deleterious effects on reproduction, growth and haematological conditions of fishes were observed when exposed to sub-lethal concentrations (0.00017-2.950 ppm) of various organochlorine pesticides (Konar, 1969). Exposure to 0.1-1.0 ppb of DDT for 7 days inhibited the growth and induced shell thinning in gastropod, *Lymnaea leuteola*.

HEAVY METALS

Since the riverine environment is continuously exposed to various types of industrial and sewage runoff containing xenobiotic substances like the heavy metals, their presence in the environment and accumulation in tissues causes serious concern. Instances of heavy metal accumulation have been detected from more and more stretches of the river Ganga and its tributaries. The main source of heavy metal pollution in the river systems is the industrial effluents. However, municipal sewage is very often accompanied by trade wastes containing heavy metals. For instance, the Calcutta city sewage shows the presence of heavy metals like Zn, Cu, Cr, Cd, Pb and Hg (Joshi, 1987).

Heavy metals, as in the case of organochloride pesticides, constitute a very serious form of pollution, as these are stable and non-biodegradable. Therefore, unlike the other forms of pollutants they not only linger on to the ecosystem, they also pass onto the living tissues in increased concentrations through bio-magnification. The ecological implications of such residues are manifold. Apart from posing a public health hazard for the fish eater, heavy metal accumulation in the tissues of fishes and other organisms causes physiological disorders such as necrosis of liver, damage of nephros in kidney, haematological abberations, decline in growth rate and fecundity, and enlargement of gall bladder.

Accumulation of Zn in ground at a high level (148.8 ppm) was found to be detrimental to fish health affecting its reproductive potential in the lung run. Fish food organisms such as *Cyclops* and *Daphnia* are more sensitive to metals like Zn, and BHC. Prawns (*Macrobrachium rude* for instance) are highly sensitive to Cu both in freshwater (90 hr. LC_{50} : 0.065 ppm and saline (2 ppt) conditions (96 hr LC_{50} : 0.060 ppm). *L. parsia* was found to be very sensitive to Zn (96 hr LC_{50} : 19.13 ppm) even at high salinity (20 ppt) and hardness (3780 ppm). Zn, Cu and Cr in combination are many times more toxic to fish compared to their individual toxicity.

Organic mercury compounds, particularly the alkyl mercury ones are extremely dangerous as these are more stable and persistent (biological half life : 86-435 days). They affect the nervous system and lead to crippling diseases in fish and ultimately man. The transport of mercury in the environment is increased by its conversion into methyl mercury. Copper brings changes in the blood chemistry and inhibits certain enzyme actions. Lethal concentration of cadmium for fish varies from 0.07 to 10.0 ppm.

THERMAL POLLUTION

Thermal power generation capacity in the country has been registering a steady growth rate of 8% per annum and by the turn of the century, the installed capacity is expected to grow to 84,000 MW (Second Citizens' Report, 1985). Various thermal plants in the country are estimated to generate 10 billion m^3 of hot water (40-52°C) and 17 million tonnes of fly ash annually. Thermal wastes are reported to be carrying heavy metals like Zn (6%), Ba (12%), Cu (1.3%), As (0.02%), V (0.08%), TI (0.02%) and Mn (0.23%) which are destined to reach the nearest river stretch on reservoir (Sundareshan *et al.*, 1983).

Chandra *et al.* (1985), while studying the impact of thermal discharge into the Rihand reservoir, reported adverse effects of heated discharge on resident aquatic organisms. They recorded mortality of fish and absence of other aquatic life within 50 m of discharge point owing to high temperature (48-52°C) of effluents.

The main ecological consequences of heated discharge into the aquatic ecosystem are increase in water temperature, alteration in chemical parameters and change in metabolism and life history of aquatic communities. The heated discharge pushes up the temperature of receiving water by 8-10°C which may cause mortality of fish and fish food organisms. Temperature also exerts a direct influence on toxicity.

Apart from the rise in temperature, power station discharges are often altered chemically during the cooling process. Davies (1966) showed that cooling tower discharge has lower ammonia level, higher concentration of nitrates and TDS, and lower levels of organic nitrogen; when cooling water is abstracted from a polluted river. The studies made by Chandra *et al.* (1985) reveal that thermal wastes of Renusagar power station were acidic, low in alkalinity and high in chloride content. Increase in water temperature is known to cause deoxygenation, but the rise of temperature within a reasonable range increases photosynthesis activities resulting in supersaturation of oxygen.

Temperature of 40°C has been reported to affect the plankton communities. Thermal effluents reduce the plankton species diversity. A lower temperature (34.5°C) encourages the growth of Myxophyceae and further lowering of temperature (30°C) increases production of Bacillariophyceae (Terembley, 1965 quoted by Langford, 1972). High temperature (37°C) shows very deleterious effects on benthic organisms too. Like plankters, only a few tolerant species survive.

Generally, fishes avoid the heated effluent discharge points by swimming away instinctively to safer places. They can also withstand a wide fluctuation of temperature (8-10°C). However, their reproduction is adversely affected due to deposition of fly-ash in the marginal areas of river/reservoir which constitute their breeding ground. Most deleterious among the impacts of thermal pollution is the blanketing effect on the river/reservoir bed. Fly ash will cover extensive areas on the river bed blanketing off the substratum resulting in retardation or total elimination of benthic communities. Thick mat of fly ash accrued on the bottom over the years may seal the nutrients away from the water phase and thereby affecting productivity.

EFFECTS OF WATER QUALITY ORGANISMS

Specific impact of a habitat constraint on the biotic communities is difficult to assess, as the different parameters that exert pressure on the ecosystem are closely interrelated. It is not even possible to prescribe safe limits in respect of any of the chemical pollutants. In fact, the practice of prescribing safe limits is no longer valid as the emphasis now is shifted to a balanced ecosystem rather than to prevent fish kills. To protect the ecosystem from gradual degradation, we must provide criteria that will protect the entire life cycle of the desirable species as well as the food chain on which these species depend.

While dealing with the multiple effluents, not only the specific knowledge on the effluents but also a knowledge of the potential chemical and physical changes is imperative. In addition, we must comprehend the potential of combined stress on aquatic life, the effect of which cannot be explained on the basis of a single contaminant. The problem involving pH and metal toxicity are common where toxicity increases due to decrease in pH values.

REFERENCES

Alderfer, R.G. and E. Lovelance, 1977 : Relationship between water quality management planning and land use planning. In : *Handbook of water quality management planning* (Ed. J.L. Pavoni) Van Nostrand, new York, p. 59.

Allen, K.R., 1969 : Distinctive aspects of the ecology of stream fishes : a review. *J. Fish Res. Bd. Canada*, 26 : 1329-1438.

Anon., 1985 : India's state of environment - Second citizen's report. *Centre for Science and Environment*, New Delhi, p. 393.

Chandra, K., D.N. Singh and R.S. Panwar, 1985 : Possible pollution problem from wastes of thermal power plants in India - A case study of Rihand reservoir, Mirzapur (U.P.). In : *Proc. Symp. Assess. Environ. Pollut.*, (Eds. R.C. Dalela and U.H. Mane) : 283-294.

Chapman, D.W., 1965 : Net production of juvenile coho salmon in three Oregon streams. *Trans. Am. Fish. Soc.*, 94 : 40-52.

Das Gupta, S.P., 1984 : Basin sub-basin inventory of water pollution - The Ganga Basin Part II (excluding Yamuna Basin). *Central Board for the Prevention and Control of Water Pollution*, New Delhi, 204.

Davies, I., 1966 : Chemical changes in cooling water towers. *Inst. J. Air Water Pollut.*, 10 : 853.

Fraser, J.C., 1972 : Regulated discharge and the stream environment. In : *River ecology and man* (Eds. T.O. Ray, C.A. Carlson and J.A. McCann). Academic Press, New York, p. 465.

Joshi, H.C., 1987 : Pesticide residues in some fish ponds of West Bengal. *IAWAPT Technical Annual*, 14 : 35-38.

Kelleberg, H., 1958 : Observations in a stream tank of territoriality and competition in juvenile salmon and trout. *Inst. Freshwater Res.*, Drottningholm, Sweden, 39 : 55-98.

Konar, S.K., 1969 : Use of control chemicals in reservoir fishery management. In : *Proceedings of the Seminar on Ecology and Fisheries of freshwater Reservoirs*, Nov. 27-29 : 169-210.

Langford, T.E., 1972 : A comparative assessment of thermal effects on some British and north American rivers. In : *River ecology and man* (Eds. Ray T.O., C.A. Carlson and J.A. McCann), Academic Press, New York.

Natarajan, K.V., 1970 : Toxicity of ammonia to marine diatoms. *J. Wat. Pollu. Cont. fed.*, 42 : R184-190.

Onodera, K., 1962 : Carrying capacity in a trout stream. *Bull. Freshwater Fisheries Research Lab.*, (Tokyo, Japan), 12 : 1-41.

Rao, K.L., 1975 : *Water wealth of India*. Orient Longman, New Delhi, p. 262.

Reymond, H.L., 1969 : Effect of John Day Reservoir on the migration rate of juvenile chinook salmon in the Columbia River. *Trans. Am. Fish. Soc.*, 98 : 513-514.

Shetty, H.P.C., 1971 : Report on fish spawn prospecting investigations, 6, Rajasthan, Uttar Pradesh, Bihar, Assam, Tamil Nadu and Mysore. *Bull. Cent. Inland Fish. Res. Inst.*, Barrackpore, 15 : 60.

Sugunan, V.V., 1991 : Changes in phytoplankton species diversity indices due to artificial impoundment in river Krishna at Nagarjunasagar. *J. Inland Fish. Soc. India*, 23 : 64-74.

Sundarshan, B.B., P.V.R. Subramaniam and A.D. Bhide, 1983 : An overview of toxic and hazardous waste in India. *Industry and Environment*, UNEP (Special issue) : 70-73.

Trembley, F.J., 1965 : Effects of cooling water discharge from stream-electric power plants on stream biota. In : *Biological Problems in Water pollution*. 3rd Seminar, August 3-17, 1962. U.S. Deptt. Health Educ. & Welfare. Public Health Service Publ., 999 - WP-25.

Wagh, S.B. and Y.K. Khillare, 1987 : Sensitivity of gills of *Barbus stigma* (Ham.) to endosulfan, malathion and sevin. In : *The First Indian Fisheries Forum*, College of Fisheries, Mangalore, December, 4-8 : 195 (Abstract).

Chapter 17

Survival of Larger Mammals in the Thar Desert

P.I. Kankane

ABSTRACT

Man has always been the single most important factor in the ecosystem and to a large extent plays deciding role as to what will survive in his sphere of activities. The dispersion of 22 largest mammals in the Thar Desert was analysed in light of above and it was found that distribution of 11 species has direct link with man or man-made activities, survival of nine species is susceptible to change in landuse pattern and five species having religious association with man have comparatively better survival prospects.

INTRODUCTION

The Thar is the eastern most extremity of the Great Palaearcatic Desert which is the cradle of old world civilization. Consequently, the land of this area has witnessed many devastating wars, overgrazing and vagaries of climate for thousands of years. The march is still going on. Undoubtedly, human beings have always been the single most important factor throughout these episodes. But, at the same time man has also been a practical biologist systematising his knowledge of plants and animals amongst which and on some of which he lived. In the process he gave some of the most important crops and domestic animals to the society. However, unfortunately, as his knowledge on environment, and natural resources expanded, he started discriminating them in terms of beneficial and non-beneficial aspects with reference to himself. Thus, human behaviour started playing a dominant role in determining the vegetational complexes as well as the range of animals, those can persist in his sphere of activities. Each animal and plant has to face this challenge in its own way. During the process, some adapt to the changed environment, others gradually shrink to their natural habitat while the remaining, either disappeared or will disappear in course of time.

In the Thar, the concentration of wild mammals is comparatively higher than outside protected areas because plenty of different categories of degraded land offer them better alternate habitats. But, outside protected areas man is an important limiting factor. This paper deals with the above phenomenon considering larger mammals of the Thar as an example. An attempt has been made to correlate distribution of free living larger mammals in the Thar Desert with that of man and his cultural, religious and professional affiliations, being the important factors determining human behaviour.

STUDY AREA

The Indian part of the Thar Desert is bounded by the Aravalli hills in the east, the international boundary with Pakistan in the west, the grand salt marsh of Kutch in the south and the districts of Haryana and Punjab in north and lies between 22°30' N and 32°05' N latitude and from 68°05' to 75°45' E longitude. The great Indian desert (India and Pakistan) covers an area of nearly half of the Arabian desert and 1/7th of Saharan desert. The Indian Thar desert occupies about 2,78,330 sq. km. (12% of geographical area of the country) of which 1,96,150 sq km (70%) lies in Rajasthan; 62,180 sq km (22%) in Gujarat and 20,000 sq km (7%) in Punjab and Haryana states. This has an elevation of about 350-450 m above the sea level at the Aravalli range in the east, about 100 m in south and west and about 20 m in Rann of Kutch.

The western Rajasthan, characterized by sparse vegetation, is neither barren nor un-inhabited, it is covered with bushes and shrubs and even small trees. It is a great sandy tract with no streams and few rocks that protrude above the lower land now covered with sand, seeming to be immobile sand dunes. The grasses on these dunes grow in clumps, indicating the availability of water just beneath the sandy soil. The desert is roving ground for camels, buffaloes and cows which are known for their strength and size.

METHODS

As far as the mammalian diversity of the Thar desert is concerned, there are 68 species belonging to nine orders (Alfred and Agarwal, 1996) inhabitating the Thar desert. This constitutes about 18% of the total mammalian fauna of the country. Their size ranges from Nilgai to a Shrew. Hence, for the purpose of convenience, the desert dwelling mammals listed under Schedule I, II and III of the Wildlife Protection Act (1972) are being considered as larger mammals, so far as the scope of this paper is concerned. Using the above criteria a total of 22 large mammals belonging to five orders (Table 17.1) are being discussed here. The author during 1993-95 had the opportunity to survey the Thar desert under two year project titled "Status Survey of Chinkara and Desert Cat in Rajasthan". While findings of this project are to be communicated elsewhere (Kankane, unpublished) records kept on the distribution pattern and man animal relations of larger mammals, including two target animals, are the base of this paper. While 17 out of total 22 animals (Table 17.1) were observed during field survey, information regarding remaining five was collected through published literature (Kankane 1995, 1996; Prakash, 1963; Rodgers and Panwar, 1988).

Table 17.1. Larger mammals of the Thar desert and their association with man

Order/Species	*Common name*	*Legal Status WL(P)A 1972*	*Free living*	*Kind of Association*			*Commercial Value*
				Human Habitation	*Agriculture Fields*	*Religious*	
Primates							
Macaca mulatta	Rhesus macaque	II	Y	Y	-	Y	-
Presbytis entellus	Hanuman langur	II	Y	Y	-	Y	-
Pholidota							
Manis crassicaudata	Indian pangolin	I	—	-	-	-	-
Carnivora							
Canis lupus	Wolf	I	Y	Y	—	—	—
Canis aureus	Jackal	II	Y	Y	—	—	—
Vulpes vulpes	Desert fox	I	Y	—	Y	—	Y
Vulpes bengalensis	Indian fox	I	Y	—	Y	—	Y
Mellivora capensis	Ratel	I	—	—	—	—	—
Lutrogale perspicillata	Otter	II	Y	—	—	—	Y
Viverricula indica	Small Indian civet	II	—	—	—	—	—
Paradoxurus hermaphroditus	Toddy cat	II	—	—	—	—	—
Hyaena hyaena	Hyaena	III	Y	—	—	—	—
Felis chaus	Jungle cat	II	—	—	—	—	—
Felis silvestris	Desert cat	I	Y	—	Y	—	Y
Felis caracal	Caracal	I	—	—	—	—	—
Panthera pardus	Leopard	I	N	—	—	—	Y
Perissodactyla							
Equus hemionus khur	Wild ass	I	—	—	—	—	—
Artiodactyla							
Sus scrofa	Wild boar	III	—	—	Y	—	—
Boselaphus tragocamelus	Nilgai	III	Y	—	Y	Y	—
Antilope cervicapra	Black buck	I	Y	Y	Y	Y	—
Gazelle bennetti	Indian gazelle	I	Y	Y	Y	Y	—
Tetraceros quadricornis	Four-horned antilope	I	—	—	—	—	—

RESULTS

While protected areas of the Thar desert constitute only 4.3% of its total geographical area, the remaining (non-protected) wildlife habitats of the Thar, in contrast to other such areas, still hold viable populations of many wild animals, which is due to various favourable factors such as arid climate, geological and socio-cultural features prevailing in the above area. The main reason for this turnout is availability of lot of degraded land (habitat for wildlife) which is either not suitable for agriculture or water (surface/underground/canal) is not available for irrigation. Hence, agricultural practices and agricultural output entirely depend on rains. Accordingly, human migration continues and pace of human settlement and human density per sq km is lowest in this area, thus, providing one of the best alternate habitats to the desert fauna outside protected areas. However, such free-living wild animals are exposed to number of extraordinary threats not faced by their cousins inside protected areas. Man-made activities and his attitude towards wildlife is the single most important factor governing survival of such populations. Their relations are just like landlord and tenants. Hence, in order to workout man-animal association, all the larger mammals of the Thar have been tabulated (Table 17.1) using following criteria

1. *Free living* : Occurrence of wild animals outside protected areas.
2. *Human habitation* : Distribution of wild animals linked with human habitations. Mostly for food.
3. *Agriculture* : Distribution of wild animals linked with agricultural fields mostly for food and shelter.
4. *Religion* : Distribution of wild animals tolerated in human settlements on religious ground.
5. *Commercial value* : Distribution of wild animals is governed by their large scale commercial exploitation.

It is evident (Table 17.1) that out of total 22 mammals, eleven species are distributed in association with man or man-made activities. Five of them, having religious association with some sects of people, are *Presbytis entellus, Macaca mulatta, Boselaphus tragocamelus, Antilope cervicapra* and *Gazella bennetti.* These five species are being tolerated in the human habitation including agricultural fields to such an extent that few of them have now acquired the status of agricultural pests. Still the free, living populations of above animals have better chances of survival in comparison to others because religion is always beyond any reasoning. Whereas, nine (including three of the above) terrestrial mammals *viz. Canis lupus, Canis aureus, Vulpes vulpes, Vulpes bengalensis, Felis silvestris* and *Hyaena hyaena* are at present inadvertently being allowed in their present habitat by the landowners but they are susceptible to any change in the landuse pattern including shift to multicrop practice on availability of water. As far as commercial exploitation of wild animals is concerned six species, *Canis auerus, Vulus vulpus, Vulpus bengalensis, Lutrogale perspicillata, Fellis silvestris* and *Panthera pardus* are facing this problem which gets further multiplied because free-living populations are easy target for poaching in comparison to population inside protected areas. Thousands of skins of Desert fox, Desert cat and Jackals caught by law enforcing authorities indicated that they are being gradually removed unnoticed.

DISCUSSION

Culturally speaking, the people of Rajasthan have over the centuries developed a dichotomous attitude towards wildlife. On the one hand there were the Rajputs - the princely caste - and their minions, besides a number of nomadic and forest dwelling tribes, who looked upon hunting of wildlife as one of the main pastimes. They killed whatever game came their way. On the other extreme of the psycho-social spectrum, there were the believers in absolute non-violence, particularly those belonging to the trading communities, *Brahmins* and practising *Jains* and some other religious sects (Prakash and Ghosh, 1980). The contribution of *Vishnoi* community towards cause of wildlife in general and Black buck in particular in the states of Rajasthan and Haryana occupies a distinctive position and needs no elaboration. In the past, during the days of the princes, it was the prerogative of the Rajas and nobles to hunt in the state forests, but poachers and defaulting commoners were severely punished for any violation of the rules. Their love of wildlife speaks for itself and some of their creations which are the best sanctuaries and national parks we see today. However, the situation gradually changed after we got independence and the commoners became the lord of the day. In the absence of any legislation, merciless, thoughtless and ruthless killing of all wildlife became the order of the day till the Parliament enacted Wildlife Protection Act in the year 1972. Nevertheless, the magnitude of impact of man and his attitude towards the wildlife has reached to such an extent that now-a-days the distribution maps of animal loving communities are becoming the distribution maps of wildlife, as far as the Thar desert is concerned.

SCOPE FOR FURTHER WORK

While we have the current state of knowledge of such species, our knowledge is still lacking on the exact distribution and population size of these animals. Effective action can only be based on accurate information. More widely shared the information, the more likely it is that concerned institutions will agree to the definition of the problem and solutions. Hence :

1. Exact distribution and population size of these animals (Table 17.1) should be immediately worked out.
2. The current land use pattern is the major promoter for occurrence of wild animals in the free living state in the Thar desert. Hence, status quo on land use should be maintained till we have details, as per the above statement. Only then change in land use be allowed after providing adequate arrangements of corridors for to and fro movement of animals from different pockets of concentration as well as protected areas.

The land use change is very much in the offing in the Thar desert as canal water has now been made available through Indira Gandhi Nahar Pariyojana (IGNP).

REFERENCES

Alfred, J.R.B. and V.C. Agarwal, 1996 : The mammal diversity of the Indian desert. In : *Faunal Diversity in the Thar Desert : Gaps in Research.* (Eds. A.K. Ghosh, Q.H. Baqri and I. Prakash), Scientific Publishers, Jodhpur, pp. 335-348.

Kankane, P.L., 1995 : Status Survey of Chinkara and Desert cat in Rajasthan. *ZSI Envis. News Letter*, 3 : 6-9.

Kankane, P.L., 1996 : Carnivores in the Thar desert. In : *Faunal Diversity in the Thar Desert : Gaps in Research.* (Eds. A.K. Ghosh, Q.H. Baqri and I. Prakash), Scientific Publishers, Jodhpur, pp. 379-388.

Prakash, I, 1963 : Taxonomic and ecological account of the mammals of Rajasthan desert. *Ann. Arid Zone*, 1 (122) : 142-162 and 2(2) : 150-160.

Prakash, I and P.K. Ghosh, 1980 : Human-animal interactions in the Rajasthan Desert. *J. Bombay Nt. Hist. Soc.*, 75 : 1259-1261.

Rodgers, W.A. and H.S. Panwar, 1988 : *Planning a Wildlife Protected Area Network in India*, Wildlife Institute of India, Dehradun.

Chapter 18

Engineered Timber Housing : An Infrastructural Need for Socio-Environmental Upgradation of Rural India

Utpal Mishra*, H.N. Mishra, Manika Mishra*****

ABSTRACT

Principles of better housing and healthy living conditions in an improved rural environment have been discussed. A sizeable section of people there under poverty line have been leading a sub-human life for decades due to want of necessary housing, health and food facilities. For a quick and economic solution of increasing housing problem, construction of engineered houses on large scale by making good use of local building materials including plantation timbers possessing amenable structural properties has been suggested. Land allotment, design and construction strategies, criteria for timber selection, its proper seasoning, chemical treatment and joining techniques for strong, durable and cost-effective housing components along with the aspect of maintenance management for old and new constructions have also been discussed in this paper.

INTRODUCTION

Rural India accounts for about eighty per cent of country's population distributed over 570000 villages. Most of them are poverty ridden and have been living a sub-human life devoid of infrastructural facilities in respect of housing, health and happiness. Housing shortage in the country has created mind boggling problem to the engineers and planners in the wake of unabated increase in population as well as soaring cost of construction materials. The shelter shortage in 1981 was shown by the National Buildings Organisation (NBO) to be to the tune of 23.8 million units including 16.3 million in the rural sector (Mishra, 1988a). In another estimate the shortage in 1991 was shown to be about 31 million housing units including 20.6 million in rural areas. The figure is likely to touch 41.0 million with break-up of 25.5 million in rural areas and 15.5 million in urban areas by 2001 A.D. (Patel, 1995). Thus, to attend to this gigantic task of urgent nature, concerted efforts should be made to chalk out the low cost housing projects to be implemented in rural and semi-urban areas by making good use of locally available timbers possessing amenable structural properties (Mishra, 1994). Making the rural area inhabitable, particularly by engineered housing systems in well planned sites is to improve the locality and living condition of the people. The paper highlights some of the factors responsible for pollution problem in rural areas, design and construction strategies of low-cost houses, scientific processing and use of available timbers, and maintenance management of such constructions including adjacent surroundings.

RURAL HOUSING AND ENVIRONMENT

The present scenario of rural housing and environment depicts a sorry figure. Factors relating to fiscal draw-back are mainly responsible for such a precarious condition. Villages full of potential human resource are aptly termed as nerve centre of a country but ironically, the villagers are termed as 'have nots' having practically no money, food and clothings. There is hardly any land under their feet and shelter over their heads. They lack in almost all types of amenities required for their substenance. Only about 5% of them are possessing good buildings, over 50% are in possession of their ancestral homes most of which have developed deterioration. The dilapidated constructions are sometimes having drooping roofs profusely leaking, wornout columns, trusses, purlins etc. posing threats of collapse, yielding walls, doors and windows revealing hazards to privacy and safety of the owners. A sizeable section of people are living in temporary huts erected mostly in disadvantageous places by using bamboo and reeds, leaves and grasses, all of which are constantly affected by vagaries of nature like rain, cyclone, water-logging and flood. Village houses constructed in intuitional way by using sub-standard materials have grown through decades in unplanned settlements. The situation there has worsened due to over-congestion of houses constructed by the increased number of inheritors in their shared land. As a result, rain water from the roof of one house wets the varanda of another and the congested structures constantly obstruct the free passage of sunrays and fresh air in the vicinity. The situation in such habitations further aggravates in rainy season when due to want of proper drainage system and sanitary arrangement the area is covered with filthy materials spreading obnoxious odour causing health hazard. People, accustomed to live in such an atmosphere, seldom attach much importance to the needful remedial measures. No repair and renovation work is taken up until warranted by any untoward happening. The causes of environmental degradation in otherwise beautiful rural areas can be attributed to :

- unhygienic sanitary condition.
- casual disposal of corpse of animals on land or in water of canal/river running through villages or even in any nearby water-logged area.
- accumulation of cow-dung, garbages etc. and non-disposal of the same at routine intervals.
- bushes and weeds growing unabated near the habitation.
- ponds, waterlogged areas containing grass, hyacinth and stagnant water suiting mosquito breeding.

- places of common use such as tube wells, dugwells remaining in unhealthy condition.
- haphazardly located cattle sheds polluting the atmosphere.

Rational approach to these aspects can eradicate the pollution hazards.

NEED FOR A CHANGE

In an area where people are thriving for better living standard equipped with modern amenities in their spacious buildings, the poor villagers comprising the major human resource of the country should not be deprived of their legitimate right of living a human life in a congenial atmosphere. In rural area there is no dearth of land. Only a change in the land allotment policy is necessary to free vast unused land from the clutches of so called zamindars and distribute the same amongst the poor for proper utilization. Better housing needs better location duly developed to contain necessary number of plots suitable to accommodate the families side by side in a colonial set up. The villagers should be made free from the bondage of age old inconvenient living condition by providing them the necessary land. The design for the houses including the layout plan should be strictly adhered to for convenience of alloting common facilities to all in respect of roads, electricity, sanitation, water supply, education and health care. The needful change in the policy should reflect the idea of better living in a better atmosphere for wider interest of the nation.

POLICY, PLANNING AND DESIGN

In rural area, cluster of houses has grown up through ages on undeveloped bank of canals, ponds and rivers, along road sides, and in-and-around places of worship, education, markets, railway stations, hospitals etc. A sudden change in the existing site plan can not be brought about to make the site free of over congestion. In such cases the old and dilapidated constructions are to be gradually removed by providing the owners new plots in the duly developed areas. Old habitations should be thinned of undesirable constructions standing in the way of proper development and the remaining houses should be repaired and renovated to improve the old environment. Further development for sites in villages depending on decided number of families should continue simultaneously. The recovered materials, mainly consisting of bricks and timber components can be re-used after due processing with considerable economy. Because of abundance and low-cost of rural land, the policy of allotment will differ from that adopted in urban areas. Here, bigger plots (at least 400 sqm) are to be earmarked to accommodate individual family units to satisfy their residential and non-residential requirements. Provision of separate bath and latrine, cattle shed, and grain storage is desirable. In the main residential block, atleast two spacious bedrooms, one reading room, one sitting-cum-dining room and one kitchen and veranda may be provided covering an area of about 60 sqm. Arrangement for facility for kitchen garden and raising plantation of some trees in individual plotsis also desirable. In termite infested areas, termite entry in building should be checked through construction of anti-termite gadgets (Anon., 1981; Mishra, 1995). Government of India is providing infrastructure facilities like electricity, pucca roads, buildings for school, college and community halls, hospitals etc. for accelerated rural upliftment. The site plan of campus should facilitate the owners deriving benefits out of such infrastructures. Target for housing project in rural area can be achieved either through active participation of the user groups, *i.e.* on self-help basis or through the agency of approved co-operative housing societies.

MATERIALS AND METHODS

Materials account for about 60 to 75 per cent of the total cost of construction and as such the success of any mass housing project depends to a great extent on the judicious and optimum use of cost effective local building materials. To comply with the deliberations of Rio-Earth Summit, a ban on felling the forest trees has been imposed and use of timber in house construction has been restricted. So, instead of some costly and conventional species like teak (*Tectona grandis*), Sal (*Shorea robusta*), deodar (*Cedrus deodara*) etc., some alternative species possessing qualities like strength, dimensional stability and durability either in the natural form or after suitable treatment with environment friendly chemicals have been suggested to be used in house construction (Mishra, 1986). The important aspect of selection of suitable timbers is perfected based on rational classification of Indian structural timbers where more than 170 species have been mentioned according to their strength characteristics, seasonal behaviour and treatability condition. Such a classification has not only permitted a proper choice of materials for different limitation of stresses and type of construction but also provided easy selection of alternative species depending on the situation (Mishra and Pruthi, 1984). Use of mixed species of timbers for structures will also ease and economise the construction (Anon., 1986). Fabrication of structural components by using suitable jointing device duly strengthened by modern fastenings like m.s. nails, bolts, wooden disc dowels, metallic rings and glue techniques deserve due attention in timber work. Timber is a biological product and as such use of it poses a problem if it is not properly processed through different stages like sawing and sizing, seasoning (Anon., 1978), chemical treatment (Anon., 1982) based on scientific methodologies. To save time and energy, some low-cost house components designed and tested by Timber Engineering Branch, FRI, may also be made use of. The components include cost reducing T-frame window which is about 43% cheaper in farming materials (Mishra, 1988b), *Eucalyptus* pole trusses, about 41% cheaper than similar trusses of sawn timber (*Pinus roxburgii*) and bamboo trusses (53% cheaper) made by species like *Dendrocalanus strictus*, *Bambusa tula* etc. duly strengthened by modern fastening devices (Mishra, 1991). Treated bamboo in round/splitted form and bamboo boards and mats may be used in componentised wall claddings and false ceilings (Mishra, 1988c; Mishra and Mishra, 1992). Use of Bamboo concrete for small span slabs suiting kitchen shelves, manhole covers, sunshades and also for making low capacity overhead water tank, fencing posts etc. can be used with considerable economy (Mishra and Mishra, 1995). Self-help construction or construction through co-operative housing societies may benefit if the systems of management adopt the above low-cost techniques suiting rural economy (Mishra, 1990 and 1994).

CONCLUSION

For ecological and economic reasons, use of conventional timbers in house construction has to be reduced by maximising the use of fast growing plantation timber like *Eucalyptus poplar* etc. after due processing. In this respect careful attention is to be paid for selection criteria of timbers based on their strength properties seasoning and chemical behaviours. Fabrication of structural components strengthened by modern fastening devices is also an important factor in timber constructions. Judicial planning of the campus devoid of congestional construction, proper orientation of the houses, provision of infrastructures and amenities for light, road, water, health care, sanitation etc. besides the tendency for togetherness on the part of the beneficiaries for a peaceful co-existence in a beautiful campus can contribute not only to an equitable living standard but also to an upgraded socio-environmental set-up in rural India.

REFERENCES

Anon., 1978 : IS : 1141 : *Code of practice for seasoning timber.* BIS, New Delhi.

Anon., 1981 : IS : 6313 : *Code of practice for anti-termite measures in buildings.* BIS, New Delhi.

Anon., 1982 : IS : 401 : *Code of practice for preservation of timber.* BIS, New Delhi.

Anon., 1986 : IS : 3629 : *Indian Standard Specification for Structural Timber in Building.* BIS, New Delhi.

Mishra, H.N. and K.S. Pruthi, 1984 : Prefabricated housing with standardised components for cost reduction through timber engineering techniques. *Proc. Int. Conf. on Low-cost Housing for Developing Countries*, CBRI Roorkee, 157-163.

Mishra, H.N., 1986 : Timber engineering in India. Lead paper for *2nd Forest Products Cont.* FRI, Dehrudun.

Mishra, H.N., 1988a : Structural use of bamboo in rural housing. *The Indian Forester*, 114 : 622-634.

Mishra, H.N., 1988b : T-frame window an-innovative technique in cost reducing construction. *J. NBO-UN Regional Housing Centre,* ESCAP, New Delhi, 33 : 19-25.

Mishra, H.N., 1988c : Know-how of bamboo house construction; bamboo current research. *Proc. Int. Bamboo Workshop*; Cochin, India, 212-249.

Mishra, H.N., 1990 : Technical aspect of timber use in housing under co-operative schemes in India. *Souvenir, 9th National Congress of Housing Co-operatives.* Panaji, Goa, 51-54.

Mishra, H.N., 1991 : Problem and prospect of structural use of bamboo and Eucalyptus Pole in Rural India; Proc. *Int. Timber Engg. Conf.*, London; 3 : 300-307.

Mishra, H.N. and Manika Mishra, 1992 : Role of bamboo in promoting rural economy. *Nat. Sem. on Socio Economic Research in Forestry.* KERI & Ford Foundation, Peechi.

Mishra, H.N., 1994 : Management for low-cost timber housing projects. *J. Non-Timber Forest Products*, 1 : 234-236.

Mishra, H.N., 1995 : Structural measures for controlling termite entry in Buildings. *Proc. Nat. Workshop on Termite Management in Buildings.* CBRI, Roorkee Pub : Tata McGraw-Hill Publishing Co. Ltd., pp. 79-86.

Mishra, H.N. and Utpal Mishra, 1995 : Managing production and utilization of bamboo crete for cost effective structural components under housing co-operatives. *11th National Congress of Housing Co-operatives.* National Co-operative Housing Federation of India, New Delhi, 103-108.

Patel, G.I., 1995 : Co-operative housing in India. Souvenir, *11th National Congress of Housing Co-operatives*, New Delhi, National Co-operative Housing Federation of India, 1-9.

Chapter 19

Quality Assessment of Groundwater Sources Adjoining Sewage Channel in Gwalior

Sanjay Sharma, R. Mathur and Asha Mathur

INTRODUCTION

Water is vital to life and about 65% of the human body is water. With the outspurt in population throughout the world, there has been great concern over the sustainability of potable water sources. Since, providing piped water supply to all is no easy task, there has been tremendous increase in the stock of private tubewells all over and particularly in India. This has resulted into a major problem of groundwater pollution. Groundwater pollution has been defined as an impairment of water quality by chemicals heat or bacteria to a degree, that does not necessarily create an actual public health hazard, but does adversely affect such waters for domestic, farm, municipal or industrial use (Walker, 1969).

Contamination of drinking water is a major source in the transmission of diseases like gastritis, enteritis, influenza, pneumonia, tuberculosis, malaria and cholera. Looking into the problems of death and incapacity caused by water-borne diseases, the decade of 1981-1990 was declared as International Drinking Water Supply and Sanitation decade. India during this decade did well by starting a Technology Mission laden with seven objectives, all mainly concerned with scientific and technological advancement and safer health. Of them the utmost priority was given to the drinking water mission which was named after the then young and dynamic Prime Minister of India, Late Shri Rajiv Gandhi. The main objectives of drinking water mission were set at to provide 40 litres/capita/day, safe water by the year 1990 at every distance of 1.6 km in 1,62,000 problem villages. Eradication of guinea worm problem, defluoridation and testing of potable waters were other important aims of the mission.

Since I.D.W.S.S. decade, there has been great awareness among the scientific community and reports are appearing from major parts of the country on the status of groundwater quality. Some recent studies in this direction are Gupta *et al.*, 1994; Mittal *et al.*, 1994; Pandey *et al.*, 1994; Rao *et al.*, 1994; Shankar and Muthukrishnan, 1994; Sharma and Mathur, 1994; Kataria, 1995; Reddy *et al.*, 1995; Rai and Sharma, 1995 and Sharma *et al.*, 1995.

THE STUDY AREA

The study area falls between latitudes 26°10' N and 26°17' N and longitude 78°10' E. The thickness of alluvium cover varies from paper thin to about 20 meters. Other geological formations exposed in the area are shale and sandstone of Gwalior series and Vindhyan series. Residents of Gwalior city derive their potable water needs from Tighra lake which was initially meant to fulfill the needs of population of 3.5-4.0 lakhs. But now when the city population has crossed ten lakhs mark, the treated water supply has gone inadequate. Inadequacy of treated water supply in the city has resulted in the exploration of underground water sources in huge numbers without looking into their locations and distance from 'point' sources of water pollution such a septic tanks or sewage disposal sites. Thus looking into this, the above study has been so framed so as to evaluate the impact of domestic sewage on raw groundwater sources of adjoining areas of Swarna Rekha river. This is a solo sewage channel which carries sewage from the heart of the city to the outskirts to Sharma Dairy Farm. The river is stretched in the city from Hanuman Bundha to Jalalpur Dam in a total stretch of 13.5 km. The study area has been shown in Fig. 19.1.

MATERIALS AND METHODS

Fifty one raw drinking water sources were identified situated adjacent to the Swarna Rekha river. These sources comprised of land pumps, dug wells and bore wells. The groundwater samples were collected between 0500-0800 h in each season during March 1989-February 1990. The physical and chemical analysis of the water samples were done in accordance with the procedures described in Standard Methods for the Examination of Water and Waste Water (APHA, 1985).

RESULTS AND DISCUSSION

In India, which is a developing country about eighty per cent of water pollution is caused by domestic wastes (Widyanto and Sorjani, 1975). In the present study, while selecting the raw drinking water sources, preference was given to the sources which were close to the sewage channel and were frequently used for drinking purpose. Distance between these sources and Swarna Rekha sewage channel varied from 2 m - 470 m. The results of the raw groundwater quality are tabulated in Table 19.1a-19.3b. As regards the distance between groundwater source and waste site, distance of 200 ft-300 ft (61 m-91.5 m) has been suggested to be satisfactory for preventing the contamination (Bedi, 1971) of groundwater. In the present study about twenty groundwater sources were found defying the standard distance range.

Turbidity value above 5 Jackson Turbidity Units (ICMR, 1975) is reported to cause gastro intestinal irritation and has been listed under primary drinking water standards of USEPA (1989). Their values crossed the standard limit in twenty one sources during summer season, seventeen sources during rainy season and nineteen sources during winter season. Hydrogen ion concentration has been categorized under secondary drinking water standard since it does not pose and health risk (USEPA, 1989). A range of 6.5-8.5 has been suggested as standard for drinking water (ISI, 1982; WHO, 1988). During

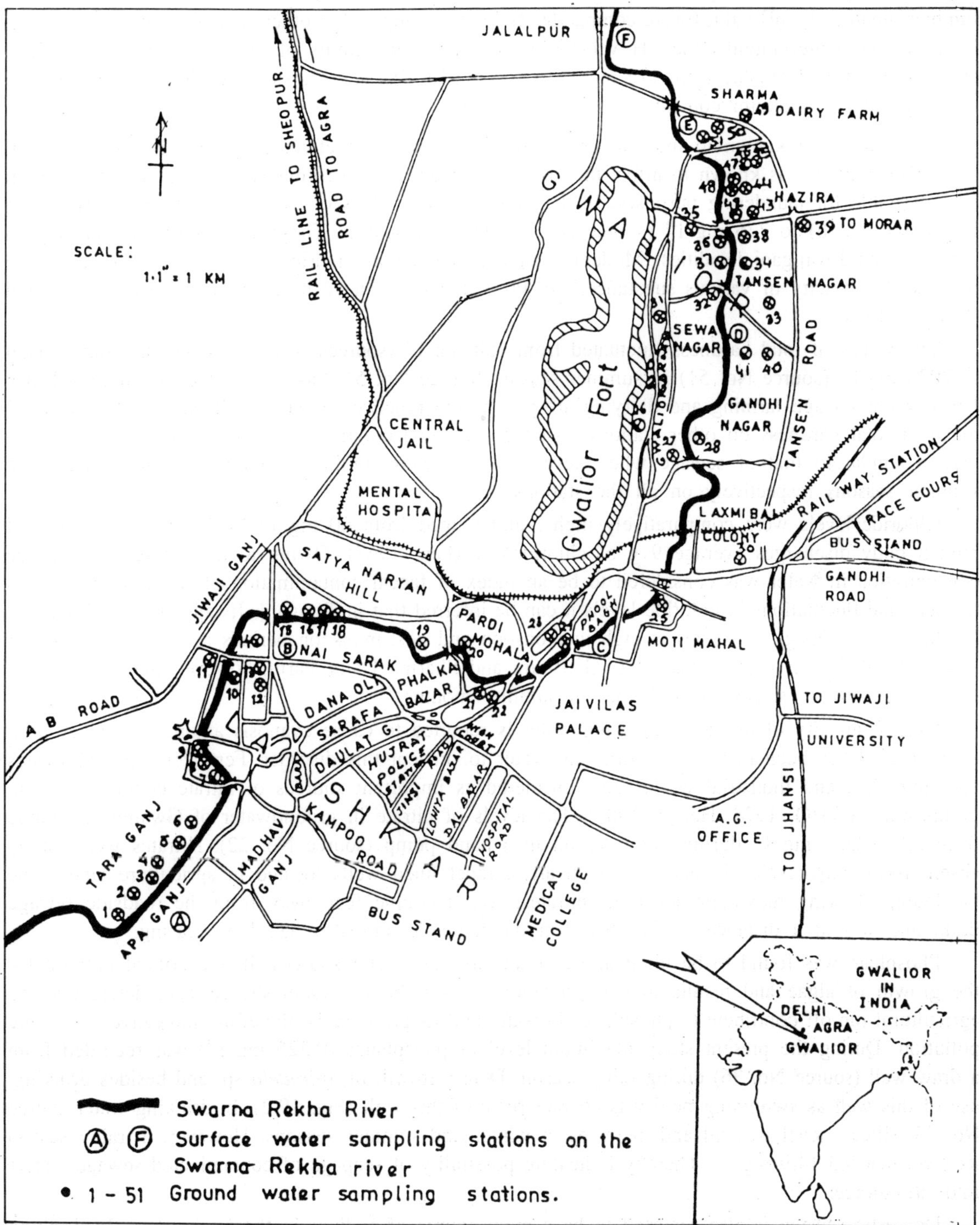

Fig. 19.1. Map showing study area and sampling sites.

summer months, the pH values were comparatively low as compared to other season. Baring few sources in rainy and winter months the pH value crossed the normal limits of drinking water. The values of total alkalinity fluctuated from 112 mg l^{-1} to 760 mg l^{-1}. The sources showing high alkalinity values also showed high sodium content.

Calcium is not known to indicate or produce any hazardous effect on human health. However, above 200 mg l^{-1} it is known to affect the taste and produce scale formation. Its higher concentration is undesirable for laundering purposes. In the present study values of calcium fluctuated from 18 mg l^{-1} (source No. 6) to 776 mg l^{-1} (source No. 10) in rainy season. These sources are respectively in Taraganj and Jiwajiganj localities. Moderate content of calcium in drinking water is rather desirable because the toxicity of various substances *i.e.* Pb, Cd, Hg etc. have been observed to be neutralized (I.I.E.E., 1984).

The values of total hardness fluctuated from 158 mg l^{-1} (sources No. 42 and 46) in rainy season to 2028 mg l^{-1} (source No. 51) in summer season. Source No. 51 was a draw well in Sharma Dairy Farm where sewage farming and dairying are year round practices. Except at Jiwajiganj, the seasonal mean of total hardness on other stations raised constantly from rainy to summer season *i.e.*, on the pattern similar to that of sewage where levels of T.H. were minimum and maximum in rainy and summer seasons respectively on all the stations.

Chloride levels were comparatively higher and ranged from 60 mg l^{-1} to 760 mg l^{-1}. A normal human body discharges average 9 g of chloride/day (Fair, 1967). Owing to this, increased chloride concentration in water was considered to be an index of faecal contamination of water. Looking into the seasonal fluctuations in chloride levels it can be inferred that in Taraganj, Jiwajiganj and Phoolbagh localities, the seepage of wastes is high in monsoon and post-monsoon periods as revealed by higher levels of chloride in winter season. At Gouspura and Sharma Dairy Farm, however, the incidence of seepage of sewage seems to be higher in summer months.

Excess quantity of nitrate in drinking water is reported to cause cyanosis among infants. A limit of 10 mg l^{-1} has been imposed on drinking water to prevent this disorder. Fertilizers, animal wastes and municipal and industrial wastes are considered as important sources of nitrate contamination in groundwater (WHO, 1978; Handa, 1983). The levels of nitrate in groundwater of Gwalior fluctuated from nil to 26.6 mg l^{-1} during rainy season in a hand pump (source No. 22). Nitrates were totally absent from sources No. 49 and 51 which are parapet lined wells located in agriculture lands. The handpump showing maximum levels of nitrates, was a couple feet away from the roadside sewage drain and was at a distance of five feet frm the manhole and also had loose casing.

Phosphate was found to be nil in most of the drinking water sources. It is a critical nutrient for the growth of algae and results in eutrophication of waterbodies. Domestic sewage, detergents and agricultural run-off contribute to phosphates in water and its presence is, therefore, indicative of organic pollution. During the present study maximum level of phosphates (0.325 mg l^{-1}) was recorded from a draw well (source No. 36) during rainy season. Dense growth of *Spirodela* sp. and besides drinking, use of this well as swimming pool was characteristic of this well (Fig. 19.2). In drinking water source No. 23, silicate level was nil and in traces in winter and summer seasons. However, in rainy season its level reached 14.9 mg l^{-1}. Thereby indicating possibility of seepage of contaminated sewage waters into the source.

Domestic sewage is also reported to be a main source of sodium in the freshwaters. In drinking water sources of Gwalior the levels of sodium varied from 10 mg l^{-1} to 400 mg l^{-1}. Most of the sources crossed the 20 mg l^{-1} of reporting level of sodium (USEPA, 1989). The levels of potassium also were very high and ranged from 1.6 mg l^{-1} to 256 mg l^{-1}. In source No. 35 the annual average level of potassium was found to be 245 mg l^{-1}.

In the present study excessive extraction of groundwater through bore wells was found to be a major cause for pollution of underground water. Various old draw wells were found to be polluted in Taraganj region due to a recently drilled bore well (source No. 5) extracting > 12 hour water/ day. Similarly in Sharma Dairy Farm locality, a bore well (source No. 45) was found to be at a distance of 1 m from the manhole. Supply of filthy water was a common problem in this area. It was thus observed that owing to lack of planning of P.H.E. department and awareness of local people, the groundwater quality was at risk. Some parameters have no health risks. Environmental Protection Agency has established standards for these parameters so as to provide an aesthetically acceptable water supply. These levels are called Secondary Maximum Contaminant levels (SMCL) (USEPA, 1991). In the present study, the levels of sulphate did not cross the SMCL limit. However chloride, pH and T.D.S. crossed the SMCL limit in many drinking water sources.

CONCLUSION

Earlier study of these groundwater sources confirmed their pollution through sewage. This study revealed poor microbial quality of those sources which were especially in the vicinity of Swarna Rekha river (Sharma and Mathur, 1994). A similar type of study was done on the groundwater sources of rural areas falling within and outside the municipal limits of Gwalior (Sharma *et al.*, 1995). The study emphasized the need for improving the sanitary conditions of the area especially near the groundwater sources so as to prevent the chances of contamination of groundwater. The present study is thus an addition on the reports of groundwater quality in Madhya Pradesh. Based on the physico-chemical characteristics of water and various field experiences, the following recommendations are suggested for attaining better environment and groundwater quality.

- Sanitary conditions in the area should be improved by regular dredging of the sewage channel so as to maintain rapid flow of the waste waters and reduce the possibilities of seepage of sewage.
- Garbage dumping in the channel should be prohibited.
- While drilling the groundwater sources all specifications pertaining to their distance from waste sites/septic tanks should be followed and all the sources in the vicinity of Swarna Rekha/manholes should be abandoned.
- Area adjoining the Swarna Rekha should be declared as "no bore" zone.
- Use of borewells adjoining sewage channel should be banned so as to avoid underground vacuum which is responsible for increased rate of percolation and ultimately pollutes the aquifer.

Table 19.1a. Raw drinking water quality in summer season with reference to the distance from the Swarna Rekha sewage channel.

Drinking water source No.	*Source Type*	*Distance from the sewage channel (Meters)*	*pH*	*Conductivity (u mhos/cm)*	*Turbidity (JTU)*	*Carbonate (mg.1-1)*	*Total alkalinity (mg.1-1)*	*Calcium hardness (mg.1-1)*	*Total hardness (mg.1-1)*	*Magnesium (mg.1-1)*	*Chloride (mg.1-1)*	*Sulfide (mg.1-1)*	*Sulfate (mg.1-1)*
I	II	III	1	2	3	4	5	6	7	8	9	10	11
							Taraganj						
1.	Draw well	10	8.2	660	22	Nil	208	108	300	46.8	70	10.2	13.0
2.	Draw well	65	8.2	726	36	Nil	236	124	338	52.1	90	8.9	41.5
3.*	Draw well	15											
4.*	Draw well	56				Nil							
5.	Bore well	27	8.1	1254	32	Nil	292	208	520	76	176	7.3	38.0
6.*	Draw well	42											
7.	Draw well	02	7.9	1254	68	Nil	760	180	700	126.4	118	9.38	66.5
8.	Draw well	08	8.3	858	01	Nil	280	140	344	49.6	83	Nil	17.5
9	Draw well	05	7.9	1056	01	Nil	408	164	380	52.5	118	Nil	37.0
							Jiwajiganj						
10.	Hand pump	70	8.5	1650	08	Nil	460	124	336	51.6	175	12.65	52.5
11.	Hand pump	25	8.7	1386	22	Nil	312	140	580	107.1	190	15.1	51.0
12.	Hand pump	450	8.4	1254	08	Nil	284	158	400	58.9	153	Nil	25.5
13.*	Hand pump	470											
14.	Bore well	83	8.0	858	01	Nil	296	176	312	33.1	61	2.45	17.0
15.	Hand pump	18	8.3	1188	36	Nil	374	192	388	47.4	124	0.816	37.0
16.	Draw well	24	8.5	1320	22	Nil	300	156	500	83.6	140	1.63	61.5
17.*	Draw well	18											
18.	Draw well	24	8.6	858	30	Nil	256	72	374	73.5	90	2.45	24.0
19.	Hand pump	30	8.2	2244	32	Nil	272	280	796	125.6	386	0.82	93.0
							Phoolbagh						
20.	Hand pump	61	8.3	1518	01	Nil	324	168	600	105.2	188	0.82	83.0
21.	Hand pump	21	8.5	1122	22	Nil	276	92	348	62.3	132	1.63	54.0
22.	Hand pump	53	8.3	759	01	Nil	300	108	372	64.3	123	1.63	13.0
23.	Bore well	106	8.8	396	01	Nil	112	88	216	31.2	72	0.82	22.5
24.	Bore well	45	8.2	1452	08	Nil	444	60	400	82.8	158	0.82	63.0

(Contd.)

I	II	III	1	2	3	4	5	6	7	8	9	10	11
25.	Hand pump	137	8.25	1122	01	Nil	292	132	400	65.3	170	2.04	21.5
26.	Draw well	360	8.2	891	04	Nil	360	92	432	82.8	158	1.22	33.0
27.	Draw well	12	7.9	1782	59	Nil	700	40	792	183.1	280	1.22	35.0
28.	Draw well	24	8.1	1320	12	Nil	450	100	568	114.0	196	1.63	54.0
29.	Draw well	240	8.3	858	04	Nil	300	68	368	73.0	176	2.04	26.0
30.	Bore well	360	8.3	1188	01	Nil	288	86	444	87.0	187	2.04	35.0
							Gouspura						
31.	Hand pump	295	8.0	2112	01	Nil	442	234	740	123.0	226	0.82	184.5
32.	Bore well	80	8.1	2376	36	Nil	370	112	820	172.1	351	2.86	142.0
33.	Bore well	160	8.1	1188	01	Nil	296	40	480	107.1	190	1.63	55.0
34.	Draw well	90	8.15	1320	08	Nil	444	80	392	76.0	170	3.3	66.5
35.	Draw well	150	8.2	1584	01	Nil	400	72	332	63.2	125	0.82	112.0
36.	Draw well	14	8.0	1947	01	Nil	528	40	540	121.5	288	3.3	112.0
37.	Draw well	15	8.05	1848	01	Nil	488	48	500	110.1	245	0.82	120.0
38.	Hand pump	70	8.1	1980	16	Nil	340	104	540	106.2	260	3.3	154.0
39.	Hand pump	280	8.1	1716	01	Nil	338	348	720	90.4	270	3.3	110.0
40.	Bore well	150	8.2	1287	01	Nil	308	132	450	77.3	200	3.3	55.0
41.	Bore well	90	8.1	1221	01	5.0	385	40	350	75.3	158	0.82	35.0
							Sharma Dairy Farm						
42.	Draw well	80	8.3	990	01	Nil	308	40	336	72.1	100	3.3	26.5
43.	Hand pump	110	8.3	1650	26	Nil	400	112	506	95.8	250	1.63	112.0
44.*	Bore well	175											
45.	Bore well	220	8.0	1980	01	Nil	368	420	800	92.3	314	1.63	174.0
46.	Draw well	195	7.9	1122	01	Nil	332	212	460	60.3	170	3.3	81.0
47.	Draw well	60	8.0	924	08	Nil	356	76	316	58.3	164	1.63	33.0
48.	Draw well	75	8.1	990	04	Nil	400	98	312	52.1	180	4.08	16.0
49.	Draw well	245	8.15	1914	01	Nil	312	348	840	119.6	350	2.45	74.0
50.	Bore well	130	8.2	1122	30	Nil	432	260	640	92.3	280	1.63	29.5
51.	Draw well	80	7.9	2904	01	Nil	272	96	2028	473.4	760	1.63	143.0

where * indicates no collection of sample due to drying up of the source.

Table 19.1b. Raw drinking water quality in summer season with reference to distance from the Swarna Rekha sewage channel.

Drinking water source No.	Source Type	Distance from the sewage channel (Meters)	Nitrite ($mg.l^{-1}$)	Nitrate ($mg.l^{-1}$)	Phosphate ($mg.l^{-1}$)	Silicate ($mg.l^{-1}$)	Sodium ($mg.l^{-1}$)	Potassium ($mg.l^{-1}$)	Chemical Oxygen Demand ($mg.l^{-1}$)	Total Solids ($mg.l^{-1}$)	Total Dissolved Solids ($mg.l^{-1}$)	Total Suspended Solids ($mg.l^{-1}$)
I	II	III	12	13	14	15	16	17	18	19	20	21
						Taraganj						
1.	Draw well	10	0.0115	11.5	Nil	7.9	25	3.2	32	640	180	460
2.	Draw well	65	0.122	1.42	0.008	7.9	30	3.2	20	760	300	460
3.*	Draw well	15										
4.*	Draw well	56										
5.	Bore well	27	0.02	2.8	Nil	9.9	45	3.2	20	1420	540	880
6.*	Draw well	42										
7.	Draw well	02	Nil	Nil	0.30	38.8	45	34.4	32	1740	540	1200
8.	Draw well	08	0.036	12.9	Nil	8.9	35	4.8	24	840	400	440
9.	Draw well	05	0.04	19.05	Nil	9.9	65	8.0	24	1180	240	940
						Jiwajiganj						
10.	Hand pump	70	0.042	20.6	0.023	10.9	260	10.0	16	1260	940	320
11.	Hand pump	25	0.0076	0.89	Nil	12.9	45	4.8	24	1220	900	320
12.	Hand pump	450	0.174	3.06	Nil	6.9	50	3.2	24	1320	780	540
13.*	Hand pump	470										
14.	Bore well	83	Traces	18.2	0.005	10.9	50	5.2	48	340	240	100
15.	Hand pump	18	0.113	14.7	0.115	6.9	70	24.8	72	580	300	280
16.	Draw well	24	0.05	2.5	0.005	10.9	75	2.8	32	760	260	500
17.*	Draw well	18										
18.	Draw well.	24	0.013	11.5	Nil	10.9	40	2.4	36	420	420	Nil
19.	Hand pump	30	0.184	0.554	Nil	7.9	55	4.0	52	2020	1620	400
						Phoolbagh						
20.	Hand pump	61	0.014	3.85	Nil	3.0	50	14.0	32	820	720	100
21.	Hand pump	21	0.025	2.66	Nil	3.0	85	1.6	20	640	340	300
22.	Hand pump	53	0.129	13.73	Nil	1.1	30	1.6	28	320	320	Nil
23.	Bore well	106	Nil	3.43	Nil	Traces	10	5.2	32	580	360	220
24.	Bore well	45	0.025	5.9	Nil	2.0	135	3.2	36	1260	580	680

(Contd.)

I	II	III	12	13	14	15	16	17	18	19	20	21
25.	Hand pump	137	0.031	3.8	Nil	2.0	55	2.8	16	640	640	Nil
26.	Draw well	360	0.0215	8.66	Nil	8.9	20	24.8	32	820	400	420
27.	Draw well	12	0.038	Nil	0.19	12.2	125	22.4	72	980	980	NIl
28.	Draw well	24	0.097	2.30	0.018	6.9	75	22.0	32	1100	800	300
29.	Draw well	240	0.354	16.6	Nil	10.9	30	26.4	12	500	500	Nil
30.	Bore well	360	0.012	1.2	Nil	4.9	50	3.2	16	3040	1080	1960
							Gouspura					
31.	Hand pump	295	0.012	Nil	Nil	6.9	225	18.0	24	3600	2000	1600
32.	Bore well	80	Traces	Nil	Nil	3.0	200	10.0	32	2120	1820	300
33.	Bore well	160	0.354	15.062	Nil	6.9	75	8.8	32	880	460	420
34.	Draw well	90	0.132	19.94	0.052	6.9	135	10.4	36	940	540	400
35.	Draw well	150	0.158	3.987	Nil	4.9	135	240.0	40	1040	700	340
36.	Draw well	14	0.021	14.2	0.115	8.9	175	120.0	48	1400	840	560
37.	Draw well	15	0.025	4.9	0.095	6.9	190	128.0	56	1480	360	1120
38.	Hand pump	70	0.0076	0.98	Nil	6.9	140	112.0	52	1620	1100	520
39.	Hand pump	280	0.047	1.44	Nil	8.9	95	3.6	48	2500	2500	Nil
40.	Bore well	150	Nil	12.2	Nil	4.9	65	4.4	40	4600	460	4140
41.	Bore well	90	0.029	14.62	Nil	4.9	11	1.6	40	6340	1680	4660
							Sharma Dairy Farm					
42.	Draw well	80	0.014	15.95	Nil	4.9	65	6.0	56	1640	640	1000
43.	Hand pump	110	0.0076	3.43	Nil	6.9	215	23.2	52	1220	840	380
44.*	Bore well	175										
45.	Bore well	220	0.194	0.77	Nil	9.9	160	18.2	08	1400	1400	Nil
46.	Draw well	195	0.047	7.09	Nil	6.9	75	19.2	12	800	800	Nil
47.	Draw well	60	0.083	8.86	0.057	6.9	65	80.0	16	640	620	20
48.	Draw well	75	0.081	11.52	0.155	8.9	75	108.0	16	760	680	80
49.	Draw well	245	0.038	Nil	0.004	6.9	135	9.2	16	1880	1880	Nil
50.	Bore well	130	0.012	16.9	Nil	5.9	75	6.4	48	1400	1360	40
51.	Draw well	80	0.035	Nil	Nil	4.9	330	4.4	40	3880	3880	Nil

Where * indicates no collection of sample due to drying up of the source.

Table 19.2a. Raw drinking water quality in rainy season with reference to the distance from the Swarna Rekha Sewage Channel

Drinking water source No.	Source Type	Distance from the sewage channel (Meters)	pH	Conductivity (u mhos/cm)	Turbidity (JTU)	Carbonate (mg.l^-1)	Total alkalinity (mg.l^-1)	Calcium hardness (mg.l^-1)	Total hardness (mg.l^-1)	Magnesium (mg.l^-1)	Chloride (mg.l^-1)	Sulfide (mg.l^-1)	Sulfate (mg.l^-1)
I	II	III	1	2	3	4	5	6	7	8	9	10	11
							Taraganj						
1.	Draw well	10	7.98	594	51	Nil	270	24	232	50.5	80	1.17	33.0
2.	Draw well	65	8.02	726	01	Nil	244	78	256	43.3	93	0.78	16.0
3.	Draw well	15	8.06	924	04	Nil	320	94	206	27.2	100	0.39	Nil
4.	Draw well	56	7.98	990	61	Nil	326	50	232	44.2	96	Nil	Nil
5.	Bore well	27	8.08	1254	22	Nil	266	80	306	54.9	168	1.17	35.0
6.	Draw well	42	7.86	1980	26	Nil	438	18	476	111.3	391	0.78	40.5
7.	Draw well	02	7.87	924	42	Nil	320	74	192	28.7	83	0.78	Traces
8.	Draw well	08	7.75	1188	01	Nil	372	110	234	30.1	95	1.17	33.0
9.	Draw well	05	7.73	1320	01	Nil	428	78	322	59.3	161	0.78	9.5
							Jiwajiganj						
10.	Hand pump	70	8.71	2904	04	Nil	520	776	978	49.1	479	0.195	111.0
11.	Hand pump	25	8.55	1056	04	Nil	310	190	476	69.5	153	0.39	37.0
12.	Hand pump	450	8.57	990	04	Nil	314	174	348	42.3	125	Nil	35.0
13.	Hand pump	470	8.59	924	12	Nil	304	78	198	29.2	91	0.195	27.5
14.	Bore well	83	8.82	990	Nil	Nil	362	78	334	62.2	84	Nil	41.5
15.	Hand pump	18	8.26	726	01	Nil	222	60	300	58.3	72	2.346	10.0
16.	Draw well	24	8.07	1452	Nil	Nil	352	172	512	82.6	175	0.780	11.0
17.	Draw well	18	8.08	528	Nil	Nil	174	82	228	35.5	70	0.780	Nil
18.	Draw well	24	7.99	990	01	Nil	234	74	382	74.8	93	0.780	30.0
19.	Hand pump	30	8.13	1980	01	Nil	260	290	622	80.7	298	1.560	96.0
							Phoolbagh						
20.	Hand pump	61	8.23	1452	01	Nil	400	156	410	61.7	151	1.170	43.5
21.	Hand pump	21	8.62	1056	12	Nil	256	66	340	66.6	100	1.170	39.0
22.	Hand pump	53	8.70	792	16	Nil	240	88	316	55.4	70	1.950	15.5
23.	Bore well	106	8.57	1320	08	Nil	324	72	388	76.8	123	1.950	49.0
24.	Bore well	45	8.42	1650	16	Nil	440	94	534	106.9	175	1.950	100.0

(Contd.)

I	II	III	1	2	3	4	5	6	7	8	9	10	11
25.	Hand pump	137	8.53	1056	Nil	Nil	302	220	388	40.8	158	0.195	42.5
26.	Draw well	360	7.68	1188	04	Nil	372	70	188	28.7	128	Nil	34.0
27.	Draw well	12	7.64	1518	16	Nil	630	324	698	90.9	279	0.391	78.0
28.	Draw well	24	7.94	1650	08	Nil	430	390	596	50.1	178	Nil	71.0
29.	Draw well	240	8.36	1056	12	Nil	266	28	318	70.5	100	0.780	43.5
30.	Bore well	360	8.57	1188	Nil	Nil	294	156	296	34.0	162	0.195	43.5
							Gouspura						
31.	Hand pump	295	8.36	2772	08	Nil	410	282	910	152.6	238	1.560	201.0
32.	Bore well	80	8.28	1980	08	Nil	268	172	550	91.8	233	1.170	114.0
33.	Bore well	160	8.78	792	08	Nil	148	34	220	45.2	85	1.560	22.5
34.	Draw well	90	8.71	1320	Nil	36	480	56	356	72.9	142	1.170	68.0
35.	Draw well	150	8.76	1584	01	Nil	382	30	280	60.8	100	1.560	96.0
36.	Draw well	14	8.83	1716	01	Nil	528	90	400	75.3	221	0.780	95.0
37.	Draw well	15	8.62	1782	01	Nil	528	36	340	73.9	157	1.560	113.0
38.	Hand pump	70	8.70	1254	22	Nil	382	42	200	38.4	86	1.560	47.5
39.	Hand pump	280	8.52	1650	01	Nil	344	60	580	126.4	226	1.170	106.0
40.	Bore well	150	8.70	1254	01	Nil	196	114	390	67.1	164	0.390	51.5
41.	Bore well	90	8.65	1254	01	Nil	212	60	384	78.7	164	0.780	56.5
							Sharma Dairy Farm						
42.	Draw well	80	8.98	594	Nil	Nil	236	24	158	32.6	60	1.560	14.5
43.	Hand pump	110	8.78	990	Nil	Nil	388	56	270	52.0	86	1.560	43.5
44.	Bore well	175	8.08	2503	Nil	Nil	290	180	800	150.7	352	0.390	183.0
45.	Bore well	220	8.54	1320	01	Nil	434	272	336	15.5	180	0.390	90.0
46.	Draw well	195	8.65	924	01	Nil	348	64	158	22.8	104	Nil	44.5
47.	Draw well	60	8.86	858	01	Nil	296	40	172	32.1	95	0.780	31.0
48.	Draw well.	75	8.65	1254	04	Nil	496	82	284	49.1	143	Nil	48.5
49.	Draw well	245	8.30	2112	01	Nil	314	578	702	30.1	311	0.390	154.0
50.	Bore well	130	8.50	1518	04	Nil	440	176	400	54.4	200	0.590	58.0
51.	Draw well	80	8.45	2772	Nil	Nil	280	176	700	127.3	576	0.195	120.0

Table 19.2b. Raw drinking water quality in rainy season with reference to the distance from the Swarna Rekha Sewage Channel.

Drinking water source No.	Source Type	Distance from the sewage channel (Meters)	Nitrite (mg.l^{-1})	Nitrate (mg.l^{-1})	Phosphate (mg.l^{-1})	Silicate (mg.l^{-1})	Sodium (mg.l^{-1})	Potassium (mg.l^{-1})	Chemmical Oxygen Demand (mg.l^{-1})	Total Solids (mg.l^{-1})	Total Dissolved Solids (mg.l^{-1})	Total Suspended Solids (mg.l^{-1})
I	II	III	12	13	14	15	16	17	18	19	20	21
						Taraganj						
1.	Draw well	10	0.018	2.76	0.018	14.9	30	3.6	40	340	340	Nil
2.	Draw well	65	0.010	12.62	0.005	12.0	30	2.4	28	420	380	40
3.	Draw well	15	0.020	1.44	0.012	14.9	35	13.2	36	360	280	80
4.	Draw well	56	Nil	1.2	0.225	29.9	50	9.2	76	520	440	80
5.	Bore well	27	Traces	9.08	0.095	12.0	50	4.0	32	600	600	Nil
6.	Draw well	42	Nil	0.77	0.011	23.2	135	23.6	64	1280	1200	80
7.	Draw well	02	Taces	8.42	0.091	9.9	40	16.0	36	460	440	20
8.	Draw well	08	Traces	8.42	0.016	11.0	50	6.4	32	700	580	120
9.	Draw well	05	0.010	11.52	0.041	17.8	65	8.4	28	700	600	100
						Jiwajiganj						
10.	Hand pump	70	0.107	Nil	0.022	14.9	400	16.4	20	2120	2120	Nil
11.	Hand pump	25	Traces	8.2	Nil	14.9	45	7.2	20	860	740	120
12.	Hand pump	450	0.232	16.61	Traces	11.0	45	2.8	12	580	580	Nil
13.	Hand pump	470	0.041	17.94	Traces	4.9	40	15.2	08	480	480	Nil
14.	Bore well	83	0.010	18.2	Nil	11.0	70	4.0	12	280	240	40
15.	Hand pump	18	0.010	4.88	0.016	14.9	35	16.0	12	280	240	40
16.	Draw well	24	0.021	3.99	0.027	16.1	75	3.6	12	720	680	40
17.	Draw well	18	0.027	3.43	0.095	9.9	15	14.8	20	160	40	120
18.	Draw well	24	Traces	10.63	0.010	16.1	30	2.4	44	180	40	140
19.	Hand pump	30	0.021	0.35	0.005	12.0	60	4.0	40	2460	220	2240
						Phoolbagh						
20.	Hand pump	61	0.072	11.08	0.010	8.9	52.5	23.2	16	780	460	320
21.	Hand pump	21	0.017	11.1	Traces	11.0	80	4.0	24	540	540	Nil
22.	Hand pump	53	0.014	26.6	0.013	11.0	20	3.2	24	380	260	120

(Contd.)

I	II	III	12	13	14	15	16	17	18	19	20	21
23.	Bore well	106	0.138	9.3	Traces	14.9	95	26.4	32	560	460	100
24.	Bore well	45	0.014	7.53	Nil	14.9	170	4.4	48	840	840	Nil
25.	Hand pump	137	0.035	8.42	Nil	9.9	45	2.8	20	860	840	20
26.	Draw well	360	0.014	15.3	0.016	24.6	20	26.0	44	560	440	120
27.	Draw well	12	0.063	16.4	0.004	14.9	125	11.2	28	600	500	100
28.	Draw well	24	0.206	3.65	0.102	16.1	85	30.8	28	880	540	340
29.	Draw well	240	0.164	19.05	Traces	22.0	35	26.4	12	500	500	Nil
30.	Bore well	360	Traces	5.76	Nil	11.0	45	2.8	16	880	680	200
							Gouspura					
31.	Hand pump	295	0.025	0.77	Nil	24.6	260	21.2	24	1980	1980	Nil
32.	Bore well	80	0.017	0.98	Traces	13.0	125	12.4	32	1220	1020	200
33.	Bore well	160	0.206	12.63	Nil	4.1	20	8.4	32	260	260	Nil
34.	Draw well	90	0.212	14.18	0.035	13.0	150	11.6	32	600	560	40
35.	Draw well	150	0.047	3.54	Nil	6.9	125	256.0	44	840	800	40
36.	Draw well	14	0.029	11.52	0.325	14.0	170	132.0	52	960	960	Nil
37.	Draw well	15	0.050	22.3	0.175	12.0	135	120.0	60	1020	940	80
38.	Hand pump	70	0.023	17.72	0.005	8.9	70	216.0	36	840	760	80
39.	Hand pump	280	0.059	2.3	Nil	7.9	80	3.6	56	1020	520	500
40.	Bore well	150	0.010	14.62	Nil	8.9	65	3.2	40	740	60	680
41.	Bore well	90	0.015	15.73	Nil	11.0	70	3.2	40	800	420	380
							Sharma Dairy Farm					
42.	Draw well	80	0.020	15.06	Traces	6.9	35	7.6	60	1620	300	1320
43.	Hand pump	110	0.015	17.72	Traces	8.9	70	28.0	52	160	160	Nil
44.	Bore well	175	0.014	0.443	Nil	9.9	225	5.6	48	2400	2300	100
45.	Bore well	220	0.090	5.32	Traces	9.9	90	3.6	08	940	940	Nil
46.	Draw well	195	0.107	19.3	0.020	8.9	45	20.8	12	480	480	Nil
47.	Draw well	60	0.035	4.65	0.025	13.0	35	39.2	24	Nil	Nil	Nil
48.	Draw well	75	0.492	15.95	0.110	17.8	80	116.0	24	560	560	Nil
49.	Draw well	245	0.059	Nil	Traces	11.0	160	8.4	24	1860	1720	140
50.	Bore well	130	0.014	21.5	Nil	9.9	80	8.8	48	1040	240	300
51	Draw well	80	0.021	Nil	Nil	8.9	290	3.6	60	2320	2200	120

Table 19.3a. Raw drinking water quality in winter season with reference to distance from the Swarna Rekha Sewage Channel.

Drinking water source No.	*Source Type*	*Distance from the sewage channel (Meters)*	*pH*	*Conductivity (u mhos/cm)*	*Turbidity (JTU)*	*Carbonate (mg.l⁻¹)*	*Total alkalinity (mg.l⁻¹)*	*Calcium hardness (mg.l⁻¹)*	*Total hardness (mg.l⁻¹)*	*Magnesium (mg.l⁻¹)*	*Chloride (mg.l⁻¹)*	*Sulfide (mg.l⁻¹)*	*Sulfate (mg.l⁻¹)*
I	II	III	1	2	3	4	5	6	7	8	9	10	11
							Taraganj						
1.	Draw well	10	8.67	528	65	Nil	266	64	226	39.4	77	1.22	35.0
2.	Draw well	65	8.35	660	16	Nil	268	68	250	44.2	95	0.82	20.5
3.	Draw well	15	8.95	858	08	16	312	72	396	78.7	118	0.41	Nil
4.*	Draw well	56											
5.	Bore well	27	8.40	1056	30	Nil	290	260	370	26.7	183	1.63	38.0
6.	Draw well	42	8.42	2112	71	Nil	496	212	532	77.8	456	Nil	36.0
7.	Draw well	02	8.58	1122	56	Nil	472	130	360	56.0	94	1.22	21.5
8.	Draw well	08	8.88	990	01	Nil	400	156	366	51.1	92	2.04	24.5
9.	Draw well	05	8.91	924	01	Nil	470	100	356	62.2	122	1.63	53.5
							Jiwajiganj						
10.	Hand pump	70	8.99	1188	08	Nil	518	220	470	60.8	390	1.22	88.0
11.	Hand pump	25	8.65	1188	16	Nil	336	160	384	54.5	130	0.82	44.5
12.	Hand pump	450	8.80	858	01	Nil	312	140	340	48.6	140	2.04	31.5
13.*	Hand pump	470											
14.	Bore well	83	8.94	1848	01	Nil	364	116	308	46.7	128	2.04	32.0
15.	Hand pump	18	8.93	1254	26	Nil	364	116	316	48.6	166	1.63	21.5
16.	Draw well	24	8.58	1518	01	12	348	150	388	56.4	180	1.22	68.0
17.*	Draw well	18											
18.	Draw well	24	8.78	990	01	Nil	246	130	348	53.0	142	1.63	27.5
19.	Hand pump	30	8.48	1584	08	Nil	342	264	468	49.6	298	1.63	96.0
							Phoolbagh						
20.	Hand pump	61	8.61	1188	01	Nil	356	224	500	67.1	178	1.22	57.5
21.	Hand pump	21	8.82	2112	16	Nil	292	96	312	52.5	146	1.63	47.5
22.	Hand pump	53	8.82	1584	01	Nil	264	68	288	53.5	152	0.41	15.0

(Contd.)

I	II	III	1	2	3	4	5	6	7	8	9	10	11
23.	Bore well	106	8.73	1122	01	Nil	362	132	372	58.3	168	2.04	35.0
24.	Bore well	45	8.61	1254	16	Nil	502	168	504	81.7	180	2.04	86.5
25.	Hand pump	137	8.57	1221	01	Nil	306	144	400	62.2	182	1.22	30.5
26.	Draw well	360	8.65	1254	01	Nil	330	104	390	69.5	144	1.22	32.0
27.	Draw well	12	8.43	1716	30	Nil	660	120	620	121.5	160	1.63	54.0
28.	Draw well	24	8.50	1518	01	Nil	420	108	448	82.6	150	1.22	64.8
29.	Draw well	240	8.40	792	08	Nil	282	60	300	58.3	170	1.63	40.5
30.	Bore well	360	8.51	1155	01	Nil	286	96	350	61.7	168	1.22	41.5
							Gouspura						
31.	Hand pump	295	8.48	1254	01	Nil	422	116	392	67.1	212	2.04	195.0
32.	Bore well	80	8.52	2310	20	Nil	312	152	780	152.6	283	2.45	127.0
33.	Bore well	160	8.48	990	01	Nil	218	38	340	73.4	191	2.45	50.0
34.	Draw well	90	8.35	1122	01	Nil	432	72	366	71.4	152	2.45	70.0
35.	Draw well	150	8.40	1551	01	Nil	372	48	300	61.2	120	2.45	103.0
36.	Draw well	14	8.48	1881	01	Nil	516	68	480	100.2	251	2.04	106.0
37.	Draw well	15	8.39	1782	01	Nil	480	40	426	93.8	216	1.63	120.0
38.	Hand pump	70	8.54	1452	08	Nil	368	76	360	69.1	140	2.45	94.0
39.	Hand pump	280	8.29	1650	01	Nil	332	272	650	91.8	241	2.04	103.0
40.	Bore well	150	8.57	1287	01	Nil	302	122	418	71.9	173	1.63	52.5
41.	Bore well	90	8.53	1254	01	Nil	346	48	356	74.8	147	1.63	42.5
							Sharma Dairy Farm						
42.	Draw well	80	8.62	726	01	Nil	254	38	280	58.8	80	3.26	21.5
43.	Hand pump	110	8.50	1188	08	Nil	390	74	356	68.5	194	2.86	66.5
44.	Bore well	175	8.00	3234	47	Nil	372	252	1024	187.6	446	0.408	156.0
45.	Bore well	220	8.32	1584	01	Nil	366	358	576	53.0	262	0.82	112.0
46.	Draw well	195	8.44	990	01	Nil	338	134	274	34.0	151	2.86	60.0
47.	Draw well	60	8.50	858	01	Nil	308	50	254	49.6	126	1.63	30.0
48.	Draw well	75	8.31	1056	01	Nil	426	88	296	50.5	158	1.22	31.0
49.	Draw well	245	8.19	1980	01	Nil	312	420	746	79.2	296	1.22	109.0
50.	Bore well	130	8.24	1254	12	Nil	424	208	552	83.6	263	0.408	39.0
51.	Draw well	80	8.25	2508	01	Nil	276	132	1428	315.0	625	0.82	128.0

Where * indicates no collection of sample due to drying up of the source.

Table 19.3b. Raw drinking water quality in winter season with reference to the distance from the Swarna Rekha Sewage Channel.

Drinking water source No.	*Source Type*	*Distance from the sewage channel (Meters)*	*Nitrite (mg.l^{-1})*	*Nitrate (mg.l^{-1})*	*Phosphate (mg.l^{-1})*	*Silicate (mg.l^{-1})*	*Sodium (mg.l^{-1})*	*Potassium (mg.l^{-1})*	*Chemmical Oxygen Demand (mg.l^{-1})*	*Total Solids (mg.l^{-1})*	*Total Dissolved Solids (mg.l^{-1})*	*Total Suspended Solids (mg.l^{-1})*
I	II	III	12	13	14	15	16	17	18	19	20	21
						Taraganj						
1.	Draw well	10	0.04	4.07	Nil	8.9	20	2.4	24	460	280	180
2.	Draw well	65	0.09	16.6	Nil	9.9	30	2.8	08	560	260	300
3.	Draw well	15	0.025	3.03	0.062	8.9	30	10.4	24	400	400	Nil
4.*	Draw well	56										
5.	Bore well	27	0.018	4.16	Nil	10.9	40	2.4	04	840	840	Nil
6.	Draw well	42	Nil	2.8	0.087	16.9	145	22.4	40	1560	1140	420
7.	Draw well	02	0.005	0.66	0.005	16.9	35	19.2	20	700	500	200
8.	Draw well	08	0.03	15.5	Nil	12.9	45	5.6	04	780	440	340
9.	Draw well	05	0.044	6.0	Nil	15.0	75	8.4	12	820	480	340
						Jiwajiganj						
10.	Hand pump	70	0.063	1.22	0.015	26.8	365	12.0	12	1880	1880	Nil
11.	Hand pump	25	0.054	7.97	Nil	15.0	40	2.4	12	460	460	Nil
12.	Hand pump	450	0.075	7.2	Nil	5.9	60	3.2	08	640	560	80
13.*	Hand pump	470										
14.	Bore well	83	0.014	12.2	Nil	4.9	70	6.8	44	680	400	280
15.	Hand pump	18	0.094	7.2	0.101	3.0	105	16.4	64	740	700	40
16.	Draw well	24	0.071	4.32	Traces	9.9	75	3.2	24	980	700	280
17.*	Draw well	18										
18.	Draw well	24	0.01	4.42	Nil	3.0	40	2.0	20	380	380	Nil
19.	Hand pump	30	0.176	2.4	Nil	2.0	60	2.8	44	1180	1180	Nil
						Phoolbagh						
20.	Hand pump	61	0.041	3.54	Nil	8.9	50	6.4	36	820	780	40
21.	Hand pump	21	0.020	4.98	Nil	5.9	75	4.4	08	660	540	120.
22.	Hand pump	53	0.067	16.6	Nil	Nil	176	24.0	16	500	400	100
23.	Bore well	106	0.068	3.0	Nil	Nil	100	4.0	20	800	800	Nil

(Contd.)

I	II	III	12	13	14	15	16	17	18	19	20	21
24.	Bore well	45	0.019	0.64	Nil	1.1	30	3.2	04	1480	720	760
25.	Hand pump	137	0.038	3.96	Nil	0.2	95	17.6	08	1140	660	480
26.	Draw well	360	0.019	4.9	Nil	4.0	25	24.4	16	1040	600	440
27.	Draw well	12	0.05	2.55	0.12	7.9	145	13.6	04	800	740	60
28.	Draw well	24	0.139	4.08	0.074	4.9	75	14.8	Nil	1160	920	240
29.	Draw well	240	0.22	18.6	0.008	6.9	35	26.0	Nil	500	500	Nil
30.	Bore well	360	0.001	3.3	Nil	8.9	50	2.8	12	1520	600	920
							Gouspura					
31.	Hand pump	295	0.021	2.3	0.004	2.0	225	20	08	2680	2000	680
32.	Bore well	80	0.01	Nil	Nil	Nil	150	10.8	24	1740	1060	680
33.	Bore well	160	0.244	0.64	Nil	9.4	45	8.4	24	580	360	220
34.	Draw well	90	0.184	16.5	0.034	11.5	135	11.2	28	820	520	300
35.	Draw well	150	0.093	4.43	Nil	5.9	125	244	36	920	660	260
36.	Draw well	14	0.025	13.3	0.17	10.5	185	128	44	1200	1000	200
37.	Draw well	15	0.039	12.2	0.16	10.9	150	140	40	1300	1180	120
38.	Hand pump	70	0.135	7.09	0.002	4.9	80	164	32	1360	1060	300
39.	Hand pump	280	0.051	1.9	Nil	9.9	75	3.6	52	940	580	360
40.	Bore well	150	Nil	13.7	Nil	5.5	65	4.0	40	1380	500	880
41.	Bore well	90	0.025	15.3	Nil	7.9	80	2.8	36	1700	760	940
							Sharma Dairy Farm					
42.	Draw well	80	0.0132	14.62	Nil	4.0	35	7.2	44	1600	560	1040
43.	Hand pump	110	0.0115	5.6	Nil	6.9	115	25.2	44	680	380	300
44.	Bore well	175	0.007	Nil	Nil	6.9	325	8.8	56	2320	2320	Nil
45.	Bore well	220	0.12	3.3	Nil	9.9	110	10.4	04	1260	1260	Nil
46.	Draw well	195	0.074	15.5	Nil	6.9	55	20.0	08	560	520	40
47.	Draw well	60	0.038	5.42	0.043	9.9	45	88.0	08	280	240	40
48.	Draw well	75	0.11	14.2	0.125	14.0	75	116.0	08	600	600	Nil
49.	Draw well	245	0.05	Nil	0.003	9.9	150	8.8	12	1840	1780	60
50.	Bore well	130	0.0105	19.05	Nil	7.9	75	7.6	44	1320	1040	280
51.	Draw well	80	0.029	Nil	Nil	6.9	300	4.0	44	2040	2000	40

Where * indicates no collection of samples due to drying up of the source.

REFERENCES

APHA, 1985 : *Standard Methods for the Examination of Water and Waste water.* American Public Health Association, APHA, AWWA, WPCF, 16th Edn., New York.

Bedi Yash Pal (Ed.), 1971 : *Hygiene and Public Health*, Anand Publishing Co., Amritsar, Eleventh Edn., p. 14.

Fair, 1967 : Cited in Occasional Monograph Series No. 84 of I.I.E.E. Delhi. *Analysis of Water Quality Chloride*, p. 10.

Gupta, M.K., V. Singh, P. Rajwanshi, S. Srivastava and S. Dass, 1994 : Fluoride in Ground Water at Agra. *Indian J. Environ. Hlth.*, 36 : 43-46.

Handa, B.K., 1983 : The effect of fertilizer use on groundwater quality in the phreatic zone with special reference to U.P. *Proc. Nat. Sem. on Assessment Development and Management of Groundwater Resources.* Central Groundwater Board, New Delhi.

ICMR, 1975 : *Manual of standards of quality for drinking water supplies* (IInd Edition). Indian Council of Medical Research. Special report series No. 44.

IIEE, 1984 : *Analysis of water quality* : Indian Institute of Ecology and Environment, New Delhi, Occasional Monograph Series No. 84.

ISI, 1982 : *Indian Standards tolerance limits for inland surface water subject to pollution.* (IInd Revision). Indian Standards Institute, IS : 2296-1982.

Kataria, H.C., 1995 : Evaluation of groundwater by Nutrients of City and Bairagarh area of Bhopal. *Indian J. Environ. Prot.*, 15 : 218-222.

Mittal, S.K., A.L.J. Rao, S. Singh and R. Kumar, 1994 : Groundwater quality of some areas in Patiala city. *Indian J. Environ. Hlth.*, 36 : 51-53.

Pandey, B.N., A.K. Jha, S.K. Mishra, R.N. Tripathi and S.K. Pandey, 1994 : Assessment of quality of drinking water of Katihar, North Bihar. *Acta Ecol.*, 16 : 144-149.

Rai, J.P. Narain and H.C. Sharma, 1995 : Bacterial contamination of groundwater in rural areas of North-West Uttar Pradesh. *Indian J. Environ. Hlth.*, 37 : 37-41.

Rao, Kaza Somsekhara, V.A. Raju, M. Singanon, B. Someswara Rao, P.V. Seshagiri Rao and K.P. Chakravarthy, 1994 : Studies on the quality of water supplied by the municipality of Kakinada and groundwaters of Kakinada town. *Indian J. Environ. Prot.*, 14 : 167-169.

Reddy, U.V.B., K.P. Reddy, V. Sudershan and B. Rajeswara Reddy, 1995 : Hydrogeochemistry of Musi river and groundwater Hyderabad city. *Indian J. Environ. Prot.*, 15 : 440-446.

Shankar, V. Bharani and N. Muthukrishnan, 1994 : Is the groundwater in Madras city potable? *Indian J. Environ. Prot.*, 14 : 176-179.

Sharma Sanjay and R. Mathur, 1994 : Bacteriological quality of groundwaters in Gwalior, India. *Indian J. Environ. Prot.*, 14 : 905-907.

Sharma Sanjay, P.C. Jain and R. Mathur, 1995 : Quality assessment of groundwater in municipal and fringe areas near Gwalior. *Indian J. Environ. Prot.*, 15 : 534-538.

USEPA, 1989 : *Is your Drinking water safe*? United States Environmental Protection Agency. Office of Water (WH-550), 570/9-89-005.

USEPA, 1991 : *Manual of Individual and Non-Public water supply systems.* United States Environmental Protection Agency EPA, 570/9-91-004.

Walker, W.H., 1969 : Illinois Groundwater Pollution, *J. Am. Wat. Wks. Assn.*, 61 : 31-40.

WHO, 1978 : *Nitrates, Nitrites and N-nitroso compounds.* World Health Organization Environmental Health Criteria-5, WHO, Geneva.

WHO, 1988 : *Assessment of Freshwater Quality. Global Environmental Monitoring Systems.* Report on the results of the WHO/UNEP program on health related environmental monitoring. World Health Organization, Geneva.

Widyanto, L. and M. Sorjani, 1975 : Paper presented at 2nd World Congress on Water Resources. Organized by I.W.R.A., p. 33.